# Graphene in Spintronics

# Graphene in Spintronics

## Fundamentals and Applications

**Jun-ichiro Inoue**
**Ai Yamakage**
**Shuta Honda**

PAN STANFORD PUBLISHING

*Published by*

Pan Stanford Publishing Pte. Ltd.
Penthouse Level, Suntec Tower 3
8 Temasek Boulevard
Singapore 038988

Email: editorial@panstanford.com
Web: www.panstanford.com

**British Library Cataloguing-in-Publication Data**
A catalogue record for this book is available from the British Library.

**Graphene in Spintronics: Fundamentals and Applications**



ISBN 978-981-4669-56-6 (Hardcover)
ISBN 978-981-4669-57-3 (eBook)

Printed in the USA

# Contents

# Preface

In these 30 to 40 years, significant progress has been achieved in solid-state physics and materials science. Not only the discovery of novel materials such as oxide superconductors with high Curie temperature, magnetic multilayers with giant magnetoresistance (GMR), and low-dimensional carbon lattices called fullerenes, nanotubes, and graphene but also the discovery of the quantum version of Hall effects and the invention of the novel scanning microscope, etc., have been made.

Modern electronics based on solid-state silicon transistors has also developed tremendously due to the development of microlithography techniques and contributed to the progress in information-based society. Silicon electronics, however, is known to be confronted with difficulties caused by the downsizing of devices. Enormous scientific and technological research has been performed to overcome the difficulties. Among them, a novel way, the so-called spintronics, which utilized both spin and charge degrees of freedom of electrons, has been developed.

Previously the spin and charge degrees of freedom of electrons have been used independently for magnetism and semiconductor technology, respectively. The discovery of GMR opened the way for spintronics because GMR is a combined phenomenon of the magnetism and transport originated from electrons. The effect of magnetoresistance has been successfully applied in the field of spintronics, and further applications to silicon technology are under investigation. As mentioned, the discovery of novel materials opens up the option for scientists and engineers to use them for spintronics applications.

Graphene is a promising material for technological applications because of its distinguished physical and chemical properties. It is, therefore, our obligation to proceed toward establishment of graphene technology. However, such an attempt is yet unsuccessful in modern electronics as well as in the field of spintronics. To bring about any scientific and technological breakthrough, an overview of the wide aspects of graphene and spintronics would be desirable. Because many excellent review articles and textbooks on each subject have already been published, it is our attempt to provide especially young scientists with an introductory view of graphene magnetoresistive junctions in relation to the present status of the field of spintronics.

The contents of this book are as follows. After a short introduction of graphene and spintronics, we present basic features of graphene in Chapter 2. The electronic structure of graphene, graphene nanoribbons, and graphene contacts are presented in Chapter 3. Transport properties relevant to graphene junctions, graphene field-effect transistors, and spin injection and magnetoresistance will be explained in Chapter 4. To give insight into the spintronics applications of magnetoresistive junctions, properties of GMR and tunnel magnetoresistance (TMR) will be explained in Chapter 5. Subsequently, theoretical results obtained by our group for magnetoresistance in realistic models will be presented in Chapter 6. In the final chapter, a summary and an overview will be given.

The authors thank A. Yamamura, T. Hiraiwa, R. Sato, and T. Kato who were master course students at the Department of Applied Physics, Nagoya University for their excellent research work in their course, as well as Dr. Itoh of Kansai University for providing us with computer program codes and computing facilities. The work was partly supported by next-generation supercomputing projects, NanoScience Program MEXT, Japan; Grants-in-Aids for scientific research in the priority area "Spin Current," MEXT, Japan; and Elements Science and Technology projects, MEXT, Japan

**Jun-ichiro Inoue**
**Ai Yamakage**
**Shuta Honda**

# Chapter 1

# Introduction

After the discovery of single-layer graphene, the science of graphene has quickly become widespread in the world. Research on technological applications of graphene to various devices is also in progress throughout the world. The discovery of new phenomena and novel materials could contribute to human society; however, many difficulties should be overcome before achieving this purpose. A successful example of such activities may be the research in spintronics developed after the discovery of giant magnetoresistance (GMR) in magnetic multilayers. Spintronics is a field of science and technology that utilizes both charge and spin degrees of freedom of electrons discovered about 100 years ago [1, 2]. Graphene could be applied to devices in the field of spintronics and also to novel ones by using characteristic features of graphene outside the field of conventional spintronics. To develop such a new field, comparison of the research in the field of graphene with that in spintronics might be effective. In this chapter we give an overview of the research on graphene and in spintronics. The overview, however, could not be comprehensive but is restricted within limited subjects because the research activities of graphene are so vast in the world.

*Graphene in Spintronics: Fundamentals and Applications*
Jun-ichiro Inoue, Ai Yamakage, and Shuta Honda

ISBN 978-981-4669-56-6 (Hardcover), 978-981-4669-57-3 (eBook)
www.panstanford.com

## 1.1 General Features of Graphene

### 1.1.1 *From Carbon Elements to Carbon Lattices*

Carbon is the sixth element in the periodic table and one of the inevitable elements in the world of life, materials, and industry. Carbon, nitrogen, and oxygen constitute most of the bodies of life in the natural world as organic materials. Among them carbon forms various lattices in solid-state materials, from zero-dimensional to three-dimensional lattices, as shown in Fig. 1.1. Diamond is a well-known material of three-dimensional lattices as the most precious stone. It has the same lattice structure as silicon and germanium, which are fundamental materials for modern electronics. Graphite has a quasi-three-dimensional lattice structure formed by stacking of two-dimensional layers. Graphite is also a well-known material that is used as lead in pencils. One- and zero-dimensional carbon lattices were discovered nearly at the end of the twentieth century, which are carbon nanotubes and fullerenes ($C_{60}$), respectively [3, 4]. The lattice structure of fullerenes had been predicted theoretically before the experimental discovery [5]. The carbon nanotube was found in the fabrication process of carbon fine particles, that is, in the soot of smoke. Furthermore, a novel structure of the

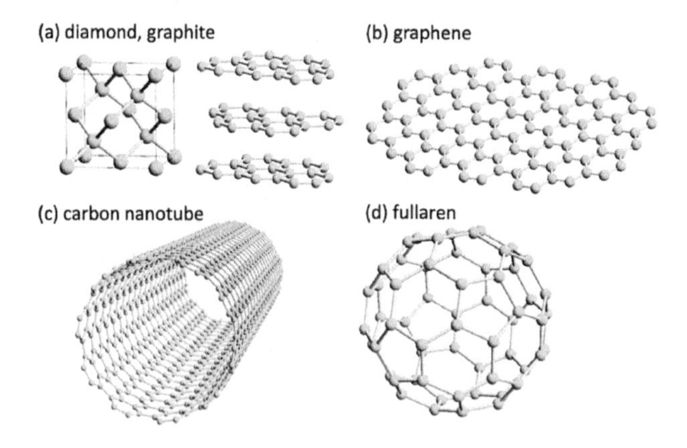


**Figure 1.1** Lattice structures of crystals made of carbon. (a) Three-dimensional diamond and graphite, (b) two-dimensional graphene, (c) one-dimensional nanotube, and (d) zero-dimensional fullerene.

carbon nanotube, which includes $C_{60}$, called "nanopeapod," has been fabricated [6].

Various types of fullerenes have been applied as, for example, compounds to strengthen plastics and as fine particles doped into lubricants. Fullerenes including metallic elements can be used in medical treatment. Carbon nanotubes can replace carbon fibers. They can be applied to solar cells, field-effect transistors (FETs), conductive transparent films, electric field emitters, and probes in scanning tunneling microscopy. Thus carbon materials are important materials for technological applications in future devices.

Graphene is nothing but an element of two-dimensional layers of graphite. Because of the simple structure of the lattice, the electronic state had already been calculated just after the Second World War by Wallace [7]. The discovery of graphene, or secession of the single sheet of carbon layer from graphite, was done in the early twenty-first century under the motivation to fabricate the thinnest carbon layer [8–10]. The fabrication technique to discover the graphene sheet was tremendously simple—to peel off the graphene sheet by using a scotch tape, as shown in Fig. 1.2. Identification of the single sheet of carbon atoms, however, was difficult because the sheet was too transparent to identify. Optical microscopy, scanning electron microscopy, and atomic force microscopy were combined to achieve it.

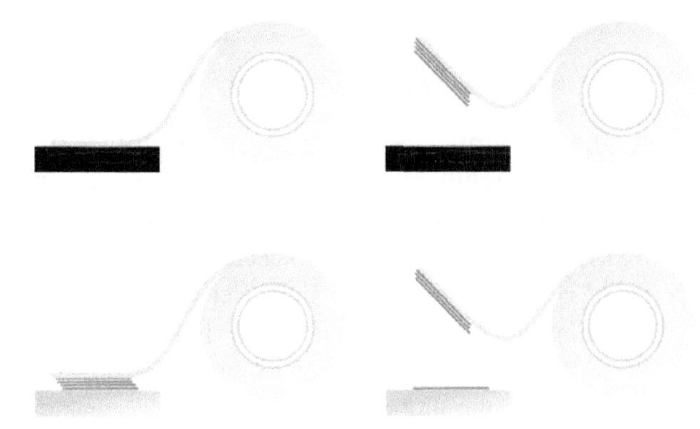

**Figure 1.2** Micromechanical cleavage of graphene sheets from graphite and attachment of a single graphene sheet on a substrate. Reprinted (figure) with permission from Ref. [16], Copyright (2011) by the American Physical Society.

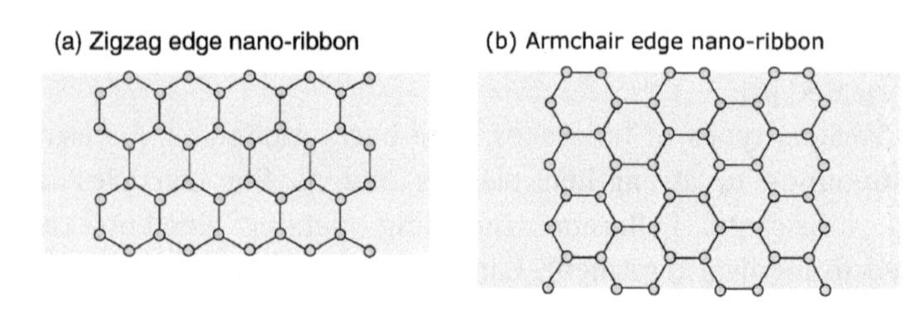


**Figure 1.3** Lattice structures of graphene nanoribbons with (a) zigzag edges and (b) armchair edges.

Graphene has a two-dimensional honeycomb lattice of carbon atoms, as shown in Fig. 1.1b. The size of the graphene sheet is usually large. Nevertheless, quasi-two-dimensional graphene sheets called graphene nanoribbons (GNRs) are extremely important. Because graphene sheets with finite size may have an edge structure, a zigzag edge or an armchair edge shown in Fig. 1.3, there are two types of GNRs, zigzag-edge GNRs and armchair-edge GNRs. In addition, there are many variations of finite-size graphene sheets, the smallest one being benzene, which could be no more a solid-state material but one of the basic organic materials. Basic features of graphene have been intensively studied, and the results have been explained in excellent review articles [11–14] and in novel lectures [15, 16].

### 1.1.2 *Fabrication Method of Graphene Sheets*

The novel study of graphene started by transcription of a single sheet peeled from graphite by an adhesive tape, as shown in Fig. 1.2. Because the method is not surely suitable to technological applications, more systematic methods of fabrication have been proposed. Occurrence of ripples in graphene sheets is also a shortcoming of the cleavage method [17]. One method is thermal decomposition of SiC [17] and the other is chemical vapor deposition (CVD) of a graphene sheet on a metallic substrate [18]. In fact, these methods were used prior to the method of cleavage of graphene sheets from graphite [19, 20]. Improvement of the fabrication methods is surely important for technological applications; however, we will not touch upon this issue in this textbook.

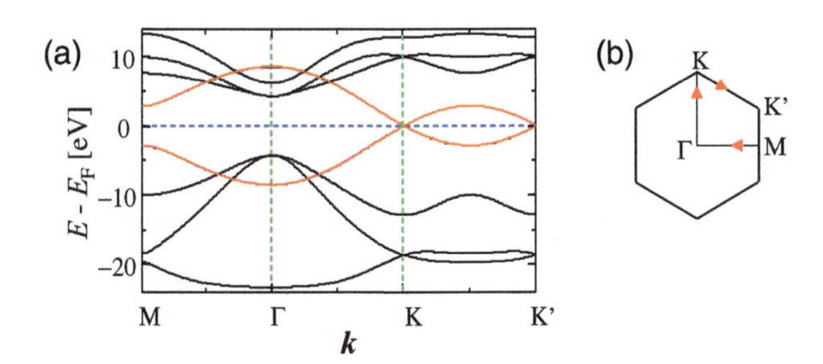


**Figure 1.4** (a) Electronic structure of graphene sheet and (b) its Brillouin zone. Red curves show energy bands responsible for various physical and chemical properties of graphene.

### 1.1.3 *Graphene Lattice and Electronic Structure*

The electron configuration of the carbon element is $(1s)^2(2s)^2(2p)^2$, among which $(2s)^2(2p)^2$ electrons form the electronic structure responsible for the solid-state properties of graphene. Because of the honeycomb lattice, a unit cell of the graphene lattice includes two atoms. The electron dispersion curve is easily calculated by using, for example, a tight-binding model with $s$ and $p$ orbitals. The results are shown in Fig. 1.4 with the Brillouin zone of the graphene sheet. The energy eigenvalues are plotted as a function of electron momentum along high-symmetry lines shown in the Brillouin zone.

The energy bands shown by black curves are formed by a linear combination of $s$ and $p_x$ and $p_y$ orbitals and are called $\sigma$ bands. The energy bands shown by red curves are formed only by the $p_z$ orbital and are called $\pi$ bands. Because carbon atoms have eight electrons in both up- and down-spin states, half of the eight energy bands are occupied by $s$- and $p$-electrons, and the Fermi level is located at the middle of the $\pi$ bands given by energy $E = 0$ (eV) in the figure.

A detail of three-dimensional picture of the energy dispersion of the $\pi$ bands is given in Chapter 2 (Fig. 2.4a). As shown in the figure, the electronic states at the Fermi energy consist of only two states, which are called $K$ and $K'$ states. These two states are called valley degrees of freedom. A schematic figure of the electronic states near $K$ and $K'$ states is depicted in Fig. 2.4b. We find that the electronic states are of cone shape and that the energy dispersion near $K$ and $K'$ is linear. The linear energy dispersion near the Fermi

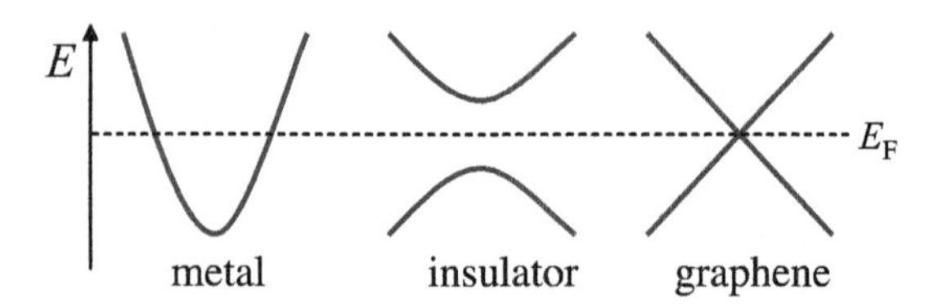


**Figure 1.5** Difference of electronic structure near the Fermi energy of metals, insulators, and a graphene sheet.

energy is much different from the energy dispersion of $k^2$ type near the Γ point in the free-electron model, for example. Because the Hamiltonian that gives a linear energy dispersion near the Fermi energy instead of $k^2$ dispersion is called Dirac Hamiltonian, the $K$ and $K'$ points are called Dirac points. The cone-shaped energy states are called Dirac cones. Although there is no energy gap in the band structure of graphene, the state at the Fermi energy is given by only two states in the Brillouin zone, and the density of states at the Fermi energy is zero. Therefore graphene is often called a zero-gap semiconductor.

Thus, the electronic structure of graphene near the Fermi level is much different from that in other materials, metals, and insulators. Figure 1.5 shows schematics of the electronic structures of metals, insulators (or undoped semiconductors), and graphene. The difference would be obvious.

## 1.1.4 *Physical Properties*

### 1.1.4.1 Electrical transport

The characteristics of graphene—the energy state with linear Energy dispersion near the Fermi energy and the material made of the carbon element—result in various interesting transport properties. Because the effective mass of electrons in metals is given by the second derivative of energy with respect to the wave vector, the linear energy dispersion gives zero effective mass ($m_{\text{eff}} = 0$) for electrons in graphene. The simplest theory for electrical transport in metals may be given by Drude's theory in which electrical conductivity is given as $\sigma = e^2 n\tau / m_{\text{eff}}$ where $e$, $n$, $\tau$, and $m_{\text{eff}}$ are electric charge, electron density, relaxation time of conduction electrons, and effective mass of conduction electrons, respectively. The zero effective mass means that the conductivity diverges (or the resistivity $\rho = 1/\sigma$ vanishes) in infinite graphene sheets even if the relaxation time is finite. On the other hand, because only two states are available at the Fermi energy, the carrier density of the graphene should be infinitesimally small, making the conductivity vanish.

Furthermore, the two available states on the Fermi level for electrical transport may give different features of electrical scattering from those in usual metals. The relaxation time $\tau$ is characterized by property of scattering of conduction electrons by impurities, phonons, etc. In ideal systems with translational invariance, the momentum conservation should be realized, and the momentum $\boldsymbol{p} = \hbar\boldsymbol{k}$ is constant in time. But in real materials, the direction of $\boldsymbol{p}$ (or $\boldsymbol{k}$) is changed by the scattering. In the quantum picture of scattering, the initial and final states of electrons in the scattering process should reside on the Fermi surface characterized by the electronic structures of the system. The characteristics of the electronic states of graphene may produce a large restriction on the process of electron scattering. Thus these conflicting features of graphene produced by the characteristic features of the electronic structure may give rise to interesting and fundamental issues on the transport property of graphene.

Experimentally, instead of the conductivity $\sigma$, values of mobility $\mu = e\tau/m_{\text{eff}}$ are often reported. Due to the argument before, it is expected that the value of $\mu$ should also be large. There are reported many results of mobility so far, for example, $\mu = 10{,}000$ measured in the electric field effect [8], $\mu = 3000$–$27000\ \text{cm}^2/\text{Vs}$ measured by using the quantum Hall effect (QHE) [10], and $\mu \approx 2000\ \text{cm}^2/\text{Vs}$ for doped graphene with carrier number $\sim 3.6\times10^{12}\ \text{cm}^{-2}$ in a four-terminal device with ferromagnetic electrodes with an Al–O tunnel barrier inserted between graphene and the electrodes [21]. Later, Novoselov et al. [22] and Bolotin et al. [23] have shown that the values of $\mu$ are $2\times10^5$ and $1.2\times10^5\ \text{cm}^2/\text{Vs}$ at 5 K and 240 K, respectively. Such high values of mobility indicate nearly vanishing of electrical scattering and therefore suggest the occurrence of quantum effects even at room temperature. Although there are some scatters of observed values of $\mu$, these values are longer than those observed for usual semiconductors. For example, the mobility of a

GaAs- or an InP-based high-electron-mobility transistor (HEMT) is about $1 \times 10^4$ cm$^2$/Vs [24].

Another important point for the transport property of graphene is that graphene consists of a light element, that is, carbon, and that the relativistic effects on the electronic properties are quite weak. This suggests that the spin diffusion length $\lambda_S$ should be quite long. First observations have been done by using nonlocal measurement in four-terminal ferromagnetic junctions with a single graphene sheet [21]. Observed values were 1.5–2 μm, which, however, may not be long enough as compared with those expected.

Graphene also shows interesting features in the Hall effect. When an external electric field $\boldsymbol{E}$ is applied to nonmagnetic metals, electrons drift to the direction given by $-\boldsymbol{E}$. When an external magnetic field $\boldsymbol{H}$ is applied in addition to the electric field, the electrons drift to the direction given by $\boldsymbol{E} \times \boldsymbol{H}$. The phenomenon is observed in any metals and is called the normal Hall effect (NHE) after the discovery of this phenomenon by E. Hall [25]. In the world of quantum mechanics, quantization of energy levels due to magnetic field occurs, that is, Landau level formation. The quantization of energy levels can be observed in the so-called QHE by using gate control of carrier density. As the carrier density is changed by gate voltage, the Fermi energy crosses the Landau levels one by one, and the Hall conductance increases stepwise by $e^2/h$.

The transverse transport properties (Hall effect) appear anomalously in graphene, as shown in Fig. 1.6. We find first that the

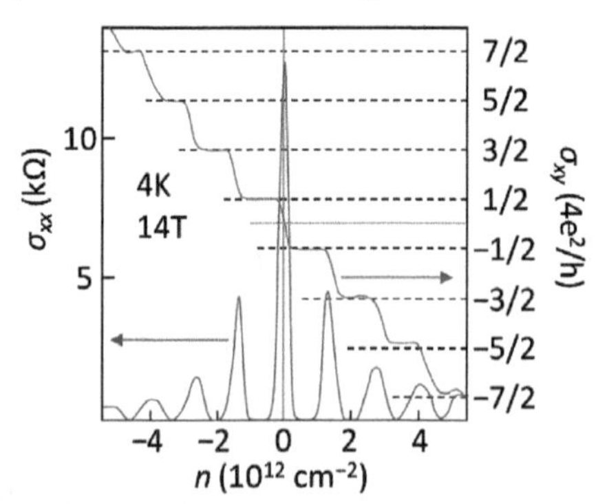


**Figure 1.6** Experimental results of the quantum Hall effect of graphene. Reprinted by permission from Macmillan Publishers Ltd: [*Nature Materials*] (Ref. [11]), copyright (2007).

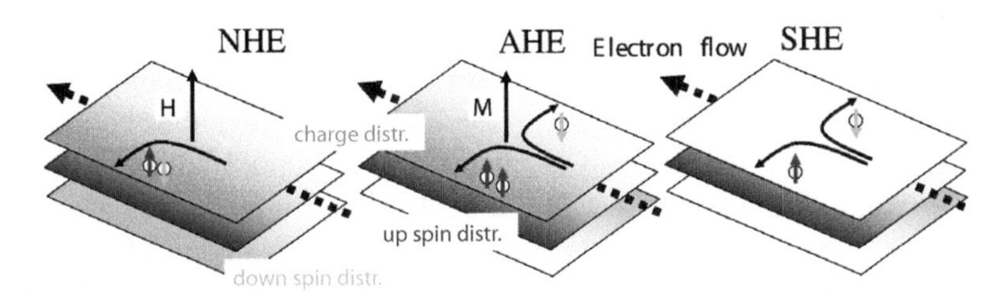


**Figure 1.7** (a) Normal Hall effect, (b) anomalous Hall effect, and (c) spin Hall effect.

conductance increases stepwise in $4e^2/h$ with the carrier density $n$, where the factor 4 indicates the spin and valley ($K$ and $K'$) degrees of freedom. Although there are stepwise increases of $4e^2/h$ in conductance, there appears no plateau at $n = 0$, and the resistivity shows a peak at this carrier density. The anomaly has been attributed to the existence of the pseudo spin degree of freedom of the valley originated to the nonequivalent two sublattices of the graphene honeycomb lattice. Thus the QHE exhibits the characteristic feature of the electronic structure of graphene.

The topological nature of the Hall effect in graphene has led to the novel concept of materials via another type of Hall effect called the spin Hall effect (SHE) in nonmagnetic metals. The SHE appears in normal metal without a magnetic field but with spin–orbit interactions. The SHE is distinct from the Hall effect in ferromagnets, called the anomalous Hall effect (AHE) [26], but predicted first by D'yakonov and Perel [27] and later by Hirsh [28] and theoretically formulated by Murakami et al. [29] and Sinova et al. [30]. The difference between the NHE, AHE, and SHE is depicted in Fig. 1.7 [31]. In the NHE, both up- and down-spin electrons drift along the same direction, as shown in Fig. 1.7a. When spin–orbit interaction exists, the up- and down-spin electrons are scattered differently, and transverse up- and down-spin currents flow toward opposite directions. Because of different carrier density of up- and down-spin electrons in ferromagnets, a net transverse current exist in AHE, as shown in Fig. 1.7b. In the SHE, on the other hand, the transverse current of up and down spin electrons in normal metals with spin–orbit interaction cancels with each other, and no transverse voltage occurs, the situation of which is shown in Fig. 1.7c.

Although the spin–orbit interaction is expected to be quite weak in graphene, it has been pointed out that an effective spin–orbit interaction appears due to the second near-neighbor hopping and/or wavy structure of graphene sheets. Kane and Mele [32] formulated the Hall effect in a model that includes a spin degree of freedom, a valley degree of freedom (the pseudo spin), and spin–orbit interaction and showed that the SHE is quantized (quantum SHE). The quantization of the SHE is related to a chirality of the electronic states and to characteristic edge states of finite-size graphene. The study has opened a novel concept of materials, that is, topological nature of insulators (see, for example, Ref. [33]).

Graphene shows also interesting magnetism. The edge state at the zigzag edges has a flat energy dispersion with a high density of states, which results in occurrence of magnetic polarization at the edge. Defects of carbon atoms in a graphene sheet also induce spin polarization around the defects. See Ref. [12]. This spin polarization might be realized only at low temperature and therefore has not yet been confirmed experimentally. Nevertheless, the magnetism of graphene attracts much interest because it could give rise to coupled phenomena of spin and charge degrees of freedom in graphene sheets.

#### 1.1.4.2 Other properties

Elastic and optical properties of graphene are also interesting. The properties are originated from the two-dimensional lattice with strong in-plane bonding. The structure makes intrinsically the out-of-plane lattice vibration large, resulting in large elastic constants and negative lattice expansion with increasing temperature [34, 37]. The large lattice oscillation also produces large thermal conductivity if about 5000 W/mK at room temperature [38].

It is easily expected that the monoatomic layer of graphene produces high optical transmittance. Actually the observed value of the optical absorption ratio is only 2.3% per layer in graphene [35]. What is the most interesting point is that the optical transmittance is given only by the fine-structure constant $\alpha = e^2/\hbar c$ [36].

### 1.1.5 *Potential Applicability of Graphene*

The characteristic features of the transport and optical properties conduct us to a novel field of graphene applications. It is quite easy for us to be reminded that the two-dimensionality, high mobility, and electron–hole symmetry of graphene are attractive features to apply the graphene sheet to high-frequency FETs in tera-Hertz frequency [39]. However, it should be noted that no energy gap in the graphene electronic state is a big obstacle to apply graphene to usual FETs. The high optical transparency may be used as transparent electrodes to replace the present indium titanium oxide (ITO) electrodes. However, it is noted that the sheet resistance of graphene is at present higher than that of ITO by 100 times [40].

Because of the long spin diffusion length originated from the weak spin–orbit interaction, graphene sheets could be applied to spin-related devices in the field of spintronics, such as spin FETs. To realize a spin FET, however, the obstacle caused by the absence of an energy gap should be overcome. The edge states and edge magnetism could also be interesting features for technological applications.

The main purpose of this chapter is to introduce possible technological applications of graphene in the field of spintronics. Prior to an explanation of spintronics applications of graphene, we first outline, in the following section of this chapter, the modern semiconductor technology and the development of the field of spintronics and give a short introduction of some graphene devices. As for spintronics, we will give a brief explanation of typical phenomena of GMR and tunnel magnetoresistance (TMR) and their technological applications.

## 1.2 Modern Electronics and Spintronics

### 1.2.1 *Semiconductor Technology and Its Limitation*

Modern electronics rests on the invention of transistors made of solid-state semiconductors. Rectification and amplification of input currents are the basic feature of the transistor. There are two types of transistors, bipolar transistors, which consist of both hole-doped

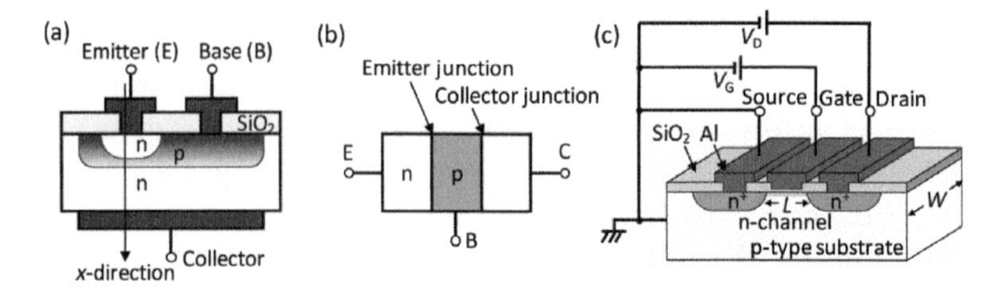


**Figure 1.8** Structures of modern silicon transistors, (a) bipolar transistor, and (b) unipolar transistor.

(*p*-type) and electron-doped (*n*-type) semiconductors, and unipolar transistors, which utilize only one (*p*- or *n*-) type of semiconductors. Device structures of bipolar and unipolar transistors are shown in Figs. 1.8a and 1.8c, respectively. The basic structure of the former consists of *n*–*p*–*n* (or *p*–*n*–*p*) junctions with three terminals: emitter, collector, and base (see Fig. 1.8b). The latter also consists of three terminal junctions with source, drain, and gate terminals. The structure shown in Fig. 1.8c is called a FET. Because the semiconducting material (usually silicon) where current flows is separated from the gate electrode by insulating oxides (usually $SiO_2$), the transistor shown in Fig. 1.8c is called a metal-oxide semiconductor FET (MOSFET).

Large-scale integration (LSI) of MOSFETs on a small tip has been quite successful for developing random access memory (RAM) for logic circuits. This type of silicon integrated circuit is often called complementary metal-oxide semiconductor (CMOS), and the silicon RAM is called dynamic RAM (DRAM). Downsizing of the CMOS structure has made tremendous contribution for developing high-speed computing machines.

The modern semiconductor (silicon) technology was developed by overcoming various difficulties associated with miniaturization and high-speed performance of computers. Very-large-scale integration (VLSI) of circuits and downsizing of MOSFETs have been successfully realized, and at present the channel length of the MOSFET is less than 20 nm. Downsizing and high-density CMOS structure, however, produce a problem of energy consumption. Leakage current is getting to be a serious problem in the MOS structure with thin oxide layers. Because DRAM data decay with time, it requires regular charging up to maintain the memory. This is also a problem of energy consumption. The leakage current could be blocked by using other oxide materials such as $HfO_2$ instead of $SiO_2$. However, the fundamental shortage, necessity of regular charging up in DRAM, could only be avoided by adopting basically a different structure for the memory device, that is, a nonvolatile memory device. Combining memory and logic circuits may also be another important progress for high-performance computing machines.

In addition to semiconductor technology, which utilizes the charge of electrons, there is an important device, the hard disc drive (HDD), which utilizes the spin of electrons. This is a permanent memory device of computing machines, made of permanent magnets. The HDD associates with magnetic sensors that are used to write data onto the disc and at the same time read data on the disc. Therefore, a high-density HDD requires also high-speed and high-sensitivity sensors.

The areal density of the HDD has been increased according to the so-called Moore's law. The density is now more than 1 $\text{Tbits/inch}^2$. The increase in the areal density of the HDD is attributed to the development of magnetoresistive sensors. The original sensors of magnetic tape memory utilize the phenomenon of electromagnetic induction. The sensor was replaced with a magnetoresistive sensor in the 1980s. The effect of magnetoresistance (MR) is a coupled phenomenon of electric current and magnetism of ferromagnetic metals. The electrical resistivity of ferromagnetic alloys, for example, Fe–Ni alloys (permalloys), changes as the direction of magnetization changes. This phenomenon is known as anisotropic magnetoresistance (AMR). The resistivity change is at most a few tens of percent. Thin films of permalloys was first used as a head material of magnetoresistive sensors. In 1986, a large MR effect was discovered for Fe/Cr multilayers [41, 42]. Because the change in the resistivity is more than 30%, the effect was called GMR. Soon after the discovery of GMR, the layered structure has been improved for application of GMR to magnetoresistive sensors. The structure is called spin valve structure [43, 44]. In 1995, an MR effect was observed in ferromagnetic tunnel junctions of Fe/Al-O/Fe junctions, in which electrical current flows perpendicular to the junction planes by a tunnel effect, and the MR effect is called TMR [45, 46]. Because the resistivity (resistance) of the tunnel junction is quite high, it is not easy to apply the TMR effect to magnetic sensors. Reduction of the resistance and increase of the MR ratio were required for such applications. Much progress in fabrication techniques of thin films and selection of suitable materials for tunnel junctions have made it possible to achieve the conditions for applications in the 2000s [47, 48].

As mentioned, the effect of MR is a coupled phenomenon of charge current of electrons and magnetism originated from spins of electrons. Therefore, the electronics based on the MR effect is called spintronics. In the next section, the representative phenomena of spintronics, GMR, and TMR are briefly explained, being followed by a section to explain an application of TMR to nonvolatile memory, magnetoresistive RAM (MRAM) and a brief introduction of applications of spintronics to semiconductor technology.

### 1.2.2 *Spintronics: GMR and TMR*

An electron has a charge $e = -1.6 \times 10^{-19}$ C and spin $S = \hbar/2$. Because spin is an angular momentum, it has two components, $S_z = \pm\hbar/2$. The electron charge is responsible for the electronics based mainly on silicon. On the other hand, the spin is responsible for magnetism, the most fundamental material being iron. Spintronics aims technological applications to utilize both charge and spin degrees of freedom of electrons on an equal footing. Here we briefly introduce the typical phenomena of GMR and TMR and their applications.

#### 1.2.2.1 GMR and TMR

GMR was first discovered for Fe/Cr/Fe trilayers. The resistivity ($\rho$) drops when an external magnetic field ($H$) is applied. The resistivity change is usually expressed by the MR ratio defined as $\rho(H = 0) - \rho(H \neq 0)/\rho(H = 0)$. The MR ratio was only a few percent in the experiment. Immediately after that, a large GMR has been observed for Fe/Cr multilayers. A schematic of the multilayer structure is depicted in Fig. 1.9a. It consists of alternate stacking of thin nonmagnetic and magnetic layers. The typical thickness of

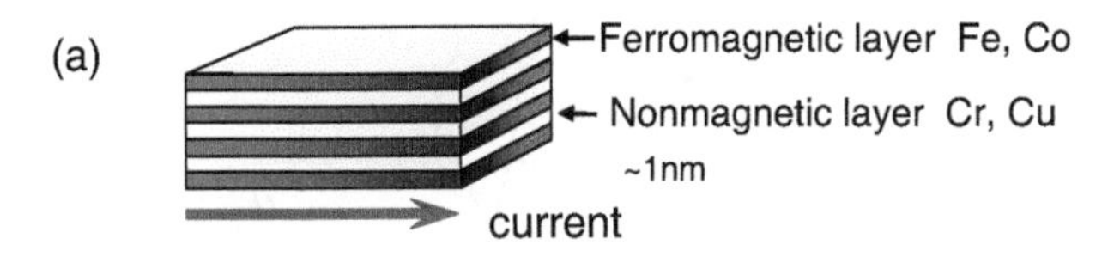


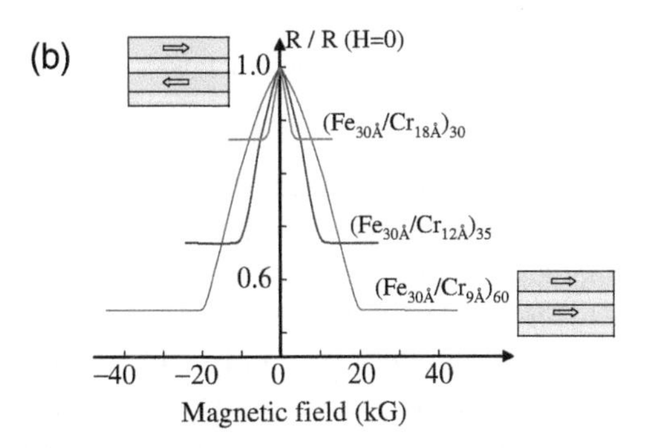


**Figure 1.9** (a) Schematic of magnetic multilayers and (b) experimental results of giant magnetoresistance (GMR) for Fe/Cr multilayers.

each layer is a few nanometers, and the number of thin layers is 10–60. The current flows in plane. The geometry of this type of GMR is called current-in-plane (CIP) GMR. There is another type of geometry in which current flows perpendicular to the layer planes. This type of GMR is called current-perpendicular-to-plane (CPP) GMR. Because the current is too tiny to observe in CPP GMR, microfabrication of samples or low temperature observation using superconducting electrodes is required [49, 50].

Figure 1.9b shows the observed results of resistivity change with an external magnetic field for Fe/Cr multilayers [42]. With increasing magnetic field, the resistivity decreases, and it becomes constant for a magnetic field higher than a certain value. The resistivity change is explained in the following way. When no magnetic field is applied, the magnetization of Fe layers aligns antiparallel. The antiparallel alignment is gradually changed to parallel alignment, and the resistivity decreases with a change in angle between magnetization on neighboring Fe layers. The resistivity decreases no more once the magnetization aligns in parallel completely.

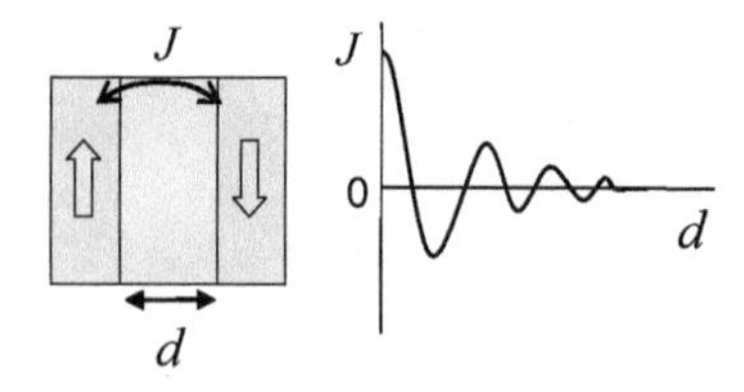


**Figure 1.10** Features of exchange coupling between magnetic layers separated by a nonmagnetic layer.

The antiparallel alignment of magnetizations at $H = 0$ is attributed to the existence of interlayer exchange coupling between magnetizations on neighboring Fe layers. Because the coupling is rather strong, a few Tesla of magnetic field is necessary to make the magnetization parallel. The antiparallel alignment of layer magnetization has been found to be intrinsic, and the alignment (parallel or antiparallel) varies with nonmagnetic layer thickness. Actually the interlayer exchange coupling oscillates with the nonmagnetic layer thickness, as shown in Fig. 1.10 [51, 52]. The character of the oscillation is similar to the well-known Rudermann–Kittel–Kasuya Yosida (RKKY) oscillation; however, the oscillation period is much longer than that of RKKY oscillation. The long oscillation period has been attributed to a discreteness of the layer thickness and to a quantum confinement of electrons within layers.

TMR is an MR effect that appears in ferromagnetic tunnel junctions made of ferromagnetic electrodes separated by an insulating barrier. A schematic of a tunnel junction is shown in Fig. 1.11a. The ferromagnetic electrodes are mainly made of iron, and the barrier materials are $Al_2O_3$ or MgO with thickness of a few nanometers. The current flows perpendicular to the layer plane: electrons transfer from one electrode to the other by tunneling. Therefore the resistance is rather high to measure.

The resistance change of the TMR effect with an external magnetic field is shown schematically in Fig. 1.11b. Without a magnetic field, the alignment of magnetization is parallel. With an increasing magnetic field, the magnetization on one of the electrodes flips and the magnetization alignment becomes antiparallel, resulting in an increase in the resistance. The magnetization of both electrodes would not flip simultaneously, because of a difference between

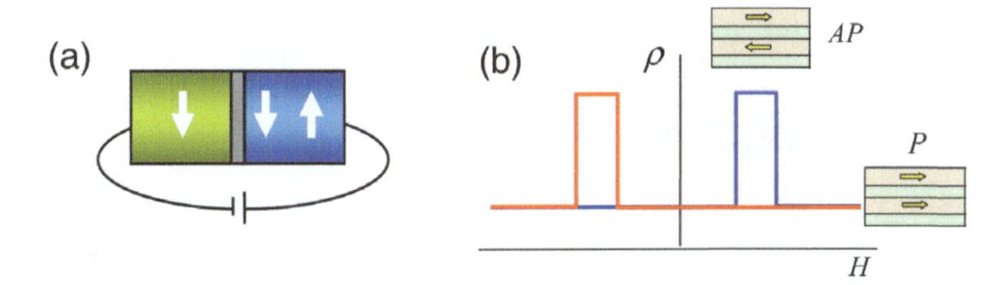


**Figure 1.11** (a) Schematic of a tunnel junction where two ferromagnetic layers are separated by a nonmagnetic insulator and (b) change in resistance with the external magnetic field depicted schematically for tunnel junctions.

magnetic anisotropy of the left and right electrodes. With a further increase in the magnetic field, the magnetization becomes parallel and the resistance decreases. The process is depicted by blue lines in Fig. 1.11b. After saturation of the magnetic field in the positive direction, the magnetic field is decreased to negative values. In this process, as shown by red lines in Fig. 1.11b, the magnetization change does not occur at the same values of the magnetic field shown by blue lines, but it occurs in the negative field because of the coercive force of the ferromagnetic electrodes. The MR ratio that characterizes the MR effect is defined to be

$$\mathrm{MR} = \frac{\rho_{\mathrm{AF}} - \rho_{\mathrm{F}}}{\rho_{\mathrm{AF}}}.$$

The definition is the same as that of the MR ratio of GMR because $\rho_{\mathrm{AF}} = \rho(H = 0)$ and $\rho_{\mathrm{F}} = \rho(H \neq 0)$ for GMR. An alternative definition in which the denominator is given by $\rho_{\mathrm{F}}$ is often used for experimental values.

The TMR is similar to CPP GMR, both of which are made of ferromagnetic trilayers; however, the ferromagnets are separated by an insulator in the TMR effect and by a nonmagnetic metal in the CPP GMR effect. The first observation of the TMR effect was reported for Co/Ge/Fe junctions where a semiconductor Ge was used to separate the ferromagnetic electrodes Co and Fe [53]. Unfortunately, the reproducibility was too poor to obtain stable results. Later a ferromagnetic junction made of Ni oxides was used to observe the TMR effect, though the MR ratio was only 2% [54]. In the 1990s a rather large TMR effect, about 30% of the MR ratio, was observed for

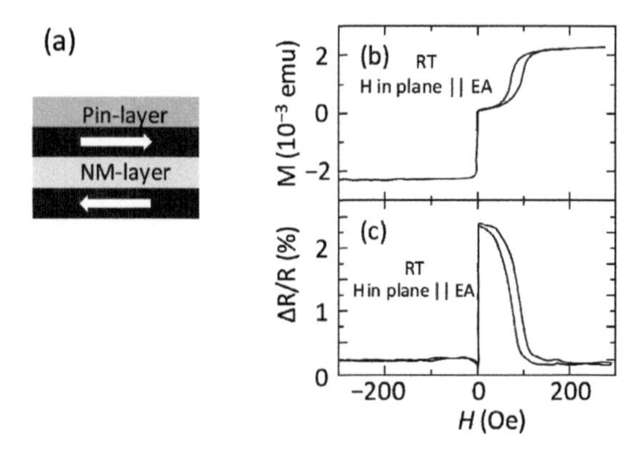


**Figure 1.12** (a) Schematic of the structure of spin valves where arrows indicate magnetizations of ferromagnetic layers and the top layer is an antiferromagnetic. (b) Observed results of the *M*–*H* curve for a spin valve and (c) corresponding resistance change. Reprinted (figure) with permission from Ref. [43], Copyright (1991) by the American Physical Society.

Fe/Al-O/Fe in which Al-O is amorphous. By successful fabrication of tunnel junctions with an epitaxial lattice structure of a MgO barrier, a large MR ratio of TMR has been realized. The success opened a gateway to technological applications of the effect of TMR.

#### 1.2.2.2 Spintronics applications of GMR and TMR

Soon after the discovery of GMR, a basic junction structure for application to sensors was proposed. Because the magnetic field to alter the magnetization alignment in magnetic multilayers such as Fe/Cr is too strong to be used for sensors, a specific structure of junctions called spin valve structure was adopted. The junction is made of ferromagnetic trilayer attached with an antiferromagnetic layer, as shown in Fig. 1.12a. The ferromagnetic layers are usually Fe–Ni alloys (permalloys) with very weak anisotropy, which are separated by a nonmagnetic Cu layer. The magnetization of one of the ferromagnetic layers is pinned by the layer-by-layer antiferromagnetism of the antiferromagnet, for example, FeMn. Because the antiferromagnet is insensitive to the external magnetic field, the ferromagnetic magnetization attached to the antiferromagnet is also nonactive to the external magnetic field. Only the magnetization of

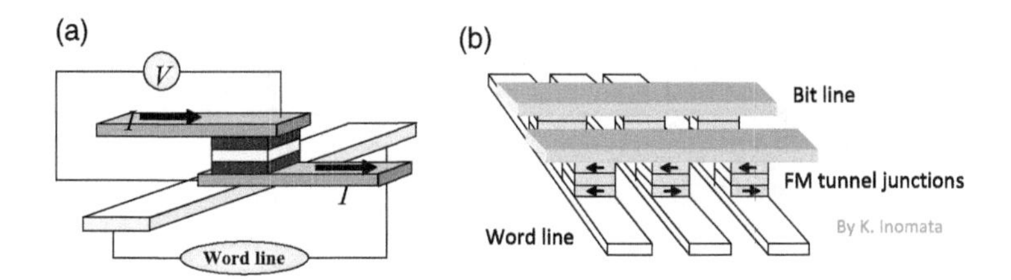


**Figure 1.13** (a) Schematic structure of a unit memory cell of magnetoresistive random access memory (MRAM) and (b) structure of MRAM.

the other ferromagnet changes its direction by the external magnetic field of a few Oe. Because the current to produce the MR effect runs in plane, the MR ratio of the spin valve structure is very small; however, an important quantity for sensors is the ratio of $\Delta\rho/\Delta H$, not the MR ratio itself. The ratio is sufficiently large, as shown in Fig. 1.12c. The spin valve MR sensors have already been commercialized as magnetic heads for HDDs.

TMR is also applicable to magnetic sensors because the basic junction structure is the same as that of the spin valve. However, because the tunneling effect is responsible for the current, the resistance of the junction is too high to apply it to sensors. Therefore, it took several years to lower the resistance by making the tunnel barrier sufficiently thin.

TMR has been used to realize nonvolatile memory for low energy consumption. The parallel and antiparallel alignments of the tunnel junctions correspond to bit memory 0 and 1, which can be identified by measuring the tunnel resistance. The basic structure of memory cells and array of cells are shown in Figs. 1.13a and 1.13b, respectively. The structure consists of ferromagnetic tunnel junctions being attached with bit and word lines. Applying an electric current to one of each of the bit and word lines, an effective magnetic field is induced at a specified tunnel junction to reverse the magnetization of the junction, as shown in Fig. 1.14a. This is writing the memory onto the tunnel junction. Reading the memory is done by applying a voltage between the corresponding bit and word lines.

Unfortunately, this type of memory cell has a shortcoming that the size of the cell cannot be sufficiently small for high-density memory because the writing current to realize magnetic field switching increases in proportion to 1/(cell volume). In spite of this

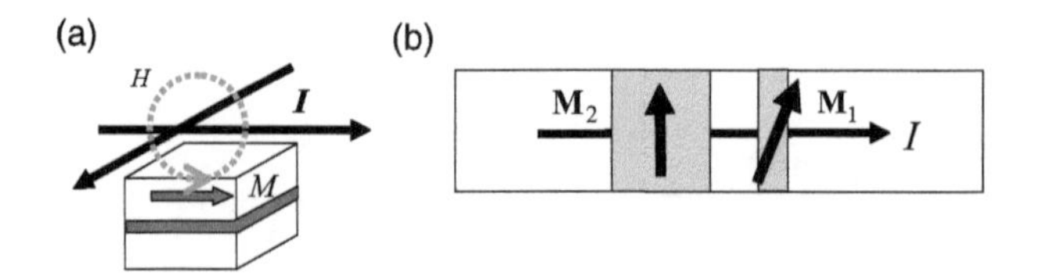


**Figure 1.14** Image of current-induced magnetization switching.

shortcoming, a low-density memory cell has been commercialized for specific purposes. To decrease the cell size, a different method to reverse the magnetization has recently been adopted. The method is called current-induced magnetization switching [55, 56]. The method utilizes a phenomenon that a spin-polarized current through junctions made of two ferromagnets with noncollinear magnetization, as shown in Fig. 1.14b, produces a torque to rotate the magnetization of one of two ferromagnets. The torque is called spin transfer torque (STT). Nonvolatile memories using STT have been developed and will be commercialized soon.

Details of the mechanism and material dependence of GMR, TMR, and STT, etc., will be explained in Chapter 5.

### 1.2.2.3 Proposal of a spin FET

The junctions explained earlier consist of two ferromagnets separated by either nonmagnetic metals or nonmagnetic insulators. Junctions including semiconductors have also been proposed. The basic structure of such junctions is shown in Fig. 1.15a. Left and

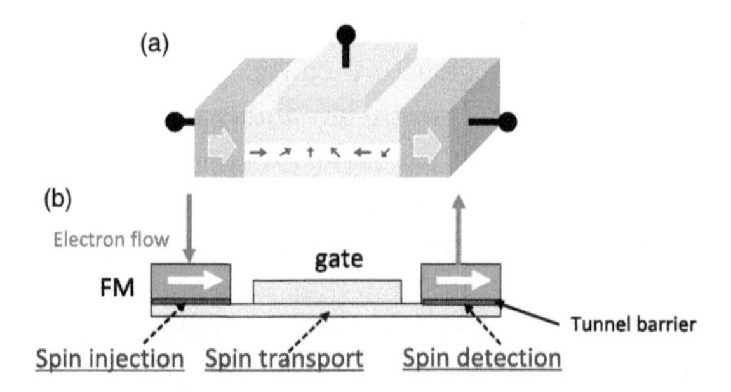


**Figure 1.15** (a) Schematic image of a spin field-effect transistor (spin FET) with a gate terminal and (b) ingredients necessary for a spin FET.

**Table 1.1** Comparison of various features in phenomena of GMR, TMR and spin FET

| Phenomena | GMR | TMR | Spin FET |
|---|---|---|---|
| Structure | Multilayers<br>Trilayers | Tunnel<br>Junctions | Lateral tunnel<br>Junctions |
| Current<br>Direction | In-plane<br>Out-of-plane | Out-of-plane | In-plane |
| Electrode<br>Materials | Fe, Co, Ni<br>Alloys | Fe, FeCoB<br>Half-metal | Fe, FeCoB<br>Fe-Ni, (GaMnAs) |
| Spacer<br>Materials | Metals:<br>Cr, Cu, Au | Insulators<br>MgO, Al-O | Semiconductors<br>GaAs, Si, Ge |
| Related<br>Phenomena | ExC<br>CIMS | CIMS<br>Torque<br>Oscillation<br>PMA | Spin injections<br>Spin<br>Accumulation<br>SHE |
| Applications | Spin valve sensor<br>(MRAM) | MRAM<br>MR sensor | |

right ferromagnetic metals are the source and the drain, respectively. A gate terminal is attached to control the current. Elements of the semiconductor may be Si, Ge, GaAs, etc. This type of junction is call spin FET because the current is spin polarized and could be controlled also by a magnetic field.

The structure of a spin FET is basically the same as that of a unipolar Transistor, except that the nonmagnetic electrodes are replaced with ferromagnetic ones. Because of the additional freedom of spin, a spin FET might be used not only for logic circuits but also for memory devices. This is because there are two alignments, parallel and antiparallel, of magnetization of the left and right electrodes, which could be used for memory bits. The magnetization alignment could be controlled by STT, and thus the spin FET has large potential applicability as a novel device. However, there are three basic issues to be solved to realize spin FETs: spin injection from the ferromagnet to the semiconductor, spin control in the semiconductor, and spin detection at the other ferromagnet. Although several concrete structures of such junctions have been proposed so far, no spin FET has yet been realized. Details of spin injection and spin FETs will also be given in Chapter 5. Finally we summarize the similarity and dissimilarity between GMR, TMR and spin FET in Table 1.1.

## 1.3 Graphene Junctions

The first research on the anomalous properties of the single-layer graphene sheet was actually performed by using an FET structure [8]. Observation of high mobility of graphene was later reported by many groups using FET structures. Nonlocal measurements have been used to confirm the long spin relaxation time in four-terminal ferromagnetic graphene junctions. MR in two-terminal ferromagnetic junctions has also been observed, as described later. More details will be given in Chapter 4.

### 1.3.1 *Proposals of Novel Graphene Devices*

Using first-principles calculation, Son et al. [57] showed that zigzag-edge GNRs show ferromagnetism at the edges when an electric field is applied across the zigzag edge. More interestingly, the ferromagnetism is half-metallic and is related to the edge state with a flat energy dispersion. Furthermore, the magnetic state can be controlled by an external magnetic field. Because only one of up- and down-spin states exists at the Fermi level in the half-metallic state, the states have large potential applicability for spintronics devices. In addition, the control of magnetism of the electric field is a coupling phenomenon of the spin and charge of electrons and therefore one of the fundamental issues of multiferroics.

A very large MR effect has been predicted by Kim and Kim [58] by using the half-metallic edge states of GNRs. They proposed a junction structure in which the left and right ferromagnetic electrodes are attached to a zigzag-edge GNR. In this junction, the zigzag edges are half-metallic and conductive. The spin polarization of the zigzag edges is controlled by the magnetization of the electrodes. When the magnetizations of the left and right electrodes are parallel, the spin polarization of the edges is either positive or negative and the edge states are metallic. When the magnetizations are antiparallel, the sign of the spin polarization near the left electrode is opposite to that of the right electrode. Because the edge state is half-metallic, no current flows when the sign of spin polarization of the left- and right-side regions of the GNR is different. This gives rise to a large

MR effect. The MR effect using edge magnetism was proposed also by Dubois et al. [59] for graphene/BN ribbon lateral junctions.

Rycerz et al. [60] have proposed a novel concept of a valley filter and a valley valve of electric currents, noting that the two Dirac points (two valleys) in the Brillouin zone give electrons a new internal degree of freedom. They considered a zigzag-edge nanoribbon, in which the zigzag edges are along the current direction, with a narrow region of a constriction. They showed that electrons in one of two valleys can transmit the region of constriction but electrons in the other valley cannot transmit the region of constriction. The degree of valley is an analog of the spin degree of freedom of electrons and thus gives rise to a novel effect. Quite recently, electronics that utilizes the degree of valley has been proposed and named *valleytronics*.

### 1.3.2 *Graphene Magnetoresistive Junctions*

Although several results of MR effects have been reported for graphene ferromagnetic junctions, the MR ratio would not be large enough for technological applications [61, 62]. A theoretical study also reported that the MR effect in two-terminal graphene junctions with ferromagnetic metal electrodes cannot be large [63]. Recently, however, a relatively large MR effect has been reported in Fe/graphene/Fe lateral junctions with an Al-O insulator between Fe and graphene [64]. The model used in the theoretical calculation could be too simple to argue the MR effect for junctions with a realistic electronic structure. It is well known that the large MR effects of GMR and TMR are closely related to realistic electronic structures of both ferromagnetic electrodes and nonmagnetic spacers, that is, nonmagnetic metals for GMR and insulators for TMR. Because the electronic state of graphene is much different from that of metals and insulators, as shown in Fig. 1.5, it is quite interesting to study how the characteristic electronic state of graphene affects the MR effect of junctions with realistic ferromagnetic electrodes.

We have so far reported theoretical results of the MR effect of junctions with ferromagnetic electrodes by using a simple model, realistic body-centered cubic (bcc) Fe/graphene/bcc Fe junctions, and realistic face-centered cubic (fcc) Ni/graphene/fcc Ni junctions and have shown that a large MR effect is realized for realistic junctions [65–67]. It is the purpose of this book to clarify the mechanism of MR, the role of the Dirac point, and conditions for high MR ratios in order to realize magnetoresistive graphene junctions. Because the electronic structures of both graphene and electrodes are important, we present several results reported for the electronic structure of graphene/metal junctions in Chapter 4, and our theoretical results of MR effects in graphene junctions with ferromagnetic electrodes will be presented in Chapter 6. Recent relevant experimental works will also be mentioned in Chapter 4. Because the topic is rapidly developing, we would not able to cover the whole results reported, but some of the details of the remaining issues will, however, be commented on in the final chapter.

## Notes

**Conductivity, drift velocity, and mobility:** In the classical Newtonian mechanics, the motion of an electron (charge $e$ and effective mass $m$) under an electric field $E$ is governed by the equation

$$m\frac{dv}{dt} + \eta v = eE,$$

where $v$ is the velocity of the electron and the second term on the left-hand side indicates resistance proportional to the velocity with a constant $\eta$. In the stational state $dv/dt = 0$, and then $v = eE/\eta$. The electrical current is defined as $J = nev$, where $n$ is the density of conduction electrons, resulting in $J = e^2nE/\eta$. Here we introduce the relaxation time $\tau = m/\eta$, and we obtain

$$J = \frac{e^2 n\tau}{m}E \equiv \sigma E.$$

The definition of the relaxation time may be understood in the following way. When the electric field is cut off at a certain time, the velocity $v$ of the electron decays to zero, which is expressed by the equation $mdv/dt + \eta v = 0$. Integrating the equation, we obtain $v = \exp(-\eta t/m)$, which is rewritten as $v = \exp(-t/\tau)$ using the relaxation time $\tau = m/\eta$.

In the derivation of conductivity, an electrical velocity $v$ was introduced. The velocity is not the actual velocity of electrons in metals and semiconductors. In these materials, electrons move fast even at zero electric field. Directions of motion of many electrons are random, and the average velocity tends to be zero. Under an external electric field, the average velocity is nonzero and contributes the electric current. Because the average velocity is realized by a balance of the acceleration by the electric field and scattering of electrons by impurities, phonons, etc., in materials, the average velocity is called drift velocity $\upsilon_D$. The velocity $\upsilon$ introduced before is the drift velocity.

The mobility indicates how electrons are accelerated by the external electric field $E$ to reach a stational velocity $\upsilon$, and thus is defined as $\upsilon = \mu eE$. The relation $\upsilon = eE/\eta$ given before and the definition of the relaxation time give $\mu = e\tau/m$.

**Half-metals:** The density of states of metallic ferromagnets is usually spin dependent. When the density of states of either an $\uparrow$ or a $\downarrow$ spin state is zero at the Fermi level, and one of two spin states is metallic and the other is insulating, the metals are referred to as half-metals. The spin polarization $P$ of these half-metals is 100%, and therefore half-metals have potential applicability as magnetoresistive devices. Many oxides, including $CrO_2$, $Fe_3O_4$, and perovskite $LaSrMnO_3$, have been shown to be half-metallic by using first-principles band calculations [68–70]. The first theoretical prediction for half-metallicity was done for Heusler alloys [68], which contain transition metal (TM) elements. Recently, it has been shown using the first-principles method that diluted magnetic semiconductors, such as(GaMn)As, may also be half-metallic. In experiments in which point contacts and tunnel junctions were used to measure spin polarization, lower values than 100% were obtained for $P$ (e.g., 90% for $CrO_2$, 70%–85% for $LaSrMnO_3$, and 60% for Heusler alloys). Recently, relatively high MR ratios have been observed in ferromagnetic tunnel junctions with Heusler alloys, suggesting that the value of $P$ is about 86%.

## References

1. J. J. Thomson, *Phil. Mag.*, **44**, 293 (1897).
2. W. Gerlach and O. Stern, *Z. Phys.*, **9**, 349 (1922).

3. S. Iijima, *Nature*, **354**, 56 (1991).
4. H. W. Kroto, J. R. Heath, S. C. O'Brien, R. F. Curl, and R. E. Smalley, *Nature*, **318**, 162 (1985).
5. E. Osawa, *Kagaku*, **25**, 854 (1970).
6. P. M. Ajayan and S. Iijima, *Nature*, **361**, 333 (1993).
7. P. R. Wallace, *Phys. Rev.*, **71**, 622 (1947).
8. K. S. Novoselov, A. K. Geim, S. V. Morozov, D. Jiang, Y. Zhang, S. V. Dubonos, I. V. Grigorieva, and A. A. Firsov, *Science*, **306**, 666 (2004).
9. K. S. Novoselov, A. K. Geim, S. V. Morozov, D. Jiang, M. I. Katsnelson, I. V. Grigorieva, S. V. Dubonos, and A. A. Firsov, *Nature*, **438**, 197 (2005).
10. Y. Zhang, Y.-W. Tan, H. L. Stormer, and P. Kim, *Nature*, **438**, 201 (2005).
11. A. K. Geim and K. S. Novoselov, *Nat. Mater.*, **6**, 183 (2007).
12. A. H. Castro Neto, F. Guinea, N. M. R. Peres, K. S. Novoselov, and A. K. Geim, *Rev. Mod. Phys.*, **81**, 109 (2009).
13. Y. H. Wu, T. Yu, and Z. X. Shen, *J. Appl. Phys.*, **108**, 072301 (2010).
14. S. Das Sarma, S. Adam, E. H. Hwang, and E. Rossi, *Rev. Mod. Phys.*, **83**, 407 (2011).
15. A. K. Geim, *Rev. Mod. Phys.*, **83**, 851 (2011).
16. K. S. Novoselov, *Rev. Mod. Phys.*, **83**, 837 (2011).
17. J. Haas, W. A. de Heer, and E. H. Conrad, *J. Phys. Condens. Matter*, **20**, 323202 (2008).
18. K. S. Kim, Y. Zhao, H. Jang, S. Y. Lee, J. M. Kim, K. S. Kim, J.-H. Ahn, P. Kim, J.-Y. Choi, and B. H. Hong, *Nature*, **457**, 706 (2009).
19. A. J. van Bommel, J. E. Crombeen, and A. van Tooren, *Surf. Sci.*, **48**, 463 (1975).
20. C. Oshima and A. Nagashima, *J. Phys. Condens. Matter*, **9**, 1 (1997).
21. N. Tombros, C. Jozsa, M. Popinciuc, H. T. Jonkman, and J. van Wees, *Nature*, **448**, 571 (2007).
22. S. V. Novoselov, K. S. Novoselov, M. I. Katsnelson, F. Schedin, D. C. Elias, J. A. Jaszczak, and A. K. Geim, *Phys. Rev. Lett.*, **100**, 016602 (2008).
23. K. I. Bolotin, K. J. Sikes, Z. Jiang, M. Klima, G. Fundenberg, J. Hone, P. Kim, and H. L. Stormer, *Solid State Commun.*, **146**, 351 (2008).
24. B. Bennett, R. Magno, J. B. Boos, W. Kruppa, and M. G. Ancona, *Solid-State Electron.*, **49**, 1875 (2005).
25. E. H. Hall, *Am. J. Math.*, **2**, 287 (1879).
26. E. H. Hall, *Philos. Mag.*, **19**, 301 (1880), A. Kundt, *Annalen Phys. Chemie*, **49**, 257 (1893).

27. M. I. D'yakonov and V. I. Perel, *ZhETF Pis. Red.*, **13**, 657 (1971).
28. J. E. Hirsh, *Phys. Rev. Lett.*, **83**, 1834 (1999).
29. S. Murakami, N. Nagaosa, and S. C. Zhang, *Science*, **301**, 1348 (2003).
30. J. Sinova, D. Culcer, Q. Niu, N. A. Sinitsyn, T. Jungwirth, and A. H. MacDonald, *Phys. Rev. Lett.*, **92**, 126603 (2004).
31. J. Inoue and H. Ohno, *Science*, **309**, 5743 (2005).
32. C. L. Kane and E. J. Mele, *Phys. Rev. Lett.*, **95**, 226801 (2006).
33. M. Z. Hasan and C. L. Kane, *Rev. Mod. Phys.*, **82**, 3045 (2010).
34. C. Lee, X. Wei, J. W. Kysar, J. Hone, *Science*, **321**, 385 (2008).
35. R. R. Nair, P. Blake, A. N. Grigorenko, K. S. Novoselov, T. J. Booth, T. Stauber, N. M. R. Peres, and A. K. Geim, *Science*, **320**, 1308 (2008).
36. A. B. Kuzmenko, E. van Heumen, F. Carbone, and D. van der Marel, *Phys. Rev. Lett.*, **100**, 117401 (2008).
37. W. Bao, F. Miao, Z. Chen, H. Zhang, W. Jang, C. Dames, and C. N. Lau, *Nat. Nanotech.*, **4**, 562 (2009).
38. A. A. Balandin, S. Ghosh, W. Bao, I. Calizo, D. Teweldebrhan, F. Miao, and C. N. Lau, *Nano Lett.*, **8**, 902 (2009).
39. Y.-M. Lin, C. Dimitrakopoulos, K. A. Jenkins, D. B. Farmer, H.-Y. Chiu, A. Grill, and Ph Avouris, *Science*, **327**, 662 (2010).
40. S. Bae, H. Kim, Y. Lee, X. Xu, J.-S. Park, Y. Zheng, J. Balakrishnan, T. Lei, H. R. Kim, Y. I. Song, Y.-J. Kim, K. S. Kim, B. Özyilmaz, J.-H. Ahn, B. H. Hong, and S. Iijima, *Nat. Nanotech.*, **5**, 574 (2010).
41. P. Grüberg, R. Schreiber, Y. Pang, M. B. Brodsky, and H. Sowers, *Phys. Rev. Lett.*, **57**, 2442 (1986).
42. M. N. Baibich, J. M. Broto, A. Fert, F. Nguyen Van Dau, F. Petroff, P. Etienna, G. Creuzet, A. Friederich, and J. Chazelas, *Phys. Rev. Lett.*, **61**, 2472 (1988).
43. B. Dieny, V. S. Sperious, S. S. P. Parkin, B. A. Gurney, D. R. Wilhoit, and D. Mauri, *Phys. Rev. B*, **43**, 1297 (1991).
44. B. Dieny, V. S. Speriosu, S. Metin, S. S. P. Parkin, B. A. Gurney, P. Baumgart, and D. R. Wilhoit, *J. Appl. Phys.*, **69**, 4774 (1991).
45. T. Miyazaki, and N. Tezuka, *J. Magn. Magn. Mater.*, **151**, 403–410 (1995).
46. J. S. Moodera, L. R. Kinder, T. M. Wong, and R. Meservey, *Phys. Rev. Lett.*, **74**, 3273 (1995).
47. S. S. P. Parkin, C. Kaiser, A. Panchula, P. M. Rice,B. Hughes, M. Samant, and S.-H. Yang, *Nat. Mater.*, **3**, 862 (2004).
48. S. Yuasa, T. Nagahama, A. Fukushima, Y. Suzuki, and K. Ando, *Nat. Mater.*, **3**, 868–871 (2004).

49. W. P. Pratt, Jr., S. F. Lee, J. M. Slaughter, R. Loloee, P. A. Schroeder, and J. Bass, *Phys. Rev. Lett.*, **66**, (1991).
50. M. A. M. Gijs, S. K. J. Lenczowski, and J. B. Giesbers, *Phys. Rev. Lett.*, **70**, 3343 (1993).
51. S. S. P. Parkin, N. More, and K. P. Roche, *Phys. Rev. Lett.*, **64**, 2304 (1990).
52. S. S. P. Parkin, *Phys. Rev. Lett.*, **67**, 3598 (1991).
53. M. Julliere, *Phys. Lett.*, **54A**, 225 (1975).
54. S. Maekawa and U. Gäfvert, *IEEE Trans. Magn.*, **18**, 707 (1982).
55. L. Berger, *J. Appl. Phys.*, **71**, 2721 (1992).
56. J. C. Slonczewski, *J. Magn. Magn. Mater.*, **159**, L1 (1996).
57. Y.-W. Son, M. L. Cohen, and S. G. Louie, *Nature*, **444**, 347 (2006).
58. W. Y. Kim and K. S. Kim, *Nat. Nanotech.*, **3**, 408 (2008).
59. S. M.-M. Dubois, X. Declerck, J.-C. Charlier, and M. C. Payne, *ACS Nano*, **7**, 4578 (2013).
60. A. Rycerz J. Tworzydlo, and C. W. J. Beenakker, *Nat. Phys.*, **3**, 172 (2007).
61. E. W. Hill, A. K. Geim, K. Novoselov, F. Schedin, and P. Blake, *IEEE Trans. Magn.*, **42**, 2694 (2006).
62. M. Nishioka and A. M. Goldman, *Appl. Phys. Lett.*, **90**, 252505 (2007).
63. L. Brey and H. A. Fertig, *Phys. Rev. B*, **76**, 205435 (2007).
64. B. Dlubak, M.-B. Martin, C. Deranlot, B. Servet, S. Xavier, R. Mattana, M. Sprinkle, C. Berger, W. A. De Heer, F. Petroff, A. Anane, and A. Fert, *Nat. Phys.*, **10**, 557 (2012).
65. A. Yamamura, S. Honda, J. Inoue, and H. Itoh, *J. Magn. Soc. Jpn*, **34**, 34 (2010).
66. S. Honda, A. Yamamura, T. Hiraiwa, R. Sato, J. Inoue, and H. Itoh, *Phys. Rev. B*, **82**, 033402 (2010).
67. T. Hiraiwa, R. Sato, A. Yamamura, J. Inoue, S. Honda, and H. Itoh, *IEEE Trans. Mag.*, **47**, 2743 (2011).
68. R. A. de Groot and K. H. J. Buschow, *J. Magn. Magn. Mater.*, **54–57**, 1377 (1986).
69. K. Schwarz, *J. Phys. F: Metal. Phys.*, **16**, L211 (1986).
70. W. E. Pickett and D. J. Singh, *Phys. Rev. B*, **53**, 1146 (1996).

# Chapter 2

# Basic Features of Graphene

## 2.1 Electronic Structure

Graphene is a monolayer crystal of carbon. Graphene was first synthesized from bulk graphite, which is regarded as stacked graphene, as illustrated in Fig. 2.1, by the exfoliation method [1]. In early times, Wallace [2] proposed that such a two-dimensional crystal has peculiar electronic states that are linear dispersions forming a cone. In this chapter, we provide a simple method dealing with electronic states on monolayer and bilayer graphene.

## 2.2 Electronic States of the $\pi$-Band in Graphene

Graphene is a two-dimensional carbon crystal with a honeycomb lattice. A carbon atom has six electrons $(1s)^2(2s)^2(2p)^2$. One $2s$- and two $2p$-electrons form $sp_2$-orbitals (aligned in the $xy$ plane in Fig. 2.2a), which play a role of bonding between neighboring carbon atoms (denoted by a stick in Fig. 2.1). The remaining $2p$-orbitals in graphene, that is, $p_z$-orbitals, can overlap with each other to move nearly freely in two-dimensional space, as illustrated in Fig. 2.2. It is important to understand the nature of the $\pi$-band since

*Graphene in Spintronics: Fundamentals and Applications*
Jun-ichiro Inoue, Ai Yamakage, and Shuta Honda

ISBN 978-981-4669-56-6 (Hardcover), 978-981-4669-57-3 (eBook)
www.panstanford.com

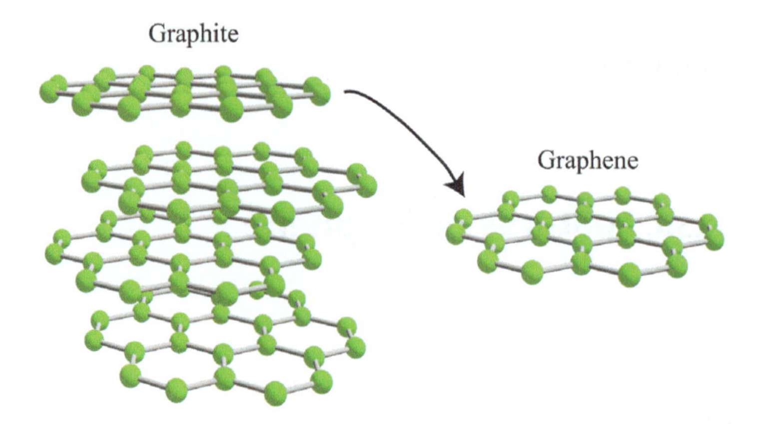


**Figure 2.1** Graphene from graphite.

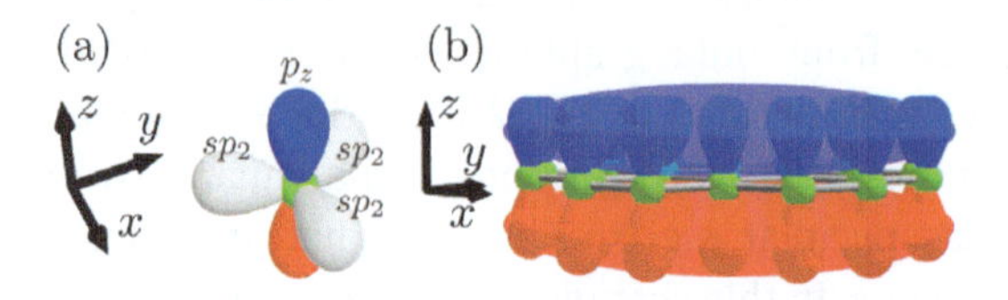


**Figure 2.2** $s$- and $p$-orbitals in a carbon atom (a) and $\pi$-orbital in graphene (b).

it dominates thermodynamic and transport properties in graphene. In the following section, we study the electronic states in graphene based on the tight-binding model.

### 2.2.1 *Graphene in Real and Reciprocal Spaces*

Before discussing the electronic states, we shall study the crystal structure of graphene in real and reciprocal spaces, quantitatively. The crystal structure of graphene is illustrated in Fig. 2.3a. A honeycomb lattice consists of A and B sublattices. Neighboring A and B sites form unit cells, which are located at $\boldsymbol{n} = n_1\boldsymbol{a}_1 + n_2\boldsymbol{a}_2$ with $n_1$ and $n_2$ being integers, $\boldsymbol{a}_1$ and $\boldsymbol{a}_2$ being the primitive translational vectors defined by

$$\boldsymbol{a}_1 = a_\text{g}\left(\frac{3}{2}, \frac{\sqrt{3}}{2}\right), \quad \boldsymbol{a}_2 = a_\text{g}\left(\frac{3}{2}, -\frac{\sqrt{3}}{2}\right), \tag{2.1}$$

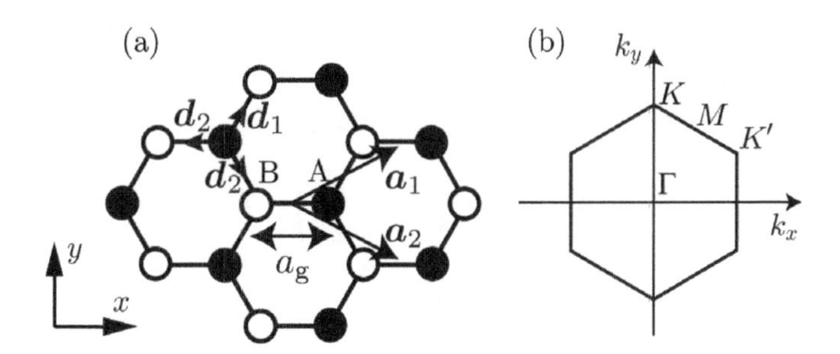


**Figure 2.3** Crystal structure of graphene (a) and the corresponding Brillouin zone (b). The closed (open) circles denote A (B) sites. $\boldsymbol{a}_1$ and $\boldsymbol{a}_2$ are the primitive translational vectors. $a_g$ is the lattice constant. The nearest-neighbor bondings are denoted by $\boldsymbol{d}_1$, $\boldsymbol{d}_2$, and $\boldsymbol{d}_3$.

with $a_g = 2.46$ Å the lattice constant of graphene. The corresponding reciprocal lattice vectors are given by

$$\boldsymbol{b}_1 = \frac{2\pi}{a_g}\left(\frac{1}{3}, \frac{1}{\sqrt{3}}\right), \ \boldsymbol{b}_2 = \frac{2\pi}{a_g}\left(\frac{1}{3}, -\frac{1}{\sqrt{3}}\right). \tag{2.2}$$

The Brillouin zone is also hexagonal, as shown in Fig. 2.3b. There are high symmetric points at

$$K\left(0, \frac{4\pi}{3\sqrt{3}a_g}\right), \ K'\left(0, -\frac{4\pi}{3\sqrt{3}a_g}\right), \ M\left(\frac{2\pi}{3a_g}, 0\right), \tag{2.3}$$

and their equivalent points.

### 2.2.2 *Tight-Binding Model of Graphene*

We first derive electronic states of the $\pi$-band in graphene, depending on the tight-binding model only with the nearest-neighbor hopping $-\gamma$ [2]. The Hamiltonian reads

$$H = -\gamma \sum_{\langle ij \rangle} c_i^\dagger c_j, \tag{2.4}$$

where $c_i$ and $c_i^\dagger$ denote annihilation and creation operators of an electron at site $i$, and the summation runs over the nearest-neighbor (A and B) sites. The value of the nearest-neighbor hopping $\gamma$ is $\gamma = 2.8$eV. More explicitly, the Hamiltonian is written as

$$H = -\gamma \sum_{\boldsymbol{n}} \sum_{\mu=1}^{3} c_{\boldsymbol{n},\mathrm{A}}^\dagger c_{\boldsymbol{n}+\boldsymbol{d}_\mu,\mathrm{B}} + \mathrm{h.c.}, \tag{2.5}$$

where $c_{n\sigma}$ is an electron annihilation operator at $\sigma = \mathrm{A}$, B sublattice in a unit cell located at $\boldsymbol{n}$. $\boldsymbol{d}_\mu$ represents the nearest-neighbor bonding shown in Fig. 2.3a, which is defined by

$$\boldsymbol{d}_1 = a_\mathrm{g}\left(\frac{1}{2}, \frac{\sqrt{3}}{2}\right), \ \boldsymbol{d}_2 = a_\mathrm{g}\left(\frac{1}{2}, -\frac{\sqrt{3}}{2}\right), \ \boldsymbol{d}_3 = a_\mathrm{g}\left(-1, 0\right). \quad (2.6)$$

Introducing the Fourier transform:

$$c_{\boldsymbol{R}\sigma} = \frac{1}{\sqrt{N}}\sum_{\boldsymbol{k}} e^{i\boldsymbol{k}\cdot\boldsymbol{R}} c_\sigma(\boldsymbol{k}), \ \frac{1}{N}\sum_{\boldsymbol{n}} e^{i\boldsymbol{k}\cdot\boldsymbol{n}} = \delta_{\boldsymbol{k},\boldsymbol{0}}, \quad (2.7)$$

with $N$ the number of the sites, the Hamiltonian is expressed as

$$H = \sum_{\boldsymbol{k}} \left(c_\mathrm{A}^\dagger(\boldsymbol{k}) \ c_\mathrm{B}^\dagger(\boldsymbol{k})\right) H(\boldsymbol{k}) \begin{pmatrix} c_\mathrm{A}(\boldsymbol{k}) \\ c_\mathrm{B}(\boldsymbol{k}) \end{pmatrix}, \quad (2.8)$$

$$H(\boldsymbol{k}) = \begin{pmatrix} 0 & H_\mathrm{AB}(\boldsymbol{k}) \\ H_\mathrm{AB}^*(\boldsymbol{k}) & 0 \end{pmatrix}, \quad (2.9)$$

$$H_\mathrm{AB}(\boldsymbol{k}) = -\gamma \sum_{\mu=1}^{3} e^{i\boldsymbol{k}\cdot\boldsymbol{d}_\mu} = -\gamma\left(e^{-ik_x a_\mathrm{g}} + 2e^{ik_x a_\mathrm{g}/2}\cos\frac{\sqrt{3}k_y a_\mathrm{g}}{2}\right). \quad (2.10)$$

Diagonalizing $H(\boldsymbol{k})$, one obtains the energy dispersion $\pm E(\boldsymbol{k})$ given by

$$E(\boldsymbol{k}) = \gamma\sqrt{1 + 4\cos\frac{\sqrt{3}k_y a_\mathrm{g}}{2}\left(\cos\frac{3k_x a_\mathrm{g}}{2} + \cos\frac{\sqrt{3}k_y a_\mathrm{g}}{2}\right)}. \quad (2.11)$$

The obtained energy band and density of states are shown in Fig. 2.4a–c. Interestingly, the band gap closes at the $K$ and $K'$ points: The low-lying excitations have linear dispersions forming a cone around the $K$ and $K'$ points, as shown in Fig. 2.4b, and the corresponding density of states is proportional to $E$ and vanishes at $E = 0$ (see Fig. 2.4d). These peculiar electronic states crucially affect the low-temperature and low-energy physics of graphene since the Fermi energy is given by $E = 0$. In the following sections, we focus on the low-energy electronic states.

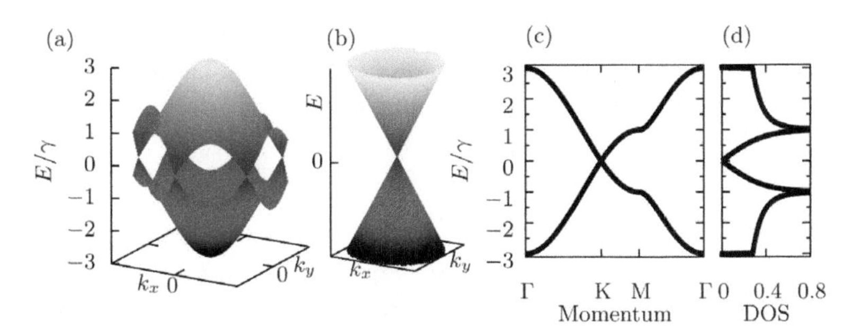


**Figure 2.4** The $\pi$-band in graphene over the whole Brillouin zone (a) and in the vicinity of the $K$ point (b). (c) The $\pi$-band along the symmetric lines. (d) The corresponding density of states.

## 2.2.3 *Dirac Fermions as Low-Lying Excitations*

As shown above, the band gap closes at two points, the $K$ and $K'$ points. Here we elucidate the low-energy electronic states in the vicinity of these points. The Hamiltonian matrix $H(\boldsymbol{k})$ is expanded with respect to $\boldsymbol{k}$ in the vicinity of the $K$ and $K'$ points as

$$H_K(\boldsymbol{k}) = v_g \begin{pmatrix} 0 & ik_x + k_y \\ -ik_x + k_y & 0 \end{pmatrix} = v_g(k_y\sigma_x - k_x\sigma_y), \qquad (2.12)$$

$$H_{K'}(\boldsymbol{k}) = v_g \begin{pmatrix} 0 & ik_x - k_y \\ -ik_x - k_y & 0 \end{pmatrix} = v_g(k_y\sigma_x + k_x\sigma_y), \qquad (2.13)$$

up to the first order, where the Fermi velocity is defined by $v_g = 3\gamma a_g/2$, which is approximately one three-hundredths of the light velocity. The shape of energy bands is a cone near $K$ and $K'$ points (see Fig. 2.4b) as

$$E(\boldsymbol{k}) = \pm v_g k = \pm v_g\sqrt{k_x^2 + k_y^2}, \qquad (2.14)$$

and the corresponding eigenvectors are

$$\boldsymbol{u}(\boldsymbol{k}) = \frac{1}{\sqrt{2}}\begin{pmatrix} 1 \\ \mp i e^{i\phi_{\boldsymbol{k}}} \end{pmatrix}, \qquad (2.15)$$

with $\tan\phi_{\boldsymbol{k}} = k_y/k_x$. This Hamiltonian is equivalent to that of massless Dirac fermions in two dimensions, which appear in high-energy physics (the sublattice spin $\sigma$ corresponds to the electron spin in the Dirac equation), and exhibits nontrivial phenomena stemming from its "relativistic" properties, as discussed later.

The low-energy Hamiltonian consists of two parts, $H_K$ and $H_{K'}$. Such a multiple minima structure of energy band is called "valley." To simplify the Hamiltonian, introducing the valley isospin $\tau$, one can rewrite the Hamiltonian as

$$H(\boldsymbol{k}) = v_{\rm g}(k_y\sigma_x - k_x\sigma_y\tau_z), \tag{2.16}$$

with $\tau_z = +1$ ($\tau_z = -1$) for $H_K$ ($H_{K'}$). These two valleys can be decoupled in the absence of atomic-scale short-range scatterers or inhomogeneities. We sometimes focus only on $H_K$ under such assumptions.

## 2.3 Electronic States of Bilayer Graphene

Not only monolayer graphene but also bilayer graphene exhibits peculiar electronic states, where the magnitude of the band gap is electrically controllable. The crystal structure of bilayer graphene is shown in Fig. 2.5. B sites of the upper layer are located above A sites of the lower layer, and there is no site above B sites of the lower layer. The Hamiltonian reads

$$H_{\rm bi} = -\gamma\sum_{\langle ij\rangle} c_i^\dagger c_j + \left(-u\sum_{\boldsymbol{n}} c^\dagger_{\boldsymbol{n},A,1}c_{\boldsymbol{n},B,2} + \text{h.c.}\right), \tag{2.17}$$

where the subscripts 1 and 2 denote the lower and upper layers, respectively. The first term of the above Hamiltonian, which has the same form of monolayer graphene (2.4), is of independent monolayer graphene, and the second term denotes interlayer

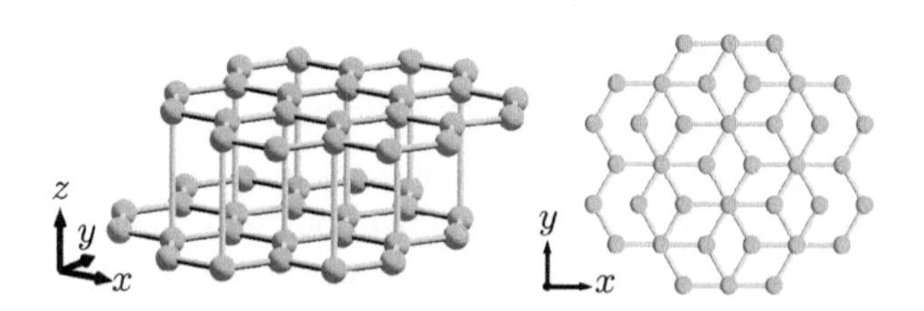


**Figure 2.5** Bilayer graphene.

hopping. The Hamiltonian is rewritten as

$$H_{\text{bi}}(\boldsymbol{k}) = \begin{pmatrix} 0 & H_{\text{AB}}(\boldsymbol{k}) & 0 & -u \\ H^*_{\text{AB}}(\boldsymbol{k}) & 0 & 0 & 0 \\ 0 & 0 & 0 & H_{\text{AB}}(\boldsymbol{k}) \\ -u & 0 & H^*_{\text{AB}}(\boldsymbol{k}) & 0 \end{pmatrix}, \tag{2.18}$$

in the basis of $(c_{A,1}(\boldsymbol{k}), c_{B,1}(\boldsymbol{k}), c_{A,2}(\boldsymbol{k}), c_{B,2}(\boldsymbol{k}))^{\text{T}}$. $H_{\text{AB}}(\boldsymbol{k})$ is given by Eq. (2.10). There four energy bands, whose energy dispersions are obtained to be

$$E_{\text{low}}(\boldsymbol{k}) = \pm\left(\sqrt{|H_{\text{AB}}(\boldsymbol{k})|^2 + \frac{u^2}{4}} - \frac{u}{2}\right), \tag{2.19}$$

$$E_{\text{high}}(\boldsymbol{k}) = \pm\left(\sqrt{|H_{\text{AB}}(\boldsymbol{k})|^2 + \frac{u^2}{4}} + \frac{u}{2}\right), \tag{2.20}$$

From the above expression, one can immediately see gapless states $E_{\text{low}}(\boldsymbol{K}) = E_{\text{low}}(\boldsymbol{K}') = 0$ at the $K$ and $K'$ points, where $H_{\text{AB}}(\boldsymbol{K}) = H_{\text{AB}}(\boldsymbol{K}') = 0$. This is the same situation as in monolayer graphene. On the other hand, the energy dispersions around the $K$ and $K'$ points become quadratic:

$$E_{\text{low}}(\boldsymbol{k}+\boldsymbol{K}) \approx \pm\frac{|H_{\text{AB}}(\boldsymbol{k}+\boldsymbol{K})|^2}{u} \approx \pm\frac{v_{\text{F}}^2}{u}k^2. \tag{2.21}$$

The energy band structure of bilayer graphene in the entire Brillouin zone is shown in Fig. 2.6. There are four nondegenerate bands, where the energy splitting is of the order of $u$.

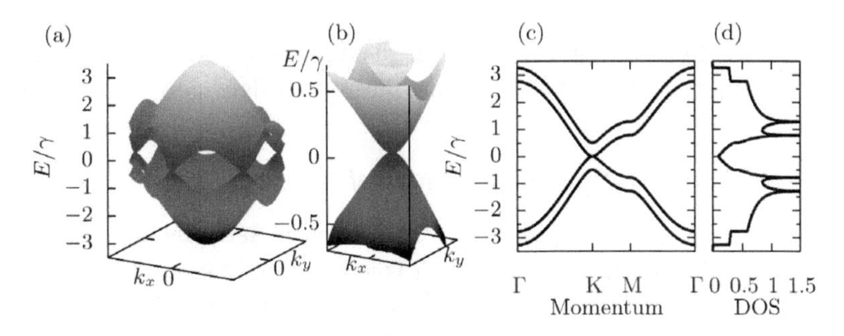


**Figure 2.6** Energy bands of bilayer graphene in the entire Brillouin zone (a and c) and in the vicinity of the $K$ point (b). The density of states is also shown in (d). The interlayer hopping is set to $u = \gamma/2$ for visibility. In the actual material, $u \approx \gamma/10$.

### 2.3.1 *Gap Opening due to the Electric Field*

Because of the application of an external electric field, the gapless states at the $K$ and $K'$ points become gapped. The electric field induces a potential difference in the bilayer as

$$H' = \sum_{\boldsymbol{n}} \sum_{\sigma=\mathrm{A,B}} m \left( c^{\dagger}_{\boldsymbol{n},\sigma,1} c_{\boldsymbol{n},\sigma,1} - c^{\dagger}_{\boldsymbol{n},\sigma,2} c_{\boldsymbol{n},\sigma,2} \right), \tag{2.22}$$

where $m$ is proportional to the applied electric field $\mathcal{E}$ and interlayer distance $\ell$. The Hamiltonian in the momentum space is explicitly represented as

$$H(\boldsymbol{k}) = \begin{pmatrix} m & H_{\mathrm{AB}}(\boldsymbol{k}) & 0 & -u \\ H^{*}_{\mathrm{AB}}(\boldsymbol{k}) & m & 0 & 0 \\ 0 & 0 & -m & H_{\mathrm{AB}}(\boldsymbol{k}) \\ -u & 0 & H^{*}_{\mathrm{AB}}(\boldsymbol{k}) & -m \end{pmatrix}, \tag{2.23}$$

and the corresponding energy eigenvalue is obtained to be

$$E_{\mathrm{low}}(\boldsymbol{k}) = \pm\sqrt{|H_{\mathrm{AB}}(\boldsymbol{k})|^2 + \frac{u^2}{2} + m^2 - \sqrt{(4m^2 + u^2)|H_{\mathrm{AB}}(\boldsymbol{k})|^2 + \frac{u^4}{4}}}, \tag{2.24}$$

$$E_{\mathrm{high}}(\boldsymbol{k}) = \pm\sqrt{|H_{\mathrm{AB}}(\boldsymbol{k})|^2 + \frac{u^2}{2} + m^2 + \sqrt{(4m^2 + u^2)|H_{\mathrm{AB}}(\boldsymbol{k})|^2 + \frac{u^4}{4}}}. \tag{2.25}$$

At the $K$ point, the energy is given by

$$E_{\mathrm{low}}(\boldsymbol{K}) = \pm m, \quad E_{\mathrm{high}}(\boldsymbol{K}) = \pm\sqrt{u^2 + m^2}. \tag{2.26}$$

Thus, the energy gap at the $K$ (and $K'$) point is induced by an electric field. It is interesting that the magnitude of the gap can be continuously controlled by tuning the electric field.

### 2.3.2 *Low-Energy Effective Model*

In the low-energy region ($v_{\mathrm{F}}k \ll u$), one can see only two gapless quadratic energy bands; hence it is useful to eliminate the higher-energy bands and derive the low-energy effective Hamiltonian $H_{\mathrm{eff}}$. In the second-order perturbation theory, the effective Hamiltonian is given by

$$(H_{\mathrm{eff}})_{ij} = \sum_{\alpha} V_{i\alpha} \frac{1}{-(H_0)_{\alpha\alpha}} V_{\alpha j},$$

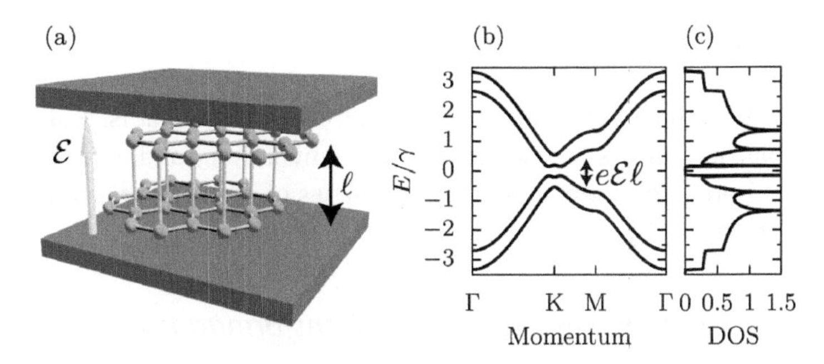


**Figure 2.7** Bilayer graphene in an electric field $\mathcal{E}$ (a), the energy band (b), and the corresponding density of states (c). The induced energy gap is of the order of $e\mathcal{E}\ell$, with $\ell$ being the interlayer distance.

for zero-energy states $[(H_0)_{ij} = 0]$, where the subscript $\alpha$ denotes the high-energy states to be eliminated. We take momentum $k$ measured from the $K$ point as a perturbation, $V = v_{\mathrm{F}}(k_y\sigma_x - k_x\sigma_y)$. The unperturbed states are given by

$$|1\rangle = (0, 1, 0, 0)^{\mathrm{T}}, \ |2\rangle = (0, 0, 1, 0)^{\mathrm{T}},$$

for $E = 0$,

$$|3\rangle = (1, 0, 0, 1)^{\mathrm{T}}/\sqrt{2},$$

for $E = -\mathrm{u}$, and

$$|4\rangle = (1, 0, 0, -1)^{\mathrm{T}}/\sqrt{2},$$

for $E = \mathrm{u}$. The matrix elements are $V_{13} = V_{14} = (-ik_x + k_y)/\sqrt{2}$, $V_{23} = (ik_x + k_y)/\sqrt{2}$, and $V_{24} = -(ik_x + k_y)/\sqrt{2}$. As a result, the effective Hamiltonian is obtained to be

$$H_{\mathrm{eff}} = \begin{pmatrix} 0 & m_{\mathrm{eff}}k_-^2 \\ m_{\mathrm{eff}}k_+^2 & 0 \end{pmatrix},$$

with $m_{\mathrm{eff}} = v_{\mathrm{F}}^2/u$, in the low-energy basis of $(|1\rangle, |2\rangle)^{\mathrm{T}}$. The form is similar to that of monolayer graphene, but the dispersion is not linear but quadratic. This means that the pseudo time reversal symmetry, which is a key for the absence of backward scattering, is broken in bilayer graphene. Therefore, transport phenomena in monolayer and bilayer graphene are completely different from each other. A typical example of this difference is *Klein tunneling*, that is, perfect transmission/reflection occurs in a monolayer/bilayer graphene bipolar junction, which will be discussed in Section 2.5 in detail.

## 2.4 Symmetries in Graphene

Graphene has several internal degrees of freedom, that is, real spin, sublattice, and valley isospins. Then many unitary and anti-unitary symmetries on them yield rich properties of transport in graphene. Here we briefly review the symmetry in graphene.

### 2.4.1 *Unitary and Anti-Unitary Transformations*

First, we review a general theory of the transformation. Wigner has shown that the transformation should be either unitary,

$$\langle\alpha|\beta\rangle = \langle\alpha'|\beta'\rangle, \tag{2.27}$$

or anti-unitary,

$$\langle\alpha|\beta\rangle = \langle\bar{\beta}|\bar{\alpha}\rangle, \tag{2.28}$$

where $|\alpha'\rangle = U|\alpha\rangle$ and $|\bar{\alpha}\rangle = \Theta|\alpha\rangle$ are the states transformed by a unitary operator $U$ and an anti-unitary one $\Theta$, respectively. For a unitary transformation, matrix elements of an arbitrary operator $\mathcal{O}$ satisfy the following relation:

$$\langle\alpha|\mathcal{O}|\beta\rangle = \langle\alpha'|U\mathcal{O}U^{\dagger}|\beta'\rangle, \tag{2.29}$$

since $U$ is a unitary operator. On the other hand, for anti-unitary transformation, one can verify

$$\langle\alpha|\mathcal{O}|\beta\rangle = \langle\bar{\beta}|\Theta\mathcal{O}^{\dagger}|\alpha\rangle = \langle\bar{\beta}|\Theta\mathcal{O}^{\dagger}\Theta^{-1}|\bar{\alpha}\rangle, \tag{2.30}$$

from Eq. 2.28.

### 2.4.2 *(Pseudo) Time Reversal Symmetry and Backward Scattering*

Time reversal is represented by the anti-unitary operator $\Theta = -is_y\mathcal{K}$, where $s_i$ is the Pauli matrix acting in the spin space and $\mathcal{K}$ is the operator of complex conjugation $\mathcal{K}f = f^*$. The Hamiltonian $H(\boldsymbol{k})$ of the system with time reversal symmetry satisfies

$$\Theta H(\boldsymbol{k})\Theta^{-1} = H(-\boldsymbol{k}). \tag{2.31}$$

Note that momentum $\boldsymbol{k}$ is reversed under time reversal. The above requires that a time reversal state $\Theta|\boldsymbol{k}\rangle$ have the same energy as that of the original state $|\boldsymbol{k}\rangle$.

Graphene has time reversal symmetry. However, $H_\tau$, which is defined only around the $K$ or $K'$ points, alone superficially breaks time reversal symmetry, since the momentum reversal $\boldsymbol{k} \to -\boldsymbol{k}$ changes the valley from $K$ to $K'$, due to $\boldsymbol{K}' = -\boldsymbol{K}$. Taking into account both valleys, in the model of Eq. 2.16, time reversal symmetry becomes preserved by defining the time reversal operator $\Theta = -is_y\tau_x\mathcal{K}$ with $\tau_x$ exchanging the valleys.

Now we can find an additional anti-unitary symmetry $\mathcal{S}$ called pseudo time reversal in $H_K$ ($H_{K'}$) as

$$\mathcal{S} = i\sigma_y\mathcal{K}, \tag{2.32}$$

satisfying

$$\mathcal{S}H_\tau(\boldsymbol{k})\mathcal{S}^{-1} = H_\tau(-\boldsymbol{k}). \tag{2.33}$$

$\mathcal{S}$ is equivalent to time reversal operator if the sublattice spin $\sigma$ is the real spin.

Time reversal symmetry guarantees the existence of doubly degenerated states, that is, Kramers pairs. Suppose that $|\alpha\rangle$ is an eigenstate of a Hamiltonian and $|\bar{\alpha}\rangle = \Theta|\alpha\rangle$ is the time reversal state. $|\alpha\rangle$ and $|\bar{\alpha}\rangle$ have the same energy owing to time reversal symmetry. From Eq. 2.28,

$$\langle\bar{\alpha}|\alpha\rangle = \langle\bar{\alpha}|\bar{\bar{\alpha}}\rangle = \Theta^2\langle\bar{\alpha}|\alpha\rangle, \tag{2.34}$$

where $\Theta^2$ is a phase factor since it is an identical operation. If $\Theta^2 \neq 1$ then $\langle\bar{\alpha}|\alpha\rangle = 0$, that is, $|\alpha\rangle$ and $|\bar{\alpha}\rangle|$ are doubly degenerated. In fact, since the pseudo time reversal (2.32) satisfies $\mathcal{S}^2 = -1$, the right-going and left-going states in a Dirac cone form a Kramers pair (see Fig. 2.8a). The pseudo time reversal symmetry protects Dirac cones, especially, the Dirac point at $E = 0$ in graphene.

Furthermore, the pseudo time reversal symmetry largely suppresses elastic scatterings on the Fermi surface. In particular, the backward scattering is completely forbidden. An intuitive explanation of the absence of backward scattering is as follows. A state $|-\boldsymbol{k}\rangle$, which is backward-scattered from $|\boldsymbol{k}\rangle$, is expressed as $|-\boldsymbol{k}\rangle = \mathcal{S}|\boldsymbol{k}\rangle$. This means that the sublattice pseudo spin of $|\boldsymbol{k}\rangle$ is opposite to that of $|-\boldsymbol{k}\rangle$. Consequently, a disorder potential varying smoothly in the atomic scale cannot "flip" the sublattice pseudo spin, that is, backward scattering never occurs by such a impurity

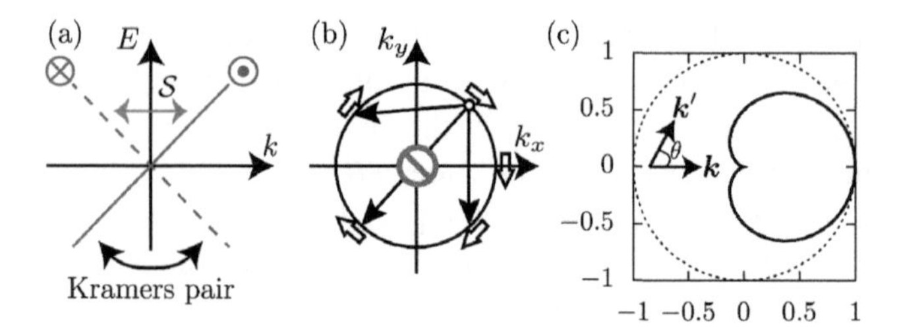


**Figure 2.8** Pseudo time reversal ($\mathcal{S}$) symmetry in a Dirac cone. (a) Energy spectrum along a certain direction. The right-going states (solid line) are Kramers partners of the left-going states (dashed line). (b) Elastic scatterings in a Dirac cone. The Fermi surface is denoted by the circle and the direction of the pseudo spin is denoted by arrows. Backward scattering is prohibited. (c) Polar plot of scattering amplitudes for conventional (dashed line) and Dirac (solid line) electrons in the presence of the delta function potential.

potential. An elastic scattering with the scattering angle in the range of $0 < \theta < \pi$, on the other hand, partially occurs since the pseudo spin of the scattered state is not completely opposite to the original one (see Fig. 2.8b).

Let us see a simple example of elastic scattering in a Dirac cone by the Delta function potential $V(\boldsymbol{x}) = V_0\delta(\boldsymbol{x})$. A state $\psi_{\boldsymbol{k}}(\boldsymbol{x}) = \boldsymbol{u}(\boldsymbol{k})e^{i\boldsymbol{k}\cdot x}$ is scattered to $\psi_{\boldsymbol{k}'}(\boldsymbol{x}) = \boldsymbol{u}(\boldsymbol{k}'^{i\boldsymbol{k}'\cdot x}$. The transition probability $f(\theta)$, with $\theta$ being the scattering angle, is proportional to

$$f(\theta) \propto \left| \int d^2x \psi^{\dagger}_{\boldsymbol{k}'}(x)\delta(\boldsymbol{x})\psi_{\boldsymbol{k}}(x) \right|^2 = \left|\boldsymbol{u}^{\dagger}(\boldsymbol{k}')\boldsymbol{u}(\boldsymbol{k})\right|^2 = \frac{1+\cos\theta}{2}. \tag{2.35}$$

in the lowest order. $f(\theta)$ is shown in Fig. 2.8c. Actually, backward scattering with $\theta = \pi$ is prohibited ($f(\pi) = 0$). In addition, $f(\theta)$ is notably suppressed for $\theta \approx \pi$. This implies high-mobility transport in graphene. It is worth mentioning that the above discussion on the absence of backward scattering is valid not only for the lowest order but also for any orders [41].

As seen above, a Dirac cone and the pseudo time reversal symmetry are important ingredients for transport. Then a natural question arises: Why does graphene respect pseudo time reversal symmetry? The answer is that pseudo time reversal symmetry emerges from the sixfold rotational symmetry and sublattice

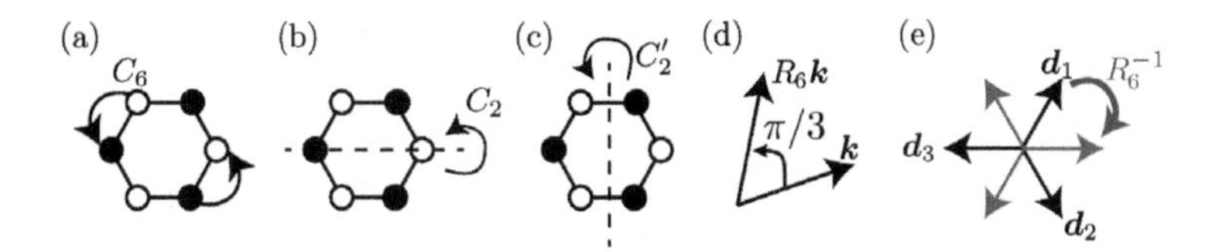


**Figure 2.9** Sixfold ($\pi/3$) rotation along the $z$ axis (a), twofold rotation along the $x$ axis (b), and twofold rotation along the $y$ axis (c) in graphene. The sixfold rotations of momentum $\boldsymbol{k}$ (d) and of the bonding vectors $\boldsymbol{d}_1$, $\boldsymbol{d}_2$, and $\boldsymbol{d}_3$ (e).

(nonsymmorphic) structure of the honeycomb lattice. The details will be discussed later.

### 2.4.3 *Rotational Symmetry and Degeneracy at the K Point*

Graphene is invariant for the sixfold rotation $C_6$ along the $z$ axis and twofold rotations $C_2$ and $C_2'$ along the $x$ and $y$ axes, respectively (Fig. 2.9a–c). The Hamiltonian satisfies

$$C_6 H(\boldsymbol{k}) C_6^\dagger = H(R_6\boldsymbol{k}), \tag{2.36}$$

$$C_2 H(\boldsymbol{k}) C_2^\dagger = H(R_2\boldsymbol{k}), \tag{2.37}$$

$$C_2' H(\boldsymbol{k}) C_2'^\dagger = H(R_2'\boldsymbol{k}). \tag{2.38}$$

$2 \times 2$ matrices $C_6$, $C_2$, and $C_2'$ are defined by

$$C_6 = \sigma_x e^{-ij_z\pi/3}, \tag{2.39}$$

$$C_2 = e^{-ij_x\pi}, \tag{2.40}$$

$$C_2' = \sigma_x e^{-ij_y\pi}, \tag{2.41}$$

$C_6$ and $C_2'$ include $\sigma_x$, which interchanges the A and B sites (see Fig. 2.9a), whereas $C_2$ does not. $\boldsymbol{j}$ denotes the total angular momentum of orbital and spin. In the present case, $j_z = 0$ for spinless electrons of the $\pi$-band ($p_z$-orbital). $C_2$ and $C_2'$ multiply $(-1)$ to wave functions of the $\pi$-band. $R_6\boldsymbol{k}$, $R_2\boldsymbol{k}$, and $R_2'\boldsymbol{k}$ are the rotated vectors defined by

$$R_6\boldsymbol{k} = \begin{pmatrix} \cos\pi/3 & -\sin\pi/3 \\ \sin\pi/3 & \cos\pi/3 \end{pmatrix} \begin{pmatrix} k_x \\ k_y \end{pmatrix}, \quad R_2\boldsymbol{k} = \begin{pmatrix} k_x \\ -k_y \end{pmatrix}, \quad R_2'\boldsymbol{k} = \begin{pmatrix} -k_x \\ k_y \end{pmatrix}, \tag{2.42}$$

as shown in Fig. 2.9d.

The relation (2.36) can be also confirmed directly from Eq. 2.10. $H_{\mathrm{AB}}$ obeys the following:

$$H_{\mathrm{AB}}(R_6\boldsymbol{k}) = \sum_{\mu=1}^{3} e^{i R_6\boldsymbol{k}\cdot\boldsymbol{d}_\mu} = \sum_{\mu=1}^{3} e^{i\boldsymbol{k}\cdot R_6^{-1}\boldsymbol{d}_\mu} = \sum_{\mu=1}^{3} e^{-i\boldsymbol{k}\cdot\boldsymbol{d}_\mu} = H_{\mathrm{AB}}^*(\boldsymbol{k}), \tag{2.43}$$

$$H_{\mathrm{AB}}(R_2\boldsymbol{k}) = H_{\mathrm{AB}}(\boldsymbol{k}), \tag{2.44}$$

$$H_{\mathrm{AB}}(R_2'\boldsymbol{k}) = H_{\mathrm{AB}}^*(\boldsymbol{k}), \tag{2.45}$$

owing to

$$R_6^{-1}\boldsymbol{d}_1 = -\boldsymbol{d}_3,\ R_6^{-1}\boldsymbol{d}_2 = -\boldsymbol{d}_1,\ R_6^{-1}\boldsymbol{d}_3 = -\boldsymbol{d}_2, \tag{2.46}$$

$$R_2^{-1}\boldsymbol{d}_1 = \boldsymbol{d}_3,\ R_2^{-1}\boldsymbol{d}_2 = \boldsymbol{d}_2,\ R_2^{-1}\boldsymbol{d}_3 = \boldsymbol{d}_1, \tag{2.47}$$

$$R_2'^{-1}\boldsymbol{d}_1 = -\boldsymbol{d}_2,\ R_2'^{-1}\boldsymbol{d}_2 = -\boldsymbol{d}_1,\ R_2'^{-1}\boldsymbol{d}_3 = -\boldsymbol{d}_3, \tag{2.48}$$

as illustrated in Fig. 2.9e. Therefore $\sigma_x H(\boldsymbol{k})\sigma_x = H(R_6\boldsymbol{k})$ and $\sigma_x H(\boldsymbol{k})\sigma_x = H(R_2'\boldsymbol{k})$ with $C_6 = C_2 = \sigma_x$, that is, Eqs. 2.36–2.38 are verified.

The rotation axis of $C_6$ introduced before is located at the $\Gamma$ point. Moreover, symmetric points on the zone boundary have some symmetries lower than that of the $\Gamma$ point, but the $K$ and $K'$ points (hereafter we focus only on the $K$ point) have threefold rotational $C_{3K}$ symmetry. Threefold rotation $C_3$ around the $\Gamma$ point is given by

$$C_3 = C_6^2 = e^{-ij_z 2\pi/3}.$$

Simultaneously, momentum $\boldsymbol{k}$ is rotated to $R_3\boldsymbol{k} = R_6^2\boldsymbol{k}$. Then the $K$ point is transferred to the other $K$ point ($R_3\boldsymbol{K} = \boldsymbol{K} - \boldsymbol{b}_1$) that differ by the reciprocal lattice vector $\boldsymbol{b}_1$. As a result, the threefold rotational symmetry requires

$$C_3 H(\boldsymbol{k} + \boldsymbol{K})C_3^\dagger = H(R_3\boldsymbol{k} + R_3\boldsymbol{K}) = H(R_3\boldsymbol{k} + \boldsymbol{K} - \boldsymbol{b}_1).$$

On the other hand, translational symmetry requires that $H(\boldsymbol{k})$ and $H(\boldsymbol{k} + \boldsymbol{G})$, where $\boldsymbol{G}$ denotes a reciprocal lattice vector, be equivalent to each other. In a honeycomb lattice, which has a sublattice structure, $H(\boldsymbol{k})$ and $H(\boldsymbol{k} + \boldsymbol{G})$ are related in a nontrivial form:

$$U_{\boldsymbol{G}} H(\boldsymbol{k}) U_{\boldsymbol{G}}^\dagger = H(\boldsymbol{k} + \boldsymbol{G}), \tag{2.49}$$

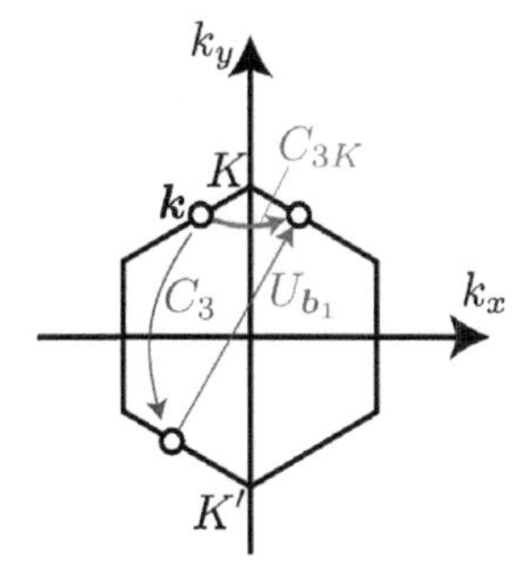


**Figure 2.10** Threefold rotational symmetry around the $K$ point.

with a unitary matrix $U_{\boldsymbol{G}}$:

$$U_{\boldsymbol{G}} = e^{-i\phi_G \sigma_z/2} = \begin{pmatrix} e^{-i\phi_G/2} & 0 \\ 0 & e^{i\phi_G/2} \end{pmatrix}.$$

For instance, the form factor for $\boldsymbol{k}+\boldsymbol{G}$ is given by $\sum_\mu e^{i(\boldsymbol{k}+\boldsymbol{G})\cdot \boldsymbol{d}_\mu}$, but $\boldsymbol{G}\cdot \boldsymbol{d}_\mu \neq 2\pi$ because $\boldsymbol{d}_\mu$ is not a translational vector in sublattice structured systems. The additional factors coming from $e^{i\boldsymbol{G}\cdot \boldsymbol{d}_\mu}$ should be a phase factor and be absorbed by a gauge transformation by $U_{\boldsymbol{G}}$. With the help of the translational symmetry, the threefold rotational $C_{3K}$ symmetry around the $K$ point is expressed as

$$C_{3K} H(\boldsymbol{k}+\boldsymbol{K}) C_{3K}^\dagger = H(R_3\boldsymbol{k}+\boldsymbol{K}),$$

with

$$C_{3K} = U_{\boldsymbol{b}_1} C_3,$$

as illustrated in Fig. 2.10.

Actually, for graphene, the form factor reduces to

$$\sum_{\mu=1}^{3} e^{i(\boldsymbol{k}+\boldsymbol{G})\cdot \boldsymbol{d}_\mu} = e^{i\boldsymbol{G}\cdot \boldsymbol{d}_3} \sum_{\mu=1}^{3} e^{i\boldsymbol{k}\cdot \boldsymbol{d}_\mu} e^{i\boldsymbol{G}\cdot(\boldsymbol{d}_\mu - \boldsymbol{d}_3)} = e^{i\boldsymbol{G}\cdot \boldsymbol{d}_3} \sum_{\mu=1}^{3} e^{i\boldsymbol{k}\cdot \boldsymbol{d}_\mu}, \quad (2.50)$$

since $\boldsymbol{d}_\mu - \boldsymbol{d}_3$ is a translational vector; $\boldsymbol{G}\cdot(\boldsymbol{d}_\mu - \boldsymbol{d}_3) = 2n\pi$. As a result, the Hamiltonian satisfies Eq. 2.49, $U_{\boldsymbol{b}_1} H(\boldsymbol{k}) U_{\boldsymbol{b}_1}^\dagger = H(\boldsymbol{k}+\boldsymbol{b}_1)$, for $\boldsymbol{G} = \boldsymbol{b}_1$ with $\phi_{\boldsymbol{b}_1} = -\boldsymbol{b}_1 \cdot \boldsymbol{d}_3 = 2\pi/3$. And then $C_{3K} H(\boldsymbol{k}+\boldsymbol{K}) C_{3K}^\dagger = H(R_3\boldsymbol{k}+\boldsymbol{K})$ also holds.

The $C_{3K}$ symmetry restricts the form of the wave function at the $K$ ($K'$) point. $C_{3K}$ commutes with the Hamiltonian at the $K$ point:

$$[C_{3K}, H(\boldsymbol{K})] = 0 \quad (2.51)$$

Hence the states at the $K$ point are eigenstates of $C_{3K}$, namely eigenstates with $\sigma_z = 1$ and $\sigma_z = -1$. The energy of the state with $\sigma_z = 1$, whose amplitude is zero at the B sublattice, is degenerated to that with $\sigma_z = -1$, whose amplitude is 0, on the contrary, at the A sublattice:

$$H(\boldsymbol{K})|\boldsymbol{K}, \pm\rangle = E_{\pm}(\boldsymbol{K})|\boldsymbol{K}, \pm\rangle, \; C_{3K}|\boldsymbol{K}, \pm\rangle = e^{\mp i\pi/3}|\boldsymbol{K}, \pm\rangle, \quad (2.52)$$

with

$$E_{+}(\boldsymbol{K}) = E_{-}(\boldsymbol{K}). \quad (2.53)$$

This is because these states $|\boldsymbol{K}, \pm\rangle$ are related by the twofold rotation $C_2'$ along the $y$ axis

$$C_2'|\boldsymbol{K}, \pm\rangle = -|\boldsymbol{K}, \mp\rangle, \quad (2.54)$$

owing to

$$C_2' C_{3K} C_2'^{-1} = C_{3K}^{-1}. \quad (2.55)$$

Since the $K$ point is invariant for $C_2'$, $[C_2', H(\boldsymbol{K})] = 0$, these states $|\boldsymbol{K}, \pm\rangle$ have the same energy. Then, they form a Dirac cone at the $K$ point.

A Dirac point, degeneracy at the $K$ and $K'$ points, is protected by the two rotational symmetries reinforced by the translational symmetry. Note that the translational symmetry (2.49) plays a role for the above discussion only at the $K$ and $K'$ points. Thus, Dirac points are located at the zone boundary, not at the zone center.

Bilayer graphene also has the $C_{3K}$ and $C_2'$ symmetries, while it does not have $C_2$ symmetry:

$$[C_{3K}, H_{\text{bi}}(\boldsymbol{K})] = [C_2', H_{\text{bi}}(\boldsymbol{K})] = 0, \quad (2.56)$$

with

$$C_{3K} = e^{-i\pi\sigma_z/3} e^{-i2\pi j_z/3}, \; C_2' = \sigma_x l_x e^{-ij_x\pi}, \quad (2.57)$$

and $C_2' C_{3K} C_2'^{-1} = C_{3K}^{-1}$, where $l_x$ denotes the Pauli matrix exchanging the layers. Thus twofold degeneracy also emerges at the $K$ and $K'$ points in bilayer graphene, protected by the above symmetries.

### 2.4.4 *Chiral Symmetry*

As we have shown in the previous section, graphene has particle–hole symmetric energy bands. This stems from chiral (sublattice) symmetry, which the systems in bipartite lattices always has.

Chiral transform is defined by $\Gamma = \sigma_z$ and satisfies

$$\Gamma H(\boldsymbol{k})\Gamma^{\dagger} = -H(\boldsymbol{k}). \tag{2.58}$$

Due to this symmetry, if there exists the eigenstate $\boldsymbol{u}(\boldsymbol{k})$ with energy $E$, there always exists the negative energy state $\Gamma\boldsymbol{u}(\boldsymbol{k})$ with energy $-E$.

## 2.5 Some Mathematics

The Pauli matrices are defined by

$$\sigma_x = \begin{pmatrix} 0 & 1 \\ 1 & 0 \end{pmatrix}, \ \sigma_y = \begin{pmatrix} 0 & -i \\ i & 0 \end{pmatrix}, \ \sigma_z = \begin{pmatrix} 1 & 0 \\ 0 & -1 \end{pmatrix}, \tag{2.59}$$

which satisfy the following commutation rule,

$$[\sigma_i, \sigma_j] = 2i\epsilon_{ijk}\sigma_{ijk}, \ \{\sigma_i, \sigma_j\} = 2\delta_{ij}, \tag{2.60}$$

with $i = 1(x), 2(y), 3(z)$. Combining these two relations, we obtain

$$\sigma_i\sigma_j = \delta_{ij} + i\epsilon_{ijk}\sigma_k. \tag{2.61}$$

The general form of the $2 \times 2$ matrix is expanded in terms of the Pauli matrices:

$$A = \sum_{\mu=0}^{3} a_{\mu}\sigma_{\mu}, \tag{2.62}$$

where $\sigma_0 = 1$ is the identity matrix. Let us consider an eigenvalue problem $A\boldsymbol{u} = \lambda\boldsymbol{u}$, namely

$$\sum_i a_i\sigma_i\boldsymbol{u} = (\lambda - a_0)\boldsymbol{u}. \tag{2.63}$$

From Eq. 2.61

$$(A - a_0\sigma_0)^2 = \sum_{i,j=1}^{3} a_i a_j \sigma_i\sigma_j = \sum_{i=1}^{3} a_i^2 \equiv a^2. \tag{2.64}$$

Therefore the eigenvalue problem

$$(A - a_0\sigma_0)^2 \boldsymbol{u} = (\lambda - a_0)^2 \boldsymbol{u} \tag{2.65}$$

is diagonalized with $(\lambda - a_0)^2 = a^2$. The eigenvalue $\lambda$ of the original problem is given by

$$\lambda_{\pm} = a_0 \pm a, \quad a = \sqrt{\sum_{i=1}^{3} a_i^2}. \tag{2.66}$$

The corresponding eigenvector is obtained as

$$\boldsymbol{u}_+ = \begin{pmatrix} \cos\dfrac{\theta}{2} \\ e^{i\phi}\sin\dfrac{\theta}{2} \end{pmatrix}, \quad \boldsymbol{u}_- = \begin{pmatrix} \sin\dfrac{\theta}{2} \\ -e^{i\phi}\cos\dfrac{\theta}{2} \end{pmatrix}, \tag{2.67}$$

where angles $\theta$ and $\phi$ are defined by

$$a_x/a = \sin\theta\cos\phi, \quad a_y/a = \sin\theta\sin\phi, \quad a_z/a = \cos\theta. \tag{2.68}$$

## 2.6 Transport Properties

Graphene is expected to be applied to a device, due to its high mobility stemming from the absence of backward scattering. One of most fundamental electronic device is the field-effect transistor (FET), as illustrated in Fig. 2.11a, where metallic leads "source" and "drain" are attached to the graphene on a insulating substrate and a

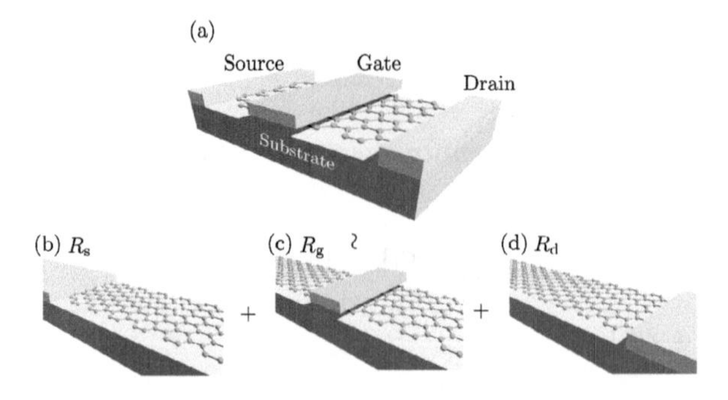


**Figure 2.11** Schematic structure of graphene field-effect transistor. For infinitely long systems, the resistance is given by $R = R_s + R_g + R_d$.

"gate" electrode is also attached but separated from the graphene by an insulating buffer layer.

The electric resistance $R$ of an ideally long FET is given by three independent terms: $R = R_s + R_g + R_d$, where $R_s$, $R_g$, and $R_d$ denote the resistances coming from the source, gate, and drain, respectively (Fig. 2.11b–d). In the following section, we discuss these basic transport properties of $R_{s/d}$ and $R_g$.

## 2.7 Bipolar Junction

Firstly, we consider the charge transport in a graphene bipolar junction illustrated in Figs. 2.11c and 2.12, which yields the resistance $R_g$ coming from the gate, where an electrostatic potential $V(x)$ is applied from the gate electrode. Suppose that the gate is infinitely long and the spatial profile of $V(x)$ is described by the step function $V(x) = V\theta(x)$, and the intervalley scattering between the $K$ and $K'$ points can be neglected, that is, the spatially varying length of $V(x)$ is much shorter than the Fermi wavelength and much longer than the lattice constant. In addition, we further simplify the problem as the gate length $L_g$ is infinitely long, where the gate resistance $R_g$ can be calculated separately: $R_g = R_l + R_r$. Also, the two resistances $R_l$ and $R_r$ are nearly the same. Thus we focus only on $R_l$ in this section.

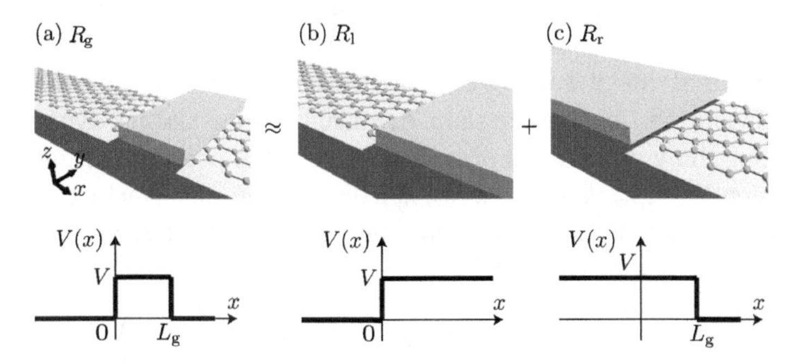


**Figure 2.12** Bipolar graphene junction with a finite length ($L_g$) of gate electrode (a). The total resistance $R_g$ can be approximately divided into two semi-infinite systems (b) and (c); $R_g \approx R_l + R_r$, for infinitely long systems. The potential profile $V(x)$ is also shown.

### 2.7.1 *Perfect Transmission in a Monolayer Graphene Junction*

Here we consider a bipolar monolayer graphene junction. The gate potential, which is illustrated in Fig. 2.12b, is given by $V(x) = V\theta(x)$. And we adapt the effective low-energy theory Eq. 2.12. The Hamiltonian reads

$$H(x) = v_g(k_y\sigma_x + i\partial_x\sigma_y) + V\theta(x). \quad (2.69)$$

Note that the momentum $k_x$ is replaced with the derivative $-i\partial_x$ since the system has no longer translational symmetry along the $x$ axis.

The scattering wave function $\psi(x)$ has the form

$$\psi(x < 0) = \boldsymbol{u}(q, k_y)e^{iqx}e^{ik_yy} + r\boldsymbol{u}(-q, k_y)e^{-iqx}e^{ik_yy}, \quad (2.70)$$

$$\psi(x > 0) = t\boldsymbol{u}'(q', k_y)e^{iq'x}e^{ik_yy}. \quad (2.71)$$

The first and second terms of Eq. 2.70 correspond to the incident wave and the reflected wave, respectively. Equation 2.71 corresponds to the transmitted wave. The momenta are expressed in terms of the incident energy $E$:

$$q = \frac{E}{v_g}\cos\theta,\ q' = \frac{E-V}{v_g}\cos\theta',\ k_y = \frac{E}{v_g}\sin\theta = \frac{E-V}{v_g}\sin\theta', \quad (2.72)$$

where $\theta$ and $\theta'$ ($|\theta|, |\theta'| < \pi/2$) denote the incident and transmission angles, and the sign of momentum is determined so that the electron comes from the left to the right, namely the group velocity is positive. Note that the $y$ components of momentum in the incident and transmitted side are the same $k_y$ since the translational symmetry along the $y$ direction is preserved even in the presence of the gate potential. $\boldsymbol{u}(k_x, k_y)$ and $\boldsymbol{u}'(k_x, k_y)$ denote the eigenvectors of the Hamiltonian in the incident and the transmitted sides, respectively, which are explicitly given by

$$\boldsymbol{u}(q, k_y) = \frac{1}{\sqrt{2}}\begin{pmatrix}1\\ -ie^{i\theta}\end{pmatrix},\ \boldsymbol{u}(-q, k_y) = \frac{1}{\sqrt{2}}\begin{pmatrix}1\\ ie^{-i\theta}\end{pmatrix}, \quad (2.73)$$

$$\boldsymbol{u}'(q', k_y) = \frac{1}{\sqrt{2}}\begin{pmatrix}1\\ -ie^{i\theta'}\end{pmatrix}, \quad (2.74)$$

where $E(k_x, k_y) = \pm v_g(k_x^2 + k_y^2)^{1/2}$.

Reflection and transmission coefficients are obtained from the boundary condition at $x = 0$, given by

$$\psi(x = -0) = \psi(x = +0). \tag{2.75}$$

Under this condition, both charge and current densities are continuous at $x = 0$. The current operator $\boldsymbol{j} = (j_x, j_y)$ is defined by

$$j_x = \frac{\partial H(\boldsymbol{k})}{\partial k_x} = -v_g\sigma_y,\ j_y = \frac{\partial H(\boldsymbol{k})}{\partial k_y} = v_g\sigma_x. \tag{2.76}$$

Therefore, the condition (2.75) immediately leads to current continuity:

$$j_x\psi(x = -0) = j_x\psi(x = +0). \tag{2.77}$$

Applying Eq. 2.75, one obtains the reflection and transmission coefficients

$$r = \frac{e^{i\theta} - e^{i\theta'}}{e^{-i\theta} + e^{i\theta'}},\ t = 1 + r. \tag{2.78}$$

The above expression tells us that the normal incidence $\theta = 0$ is special. In this case, the transmitted angle is also $\theta' = 0$, that is, $r = 0$ and $t = 1$ are satisfied for an arbitrary gate potential $V$phenomenon was originally discussed by Klein in the context of relativistic quantum mechanics, the tunneling problem of a Dirac fermion called Klein tunneling. The perfect transmission is one of the realizations of the absence of backward scattering. Namely, the incident and reflected waves have the pseudo spin $\sigma$ opposite to each other. This prohibits the reflection and the electron have to go forward through the potential barrier, even if the barrier height is infinitely large. Transmission probability $T = 1 - |r|^2$, on the other hand, decreases as $\theta$ changes from $\theta = 0$ to $\theta = \pm\pi/2$.

The most direct observable of transport is charge conductance, which is obtained by summing up the transmission probability $T = 1 - |r|^2$ over all the possible incident channels:

$$G = \frac{e^2}{h}\frac{W}{2\pi}\int_{-k_F}^{k_F} dk_y T = G_0\int_{-\pi/2}^{\pi/2} d\theta\frac{\cos\theta}{2}T, \tag{2.79}$$

with $W$ being the width of the junction and $G_0 = e^2/h \times WE/(\pi\hbar v_g)$ the maximum value of the conductance. $G/G_0$ as a function of $V/E$, which is shown in Fig. 2.13c, is suppressed as one applies gate

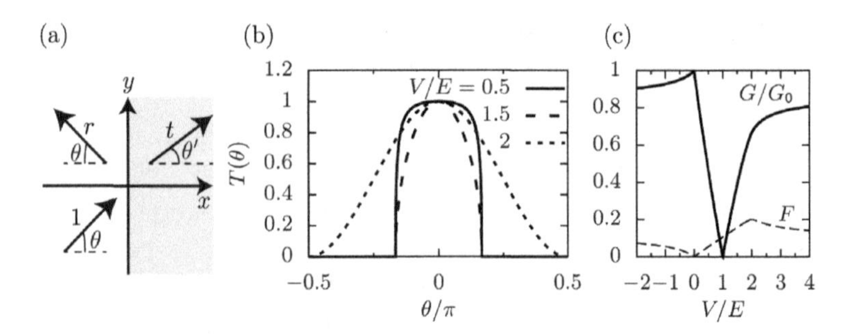


**Figure 2.13** Transport in a bipolar junction of monolayer graphene (a). The angle-resolved transmission probability $T(\theta)$ (b), charge conductance $G$, and Fano factor $F$ as a function of the gate voltage normalized by the incident energy $V/E$ are shown (c).

potential $V$, and vanishes at $V = E$, where the Fermi energy in the transmitted side is located at the Dirac point and the density of states vanishes. Note that the conductance takes a finite value except for $V = E$ since graphene has no band gap. If one wants to apply graphene to a FET, one has to make somehow a finite size of the band gap for future applications of graphene transistors.

From the transmission probability, we can derive the shot noise of charge current. The noise–signal ratio, the Fano factor $F$, is given by

$$F = \frac{\sum_{k_y} T(1-T)}{\sum_{k_y} T}. \tag{2.80}$$

By definition, $F$ takes $0 \leq F \leq 1$.

## 2.7.2 *Perfect Reflection in Bilayer Graphene Junction*

Replacing the monolayer graphene with bilayer graphene in the junction, the bipolar junction exhibits perfect reflection instead of perfect transmission as in the case of monolayer graphene. We start with the low-energy model of bilayer graphene (Eq. 2.81):

$$H = \begin{pmatrix} 0 & \dfrac{(k_x - ik_y)^2}{2m} \\ \dfrac{(k_x + ik_y)^2}{2m} & 0 \end{pmatrix}. \tag{2.81}$$

The energy eigenvalue $E(\boldsymbol{k})$ and the corresponding eigenvector $\boldsymbol{u}(\boldsymbol{k})$ are obtained to be

$$E(\boldsymbol{k}) = \pm\frac{k^2}{2m}, \ \boldsymbol{u}(\boldsymbol{k}) = \frac{1}{\sqrt{2}}\begin{pmatrix} 1 \\ \pm\dfrac{k^2}{(k_x - ik_y)^2} \end{pmatrix} = \frac{1}{\sqrt{2}}\begin{pmatrix} 1 \\ \pm e^{i2\theta_k} \end{pmatrix}, \tag{2.82}$$

with $k = (k_x^2 + k_y^2)^{1/2}$. In the junction, the Hamiltonian is replaced with

$$H(x) = \begin{pmatrix} V\theta(x) & \dfrac{(-i\partial_x - ik_y)^2}{2m} \\ \dfrac{(-i\partial_x + ik_y)^2}{2m} & V\theta(x) \end{pmatrix}. \tag{2.83}$$

For a definite energy $E$, there are two solutions of momentum $k = \sqrt{\pm 2mE}$, that is, a real solution for propagating modes and a pure imaginary solution for evanescent modes. Therefore, the scattering wave function has the form

$$\psi(x < 0) = \boldsymbol{u}(q, k_y)e^{iqx} + r_\mathrm{p}\boldsymbol{u}(-q, k_y)e^{-iqx} + r_\mathrm{e}\boldsymbol{u}(-i\kappa, k_y)e^{\kappa x}, \tag{2.84}$$

$$\psi(x > 0) = t_\mathrm{p}\boldsymbol{u}'(q', k_y)e^{iq'x} + t_\mathrm{e}\boldsymbol{u}'(i\kappa', k_y)e^{-\kappa' x}, \tag{2.85}$$

with

$$q = s\sqrt{2m|E|}\cos\theta, \ \kappa = \sqrt{2m|E| + k_y^2} = \sqrt{2m|E|}\sqrt{1 + \sin^2\theta}, \tag{2.86}$$

$$q' = s'\sqrt{2m|E - V|}\cos\theta', \tag{2.87}$$

$$\kappa' = \sqrt{2m|E - V| + k_y^2} = \sqrt{2m|E - V|}\sqrt{1 + \sin^2\theta'}, \tag{2.88}$$

$$k_y = s\sqrt{2m|E|}\sin\theta = s'\sqrt{2m|E - V|}\sin\theta', \tag{2.89}$$

and $s' = \mathrm{sign}(E - V)$. The eigenvectors for the propagating and evanescent modes are expressed as

$$\boldsymbol{u}(q, k_y) = \frac{1}{\sqrt{2}}\begin{pmatrix} 1 \\ se^{i2\theta} \end{pmatrix}, \ \boldsymbol{u}'(q', k_y) = \frac{1}{\sqrt{2}}\begin{pmatrix} 1 \\ s'^{i2\theta'} \end{pmatrix}, \tag{2.90}$$

$$\boldsymbol{u}(i\kappa, k_y) = \frac{1}{\sqrt{2}}\begin{pmatrix} 1 \\ -s\dfrac{\kappa + k_y}{\kappa - k_y} \end{pmatrix}, \ \boldsymbol{u}'(i\kappa', k_y) = \frac{1}{\sqrt{2}}\begin{pmatrix} 1 \\ -s'\dfrac{\kappa' + k_y}{\kappa' - k_y} \end{pmatrix}. \tag{2.91}$$

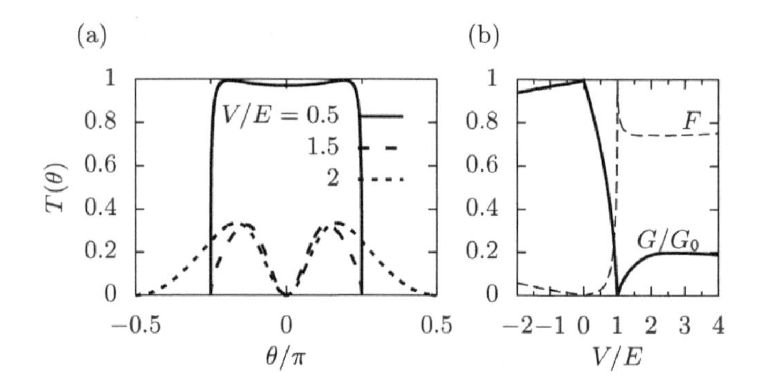


**Figure 2.14** Transport in a bipolar junction of bilayer graphene.

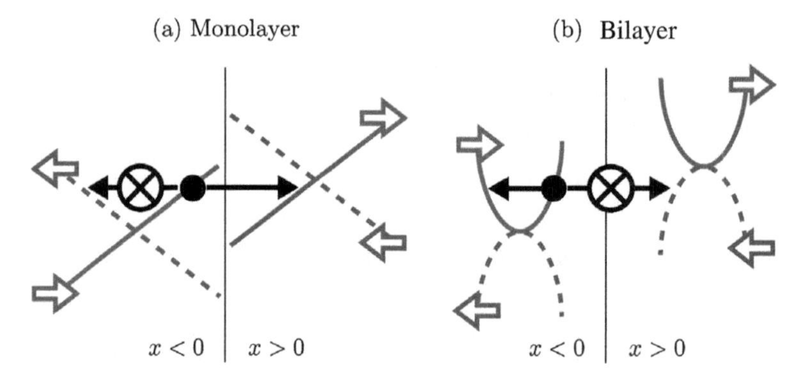


**Figure 2.15** Energy bands and pseudo spin in monolayer and bilayer graphenes. The pseudo spins in the energy bands denoted by the solid line and by the dashed line have opposite directions.

The boundary condition at $x = 0$ is the continuities of the wave function $\psi(-0) = \psi(+0)$ and of the derivative of the wave function $\psi'(-0) = \psi'(+0)$.

The solution is easily obtained in the normal incident case ($\theta = \theta' = 0$):

$$r_{\rm p} = \frac{1-\eta}{1+\eta}, \; r_{\rm e} = 0, \; t_{\rm p} = \frac{2}{1+\eta}, \; t_{\rm e} = 0, \tag{2.92}$$

for $ss' = 1$, and

$$r_{\rm p} = \frac{1-i\eta}{1+i\eta}, \; r_{\rm e} = 0, \; t_{\rm p} = 0, \; t_{\rm e} = \frac{2}{1+i\eta}, \tag{2.93}$$

for $ss' = -1$ and $\eta = \sqrt{|E-V|/|E|}$. Namely, in the case of *pn* junction ($ss' = -1$), the reflection probability becomes unity $R = |r_{\rm p}|^2 = 1$ for arbitrary $\eta$. This perfect reflection can be also understood in the viewpoint of the orthogonality of the pseudo spin; the positive and negative energy bands have the opposite pseudo spin. This orthogonality prohibits the interband tunneling, which corresponds to the *pn* junction case.

## 2.8 Metal/Graphene Junction

We consider the electron transport in a metal/graphene junction shown in Fig. 2.16. Let us suppose a simple metal described as

$$H_{\rm s} = -\epsilon_0 \sum_n c_n^\dagger c_n + \left[ -t_{\rm s} \sum_n \left( c_n^\dagger c_{n+a_x} + c_n^\dagger c_{n+a_y} \right) + \text{h.c.} \right], \quad (2.94)$$

in the left side $x < 0$. Without loss of generality, we set $t_{\rm s} > 0$ in the following. Here, the primitive translational vectors in the square lattice are given by $\boldsymbol{a}_x = (a_{\rm s}, 0)$, $\boldsymbol{a}_y = (0, a_{\rm s})$, where $a_{\rm s} = \sqrt{3}a_{\rm g}$ is the lattice constant of square lattice matching that in the honeycomb lattice. The energy spectrum of the metal is obtained as $E_{\rm s}(\boldsymbol{k}) = -2t_{\rm s}(\cos k_x a_{\rm s} + \cos k_y a_{\rm s}) - \epsilon_0$. For simplicity, we assume the half-filling case $\epsilon_0 = 0$

Now, we solve the scattering problem where an electron is injected from the left side. And we apply translational symmetry along the $y$ axis, and then the $y$component of momentum $k_y$ becomes

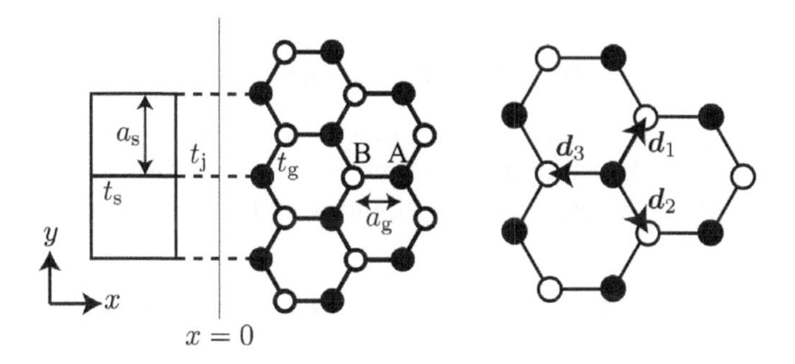


**Figure 2.16** Metal with a square lattice/graphene junction.

a good quantum number. We focus on the low-energy ($E \approx 0$) transport in the vicinity of the $K$ and $K'$ points, where Dirac fermions in graphene play an important role. Note that the contribution from the $K'$ point is the same as that from the $K$ point.

On the left side, the scattering wave function has the form

$$\psi(x < 0) = \left(e^{ik_s x} + re^{-ik_s x}\right) e^{ik_y y}, \tag{2.95}$$

where the momentum $\boldsymbol{k}_s = (k_s, k_y)$, $k_s > 0$ satisfies the energy–momentum relation $E = E_s(k_s, k_y)$. On the other hand, on the right side, the wave function is given by

$$\psi(x > 0) = t \begin{pmatrix} u(\boldsymbol{k}_g) \\ v(\boldsymbol{k}_g) \end{pmatrix} e^{i\boldsymbol{k}_g \cdot \boldsymbol{x}}, \tag{2.96}$$

where $t$ is the transmission coefficient and $\boldsymbol{k}_g = (q_g, k_y)$ satisfying $E = E_g(\boldsymbol{k}_g)$, and

$$\frac{v(\boldsymbol{k}_g)}{u(\boldsymbol{k}_g)} = \frac{E}{H_{AB}(\boldsymbol{k}_g)} \approx -ie^{i\theta}, \tag{2.97}$$

where the transmitted angle is defined as follows:

$$q_g = \frac{E}{v_g}\cos\theta, \ k_y - K = \frac{E}{v_g}\sin\theta, \tag{2.98}$$

in the vicinity of the $K$ point (see Section 2.7.1). The reflection and transmission coefficients $r$ and $t$ are determined by the boundary condition at the interface located at $x = 0$, which is derived from the Schrödinger equation of the tight-binding model [3]:

$$E\phi(\boldsymbol{0}) = -t_j\psi_A(\boldsymbol{0}) - \epsilon_0\phi(\boldsymbol{0}) - t_s\phi(\boldsymbol{a}_y) - t_s\phi(-\boldsymbol{a}_y) - t_s\phi(-\boldsymbol{a}_x), \tag{2.99}$$

$$E\psi_A(\boldsymbol{0}) = -t_j\phi(\boldsymbol{0}) - t_g\psi_B(\boldsymbol{d}_1) - t_g\psi_B(\boldsymbol{d}_2), \tag{2.100}$$

where $\phi = \psi(x < 0)$ is the wave function in the metal and $(\psi_A, \psi_B)^{\mathrm{T}} = \psi(x > 0)$ is that in the graphene. $\phi$ and $\psi_\sigma$ are also a solution of the Schrödinger equation of the uniform systems, then the following hold:

$$E\phi(\boldsymbol{0}) = -t_s\phi(\boldsymbol{a}_x) - \epsilon_0\phi(\boldsymbol{0}) - t_s\phi(\boldsymbol{a}_y) - t_s\phi(-\boldsymbol{a}_y) - t_s\phi(-\boldsymbol{a}_x), \tag{2.101}$$

$$E\psi_A(\boldsymbol{0}) = -t_g\psi_B(\boldsymbol{d}_3) - t_g\psi_B(\boldsymbol{d}_1) - t_g\psi_B(\boldsymbol{d}_2), \tag{2.102}$$

Equations 2.99–2.102 reduce to

$$t_j \psi_A(\mathbf{0}) = t_s \phi(\boldsymbol{a}_x), \;\; t_j \phi(\mathbf{0}) = t_g \psi_B(\boldsymbol{d}_3). \tag{2.103}$$

Substituting Eqs. 2.95 and 2.96 in the above conditions, as a result, the reflection probability $R = |r|^2$ is obtained to be

$$R = |r|^2 = \frac{\beta + \beta^{-1} + 2\sin(k_g a_g - \theta - k_s a_s)}{\beta + \beta^{-1} + 2\sin(k_g a_g - \theta + k_s a_s)}, \tag{2.104}$$

with $\beta = t_j^2/(t_g t_s)$. Due to the incident electron is a right-going state with $k_s > 0$, the reflection probability takes a value in the range of $0 < R < 1$. An opaque junction ($R \to 1$) is realized for $\beta \to 0$ or $\beta \to \pm\infty$. The most transparent junction, on the other hand, is realized for $t_j^2 = |t_g t_s|$, that is, $\beta = 1$.

Let us consider the low-energy ($E \approx 0$) transport, where states near the $K$ and $K'$ points can contribute to the transport. The transmitted momentum is evaluated as $k_g a_g \ll k_s a_s$. The transmitted angle $\theta$ is given by $k_y = \text{sign}(E) k_{\text{F g}} \sin\theta$ with $k_{\text{F g}} = |E|/v_g$. As a result, the conductance $G$ from the low-energy states is derived as

$$G = \frac{e^2}{h}\frac{W}{2\pi}\int_{-k_{\text{F g}}}^{k_{\text{F g}}} dk_y (1 - R) \approx \frac{e^2}{h}\frac{W}{\pi}\frac{|E|}{v_g}\int_{-\pi/2}^{\pi/2}\frac{d\theta}{2}\cos\theta(1 - R), \tag{2.105}$$

where $R$ is approximately independent of $E$:

$$R \approx \frac{\beta + \beta^{-1} - 2\sin(k_s a_s + \theta)}{\beta + \beta^{-1} + 2\sin(k_s a_s - \theta)}. \tag{2.106}$$

This leads to $G \propto |E|$, which is the same as the density of states. The whole structure of $G$, which is calculated from Eq. 2.104, is shown in Fig. 2.17a. $G$ becomes larger, approaching $\beta = 1$. The asymmetric structure of $G$ with respect to $E$ is suppressed as one decreases $\beta$. And $G$ has the form similar to the density of states shown in Fig. 2.4d for $\beta \ll 1$. A peak appears at $E/t_g = \pm 1$, where the van Hove singularity in the density of states shows up. Figure 2.17b shows the Fano factor $F$. Again, $F$ takes a finite value at the Dirac point ($E = 0$), while the conductance vanishes, similarly to the case of a bipolar junction of graphene (Fig. 2.13c).

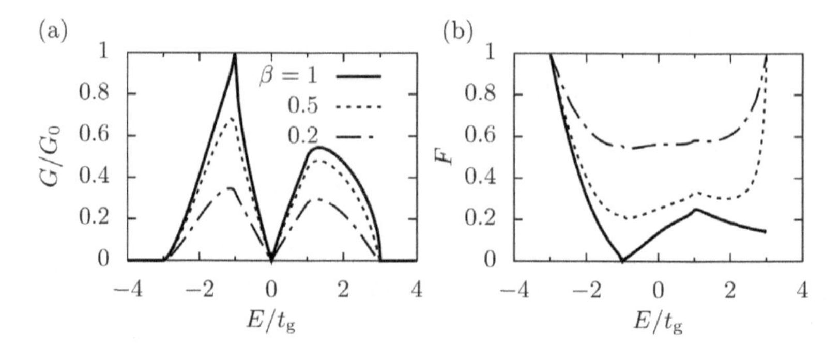


**Figure 2.17** Conductance (a) and Fano factor (b) of a metal/graphene junction for $t_s \gg t_g$. The Fano factor is defined only for $|E| < 3t_g$ since the conductance vanishes for $|E| > 3t_g$. $G_0$ is defined by $G_0 = Wt_g/\pi v_g \times e^2/h$.

## 2.9 Minimal Conductivity

Transport at Dirac points is one of the central issues in graphene. One may naively expect that no transport occurs at Dirac points since the density of states vanishes, as shown in Fig. 2.18a. Contradictory to the naive expectation, nontrivial behaviors, finite conductivity $\sigma_{\min}$ and its nonuniversal values, and metal–insulator transition have been observed. Here we briefly summarize the history and present status of transport at Dirac points.

The early experiments [4, 5] observed a finite value of charge conductivity $\sigma_{\min} \approx 4e^2/h$ at the Dirac point $E = 0$, as shown in Fig. 2.18b; nevertheless the density of states vanishes for $E = 0$.

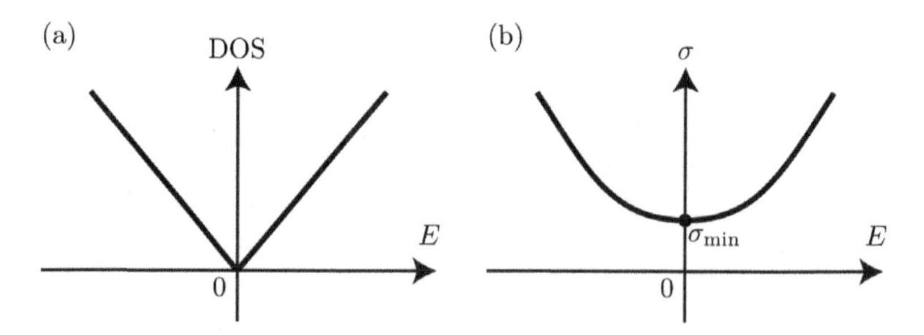


**Figure 2.18** Density of states (a) and typical behavior of conductivity (b) as a function of the Fermi energy in graphene. $\sigma_{\min}$ denotes the value of conductivity for $E = 0$.

Extensive studies based on the Kubo formula have been done to elucidate why the conductivity takes a finite value and succeeded to reproduce a finite value of conductivity. The predicted value $\sigma_{\min} = 4e^2/\pi h$ is, however, about three times as small as that observed in the experiments $\sigma_{\min} \approx 4e^2/h$ [4]. This contradiction between theories and experiments has been called "missing pi." Moreover, the conductivity derived from the Kubo formula depends highly on the frequency $\omega$ and impurity-scattering rate $\eta$, while both $\omega$ and $\eta$ are taken to be zero in the direct current (DC) conductivity for the clean limit [6, 7], namely

$$\lim_{\eta \to 0} \lim_{\omega \to 0} \sigma_{\min} = \frac{4}{\pi} \frac{e^2}{h}, \tag{2.107}$$

$$\lim_{\omega \to 0} \lim_{\eta \to 0} \sigma_{\min} = \frac{\pi}{2} \frac{e^2}{h}. \tag{2.108}$$

In addition, an alternative approach from the Landauer formula predicts conductivity, depending on the aspect ratio of the system, and $\sigma_{\min} = 4e^2/\pi h$ in the limit of $L \ll W$ with $L$ and $W$ being the length and width of the system, respectively [8, 9]. The sensitivity and ambiguity of conductivity indicate that another mechanism dominates the observed minimal conductivity $\sigma_{\min} \approx 4e^2/h$.

The mystery of finite conductivity has been partially solved from a different point of view, that is, the disorder effect. Several studies have demonstrated that in the presence of the long-range disorder potential by charged impurities in graphene, minimal conductivity is largely enhanced to reach $\sigma_{\min} \approx 4e^2/h$, which is comparable to the experimental value [10–12]. The disorder effect scenario has been supported in subsequent experimental results on disordered graphene. In the devices with various strengths of disorder, various values of minimal conductivity from $\sigma_{\min} \approx e^2/h$ to $\sigma_{\min} \approx 10e^2/h$ have been obtained [13]. Furthermore, details of the charged impurity have been studied. Experiments [14, 17] have observed charge inhomogeneity, that is, electron-rich and hole-rich regions, called electron–hole puddles. And theories [15, 16] based on such inhomogeneity have well explained and fitted experimental results. So far, the charge transport at the Dirac points has been understood to be described by disorder effects induced by the electron–hole puddles.

Recently, on the other hand, the metal–insulator transition has been observed in cleaner graphene samples [18, 19]. As discussed above, conductivity takes $\sigma_{\text{min}} = 4e^2/\pi h$ in the clean limit. Therefore, the insulating behavior has suggested that the Dirac electrons in graphene somehow become gapped in the clean limit. This is a remaining issue to be solved in future experimental and theoretical studies.

## 2.10 Half-Integer Quantized Hall Effect

One of the characteristic behaviors intrinsic to graphene emerges under a magnetic field, that is, Landau level and quantum Hall effect. Here we briefly review the conventional quantum Hall effect and develop it for the peculiar quantum Hall effect in graphene.

### 2.10.1 *Landau Level in Conventional Electron Systems*

First, we review the Landau level in conventional electron systems described by the Hamiltonian

$$H = \frac{\pi_x^2 + \pi_y^2}{2m}, \tag{2.109}$$

where $\pi_i = p_i + eA_i/c$ the covariant momentum. Since $\pi_x$ and $\pi_y$ do not commute with each other, one can define creation $a^\dagger$ and annihilation $a$ operators, which satisfy $[a, a^\dagger] = 1$, via $\pi_x$ and $\pi_y$:

$$a = \frac{l_B}{\sqrt{2}\hbar}\pi_-, \; a^\dagger = \frac{l_B}{\sqrt{2}\hbar}\pi_+, \tag{2.110}$$

with $l_B = \sqrt{\hbar c/e|B|}$ the magnetic length.

Eigenstates of the number operator $a^\dagger a$ are given by

$$|n\rangle = \frac{(a^\dagger)^n}{\sqrt{n!}}|0\rangle, \; a^\dagger a|n\rangle = n|n\rangle, \; n = 0, 1, \cdots, \tag{2.111}$$

with

$$a|n\rangle = \sqrt{n}|n-1\rangle, \; a^\dagger|n\rangle = \sqrt{n+1}|n+1\rangle. \tag{2.112}$$

The Hamiltonian is rewritten as

$$H = \hbar\omega_c\left(a^\dagger a + \frac{1}{2}\right), \tag{2.113}$$

with $\omega_c = e|B|/mc$ the cyclotron frequency. As a result, the energy eigenvalue is obtained as

$$E = \hbar\omega_c \left(n + \frac{1}{2}\right), \quad n = 0, 1, \cdots, \tag{2.114}$$

and the eigenvector is given by $|n\rangle$.

The wave function $\psi_0(\boldsymbol{x})$ of the ground state obeys the following equation:

$$a|0\rangle = 0, \text{ i.e., } \pi_- \psi_0(\boldsymbol{x}) = 0. \tag{2.115}$$

In the Landau gauge $\boldsymbol{A} = (0, Bx, 0)$, the above equation is rewritten as

$$\frac{\partial \phi_0}{\partial x} = -\left(k_y + \frac{eB}{c}x\right)\phi_0, \tag{2.116}$$

with $\psi_0(\boldsymbol{x}) = e^{ik_y y}\phi_0(x)$. The solution of the above differential equation is given by

$$\phi(x) = \frac{1}{\pi^{1/4}\sqrt{l_B}} \exp\left[-\frac{(x + l_B^2 k_y)^2}{2l_B^2}\right]. \tag{2.117}$$

Assume a system whose length and width are given by $L_x$ and $L_y$. The momentum takes $k_y = 2q\pi/L_y$, $q = 0, \pm 1, \cdots$. The guiding center of the wave function $X \equiv -l_B^2 k_y$ must be in the range of $0 < X < L_x$, namely $0 < q < \Phi/\Phi_0$, with $\Phi = |B|L_x L_y$ being the magnetic flux and $\Phi_0 = hc/e$ the flux quantum. This condition means that each Landau level has $\Phi/\Phi_0$-fold degenerated states.

## 2.10.2 *Semiclassical Approach for the Quantum Hall Effect*

Here, the quantum Hall effect is analyzed by a semiclassical approach. Assume a Hall bar, as shown in Fig. 2.19, where a bias voltage and magnetic field are applied along the $x$ and $z$ directions, respectively. An electron injected into the Hall bar is accelerated by the Lorentz force and is accumulated on the edge of the system. Then the Hall voltage along the $y$ direction is induced so that the net force along the $y$ direction vanishes:

$$\frac{e}{c}v_x B - eE_y = 0. \tag{2.118}$$

As a result, charge current along the $x$ direction $j_x$ is given by

$$j_x = -e\rho v_x = -\frac{ec\rho}{B}E_y, \tag{2.119}$$

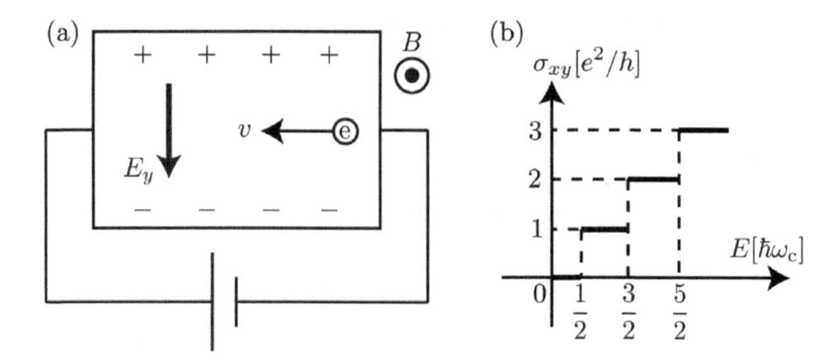


**Figure 2.19** Quantum Hall effect. (a) Hall bar and (b) Hall conductance.

where the electron density $\rho$ is given by

$$\rho = \nu \frac{1}{L_x L_y} \frac{\Phi}{\Phi_0} = \nu \frac{eB}{hc}, \tag{2.120}$$

with $\nu$ being the number of the occupied Landau levels. Therefore, the Hall conductance $\sigma_{xy} = j_x/E_y$ is obtained to be

$$\sigma_{xy} = \nu \frac{e^2}{h}, \tag{2.121}$$

with $\nu = 0, 1, \cdots$. Namely, the Hall conductance is quantized in the unit of $e^2/h$.

This semiclassical approach is a simple explanation of the quantum Hall effect. Here, many perturbations such as electron correlation and disorder potential are neglected. But, even if such a perturbation is taken into account, the quantum Hall effect still survives. Actually, the quantization is experimentally confirmed with awesome precision. This fact suggests that the quantum Hall effect can be understood in a different way, that is, *topology*. Thouless, Kohmoto, Nightingale, and den Nijs have shown that the Hall conductance in two-dimensional systems can be expressed by the topological number, which has been known as the Chern number in topology. Topological numbers are invariant for continuous change of manifolds. In physics, this corresponds to the robustness of quantized Hall conductance against perturbations. The quantum Hall effect is a milestone in a sense that it has established the usefulness of topology in physics. Recently, such a topological concept has been widely extended to other systems including three-dimensional insulators, zero-gap semiconductors, and superconductors, in which a kind of quantization occurs, guaranteed by topological properties of wave functions. These materials are called topological insulators and superconductors and have attracted much attention.

### 2.10.3 *Landau Level in Graphene*

Now we turn to graphene. Under a magnetic field, the Hamiltonian (2.12) is modified as

$$H = v_g \begin{pmatrix} 0 & i\pi_- \\ -i\pi_+ & 0 \end{pmatrix} = \hbar\omega_c \begin{pmatrix} 0 & ia \\ -ia^\dagger & 0 \end{pmatrix}, \tag{2.122}$$

with $\omega_c = v_g/l_B$. The above Hamiltonian has the zero-energy state $\boldsymbol{u}_0 = (0, |0\rangle)^T$, and finite-energy states $\boldsymbol{u}_n = (a_n|n-1\rangle, b_n|n\rangle)^T$, $n \geq 1$, with

$$\hbar\omega_c \begin{pmatrix} 0 & i\sqrt{n} \\ -i\sqrt{n} & 0 \end{pmatrix} \begin{pmatrix} a_n \\ b_n \end{pmatrix} = E_n \begin{pmatrix} a_n \\ b_n \end{pmatrix}. \tag{2.123}$$

Here, $|n\rangle$ is the $n$-th eigenstate of a harmonic oscillator given by Eqs. 2.111 and 2.112. Therefore, one obtains the energy of the $n$-th Landau level in graphene:

$$E_{\pm n} = \pm\hbar\omega_c\sqrt{n}. \tag{2.124}$$

The eigenvectors are given by $(a_n, b_n) = (1, -i)/\sqrt{2}$ for positive-energy states and by $(a_n, b_n) = (1, i)/\sqrt{2}$ for negative-energy states. The energy of Landau levels is proportional to $\sqrt{n}$, which is completely different from those for conventional electron systems ($\propto n$) given by Eq. 2.114.

### 2.10.4 *Half-Quantized Hall Conductance in Graphene*

Hall conductance in graphene is also understood in a semiclassical approach discussed in Section 2.10.2, namely $\sigma_{xy} = -ec\rho/B$. Note that $\rho$ in the expression of Hall conductance is reinterpreted as a carrier density with $\rho > 0$ for $n$-type and $\rho < 0$ for $p$-type semiconductors. Each Landau level with a positive/negative energy supplies $\Phi/\Phi_0$ electrons/holes. The zero-energy Landau level peculiar to graphene, on the other hand, supplies $1/2 \times \Phi/\Phi_0$

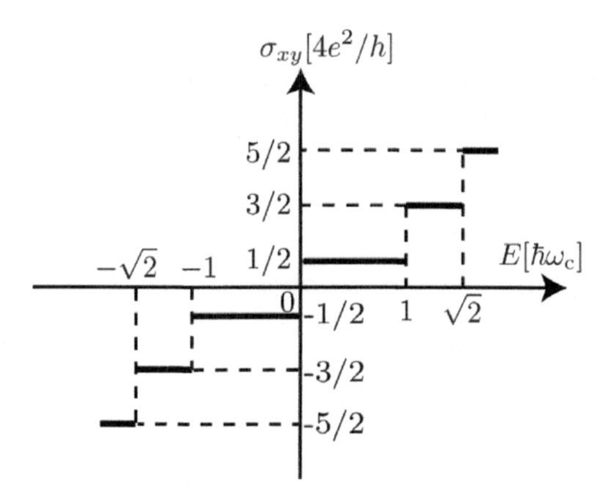


**Figure 2.20** Half-quantized Hall effect.

electrons/holes for a positive/negative Fermi energy in order to satisfy the charge neutrality of zero energy. As a result, the Hall conductance in graphene is given by

$$\sigma_{xy} = \frac{4e^2}{h}\left(\nu + \frac{1}{2}\right), \quad \nu = 0, \pm 1, \pm 2, \cdots, \tag{2.125}$$

where factor 4 comes from spin (up and down spins) and valley ($K$ and $K'$ points) degeneracies. It is emphasized that the offset 1/2 is a consequence of the zero-energy Landau level. Interestingly, this half-quantized Hall conductance has been clearly observed in experiments.

## 2.11 From Graphene to Topological Insulators

We have shown that graphene has massless Dirac fermions, as low-lying excitations, which are protected by symmetry. Symmetry-breaking perturbations open a gap and make the Dirac fermions massive. This gapped state is also interesting because massive Dirac fermions can exhibit a kind of quantum Hall effect. Developing this idea, one can reach to a new concept of matter, that is, topological insulators. In this section, we briefly review the properties of massive Dirac fermions and various types of the quantum Hall effects and finally introduce an idea of topological insulators.

### 2.11.1 *Massive Dirac Fermions*

Firstly, we introduce a massive Dirac fermion described by the following Hamiltonian:

$$H = m\sigma_z + v(k_y\sigma_x - \tau k_x\sigma_y), \ \tau = \pm 1, \tag{2.126}$$

where mass $m$ corresponds to the energy gap. This model shows the half-quantized Hall effect. The Hall conductivity $\sigma_{xy}$ of $2 \times 2$ Hamiltonian is calculated by the formula [20]

$$\sigma_{xy} = -\frac{e^2}{h}\int \frac{d^2k}{4\pi}\hat{\boldsymbol{d}}(\boldsymbol{k}) \cdot \frac{\partial \hat{\boldsymbol{d}}(\boldsymbol{k})}{\partial k_x} \times \frac{\partial \hat{\boldsymbol{d}}(\boldsymbol{k})}{\partial k_y}, \tag{2.127}$$

where $\hat{\boldsymbol{d}}(\boldsymbol{k})$ is the $d$-vector associated to $\sigma$ defined by $\hat{\boldsymbol{d}}(\boldsymbol{k}) = \boldsymbol{d}(\boldsymbol{k})/d(\boldsymbol{k})$ with

$$d_x(\boldsymbol{k}) = vk_y, \ d_y(\boldsymbol{k}) = -\tau v k_x, \ d_z(\boldsymbol{k}) = m, \tag{2.128}$$

and

$$d(\boldsymbol{k}) = \sqrt{d_x^2(\boldsymbol{k}) + d_y^2(\boldsymbol{k}) + d_z^2(\boldsymbol{k})} = \sqrt{m^2 + v^2(k_x^2 + k_y^2)}. \tag{2.129}$$

Using this formula, $\sigma_{xy}$ for a massive Dirac fermion is obtained to be

$$\sigma_{xy} = -\frac{\tau}{2}\frac{e^2}{h}\text{sign}(m). \tag{2.130}$$

Surprisingly, as long as $m$ is finite, the system shows a half-quantized quantum Hall effect. Though graphene has two Dirac fermions located at the $K$ and $K'$ points, $\sigma_{xy}$ becomes an integer multiplied by $e^2/h$, including $\sigma_{xy} = 0$. The existence of a finite value for $\sigma_{xy}$ depends on the symmetry of the mass term of Dirac fermions. In the following section, we discuss several cases of finite quantum Hall conductance from massive Dirac fermions in graphene with symmetry-breaking perturbations.

### 2.11.2 *Quantum Valley Hall Effect*

A simple mass term is induced by an inequivalent sublattice structure, which is realized in boron nitride with the honeycomb lattice. The Hamiltonian is given by $H = H_{\mathrm{g}} + H_{\mathrm{BN}}$ with $H_{\mathrm{g}}$ being the Hamiltonian of graphene and

$$H_{\mathrm{BN}} = \sum_n m_{\mathrm{BN}} \left( c_{n\mathrm{A}}^\dagger c_{n\mathrm{A}} - c_{n\mathrm{B}}^\dagger c_{n\mathrm{B}} \right). \tag{2.131}$$

The resulting low-energy Hamiltonians $H_\tau$, $\tau = K, K'$, in the vicinity of the $K$ and $K'$ points, are given by

$$H_\tau = m\sigma_z + v(k_y\sigma_x - \tau k_x\sigma_y). \quad (2.132)$$

The Hall conductance $\sigma^\tau_{xy}$ from the $K$ or $K'$ points is given by Eq. 2.130, $\sigma^\tau_{xy} = -\tau e^2/2h$. The total (charge) Hall conductance $\sigma^c_{xy}$ is obtained to be the summation of those contributed from the $K$ and $K'$ points, but these are canceled out:

$$\sigma^c_{xy} = \sigma^K_{xy} + \sigma^{K'}_{xy} = 0 \quad (2.133)$$

On the other hand, if one examines the difference, not the summation, a finite value of Hall conductance, so-called valley Hall conductance $\sigma^v_{xy}$, is obtained:

$$\sigma^v_{xy} = \sigma^K_{xy} - \sigma^{K'}_{xy} = -\frac{e^2}{h}. \quad (2.134)$$

This is a new type of quantum Hall effect [21]. The valley Hall conductance, however, is hard to detect since the valley is an object not in the real space but in the reciprocal space. Nonetheless, the above discussion serves as a good example leading to a novel state of matter, that is, a kind of quantum Hall effect can occur in a system effectively described by massive Dirac fermions. In the subsequent section, we shall see a system with massive Dirac fermions that is more directly related to actual materials and experiments.

### 2.11.3 *Quantum Spin Hall Effect*

In 2005, Kane and Mele found that spin–orbit interaction in graphene opens an energy gap and yields the quantum spin Hall effect [22]. The spin–orbit interaction $H_{SO}$ for low-energy states in graphene is expressed as

$$H_{SO} = \Delta\sigma_z s_z \tau, \quad (2.135)$$

where $s_i$ denotes the Pauli matrix representing spin. The Hall conductance of a massive Dirac fermion for spin $s = \uparrow, \downarrow$ in the vicinity of the $\tau = K$ and $\tau = K'$ points is given by

$$\sigma^{\tau s}_{xy} = -\frac{\tau}{2}\frac{e^2}{h}\mathrm{sign}(s\tau). \quad (2.136)$$

From this expression, one finds that the charge and valley Hall conductances vanish

$$\sigma_{xy}^{\mathrm{c}} = \sum_{\tau s} \sigma_{xy}^{\tau s} = 0, \; \sigma_{xy}^{\mathrm{v}} = \sum_{\tau s} \tau \sigma_{xy}^{\tau s} = 0, \tag{2.137}$$

while the spin Hall conductance becomes finite

$$\sigma_{xy}^{\mathrm{s}} = \sum_{\tau s} s \sigma_{xy}^{\tau s} = -\frac{e^2}{h}. \tag{2.138}$$

The spin Hall conductance is a well-defined quantity if spin–orbit interaction is weak, and can be detected experimentally. Unfortunately, spin–orbit interaction in graphene is quite weak since carbon is a light atom, and therefore it is difficult to experimentally detect the spin Hall conductance. After the proposal by Kane and Mele [22], a promising material candidate, HgTe/CdTe quantum well, which has strong spin–orbit interaction generated by heavy atoms Hg, Cd, and Te, was proposed [23]. Soon after that, the experiments revealing quantum spin Hall effect has been reported [24]. Also, another material InAs/GaSb quantum well exhibiting the quantum spin Hall effect has been found [25–28]. Thus, the quantum spin Hall effect has been established.

### 2.11.4 $\mathbb{Z}_2$ *Topological Insulator*

In the previous section, we focused on the quantization of spin Hall conductance. However, this is not sufficient to define a stable nontrivial phase, since the spin is not conserved in the presence of strong spin–orbit interaction. In fact, the concept of quantization of spin Hall conductance can be generalized even for strongly spin–orbital coupled systems. The key is time reversal symmetry [29]. Spin up $|\uparrow\rangle$ and down $|\downarrow\rangle$ form a Kramers pair:

$$\Theta \left|\uparrow\right\rangle = \left|\downarrow\right\rangle, \; \Theta \left|\downarrow\right\rangle = -\left|\uparrow\right\rangle, \tag{2.139}$$

where $\Theta = -is_y\mathcal{K}$ denotes the time reversal operator. Spin is not a good quantum number in the presence of spin–orbit interaction, whereas the above relation still holds as

$$\Theta \left|\boldsymbol{k}, \alpha, \mathrm{I}\right\rangle = \left|-\boldsymbol{k}, \alpha, \mathrm{II}\right\rangle, \; \Theta \left|\boldsymbol{k}, \alpha, \mathrm{II}\right\rangle = -\left|-\boldsymbol{k}, \alpha, \mathrm{I}\right\rangle, \tag{2.140}$$

in time reversal invariant systems, where $|\boldsymbol{k}, \alpha, \mathrm{I}/\mathrm{II}\rangle$ denotes an eigenstate of Hamiltonian, $\alpha$ is the band index, and I(II), which corresponds to spin in the absence of spin–orbit interaction, labels a Kramers partner. As a result, one can define an operator $\hat{s}_\Theta$ alternative to spin by

$$\hat{s}_\Theta = \sum_{\boldsymbol{k},\alpha} \left( |\boldsymbol{k}, \alpha, \mathrm{I}\,\rangle\langle \boldsymbol{k}, \alpha, \mathrm{I}| - |-\boldsymbol{k}, \alpha, \mathrm{II}\rangle \langle -\boldsymbol{k}, \alpha, \mathrm{II}| \right), \qquad (2.141)$$

with $[\hat{s}_\Theta, H(\boldsymbol{k})] = 0$, and would expect the quantum Hall effect with respect to $\hat{s}_\Theta$. However, $\hat{s}_\Theta$ itself has an ambiguity: $|\boldsymbol{k}, \alpha, \mathrm{I}\,\rangle$ cannot be uniquely defined, although if $|\boldsymbol{k}, \alpha, \mathrm{I}\,\rangle$ is defined $|\boldsymbol{k}, \alpha, \mathrm{II}\rangle$ and $\hat{s}_\Theta$ are defined. In spite of this, the $\mathbb{Z}_2$ part (even or odd) of the Hall conductance of $\hat{s}_\Theta$ becomes free from the ambiguity for the gauge choice of $|\boldsymbol{k}, \alpha, \mathrm{I}\,\rangle$ [29, 30]. In this way, time reversal invariant insulators are characterized by the $\mathbb{Z}_2$ invariant, instead of $\mathbb{Z}$ from the quantum Hall conductance. This $\mathbb{Z}_2$ classification has opened a new way to the state of matter. Namely, gapped states, including the integer quantum Hall effect, can be classified by topological numbers such as $\mathbb{Z}$ and $\mathbb{Z}_2$ invariants [34–36]. Furthermore, the concept of $\mathbb{Z}_2$ topological insulators, which has the nontrivial $\mathbb{Z}_2$ invariant, can be straightforwardly extended to three-dimensional insulators [31–33]. More interestingly, superconductors are also characterized by such topological numbers (topological superconductors). Topological insulators and superconductors have opened a novel field of condensed matter physics [37–40].

## References

1. K. S. Novoselov, A. K. Geim, S. V. Morozov, D. Jiang, Y. Zhang, S. V. Dubonos, I. V. Grigorieva, and A. A. Firsov, *Science*, **306**, 666 (2004).
2. P. R. Wallace, *Phys. Rev.*, **71**, 622 (1947).
3. Y. M. Blanter and I. Martin, *Phys. Rev. B*, **76**, 155433 (2007).
4. K. S. Novoselov, A. K. Geim, S. V Morozov, D. Jiang, M. I. Katsnelson, I. V Grigorieva, S. V Dubonos, and A. A. Firsov, *Nature*, **438**, 197 (2005).
5. Y. Zhang, Y.-W. Tan, H. L. Stormer, and P. Kim, *Nature*, **438**, 201 (2005).
6. A. W. W. Ludwig, M. Fisher, R. Shankar, and G. Grinstein, *Phys. Rev. B*, **50**, 7526 (1994).
7. K. Ziegler, *Phys. Rev. B*, **75**, 233407 (2007).

8. J. Tworzydło, B. Trauzettel, M. Titov, A. Rycerz, and C. W. J. Beenakker, *Phys. Rev. Lett.*, **96** (2006).
9. M. I. Katsnelson, *Eur. Phys. J. B*, **51**, 157 (2006).
10. H. Kumazaki and D. S. Hirashima, *J. Phys. Soc. Jpn.*, **75**, 053707 (2006).
11. T. Ando, *J. Phys. Soc. Jpn.*, **75**, 074716 (2006).
12. K. Nomura and A. H. MacDonald, *Phys. Rev. Lett.*, **98**, 076602 (2007).
13. Y. W. Tan, Y. Zhang, K. Bolotin, Y. Zhao, S. Adam, E. H. Hwang, S. Das Sarma, H. L. Stormer, and P. Kim, *Phys. Rev. Lett.*, **99**, 246803 (2007).
14. A. K. Geim and K. S. Novoselov, *Nat. Mater.*, **6**, 183 (2007).
15. E. H. Hwang, S. Adam, and S. Das Sarma, *Phys. Rev. Lett.*, **98**, 186806 (2007).
16. V. V. Cheianov, V. I. Fal'ko, B. L. Altshuler, and I. L. Aleiner, *Phys. Rev. Lett.*, **99**, 176801 (2007).
17. J. Martin, N. Akerman, G. Ulbricht, T. Lohmann, J. H. Smet, K. von Klitzing, and A. Yacoby, *Nat. Phys.*, **4**, 144 (2007).
18. L. A. Ponomarenko, A. K. Geim, A. A. Zhukov, R. Jalil, S. V. Morozov, K. S. Novoselov, I. V. Grigorieva, E. H. Hill, V. Cheianov, V. Falko, K. Watanabe, T. Taniguchi, and R. V. Gorbachev, *Nat. Phys.*, **7**, 958 (2011).
19. F. Amet, J. R. Williams, K. Watanabe, T. Taniguchi, and D. Goldhaber-Gordon, *Phys. Rev. Lett.*, **110**, 216601 (2013).
20. X.-L. Qi, Y.-S. Wu, and S.-C. Zhang, *Phys. Rev. B*, **74**, 085308 (2006).
21. G. W. Semenoff, *Phys. Rev. Lett.*, **53**, 2449 (1984).
22. C. L. Kane and E. J. Mele, *Phys. Rev. Lett.*, **95**, 226801 (2005).
23. B. A. Bernevig, T. L. Hughes, and S.-C. Zhang, *Science*, **314**, 1757 (2006).
24. M. König, S. Wiedmann, C. Brune, A. Roth, H. Buhmann, L. W. Molenkamp, X.-L. Qi, and S.-C. Zhang, *Science*, **318**, 766 (2007).
25. C. Liu, T. L. Hughes, X.-L. Qi, K. Wang, and S.-C. Zhang, *Phys. Rev. Lett.*, **100**, 236601 (2008).
26. I. Knez, R.-R. Du, and G. Sullivan, *Phys. Rev. B*, **81**, 201301 (2010).
27. I. Knez, R.-R. Du, and G. Sullivan, *Phys. Rev. Lett.*, **107**, 136603 (2011).
28. K. Suzuki, Y. Harada, K. Onomitsu, and K. Muraki, *Phys. Rev. B*, **87**, 235311 (2013).
29. C. L. Kane and E. J. Mele, *Phys. Rev. Lett.*, **95**, 146802 (2005).
30. L. Fu and C. L. Kane, *Phys. Rev. B*, **74**, 195312 (2006).
31. L. Fu, C. L. Kane, and E. J. Mele, *Phys. Rev. Lett.*, **98**, 106803 (2007).
32. J. E. Moore and L. Balents, *Phys. Rev. B*, **75**, 121306(R) (2007).
33. R. Roy, *Phys. Rev. B*, **79**, 195322 (2009).

34. A. P. Schnyder, S. Ryu, A. Furusaki, and A. W. W. Ludwig, *Phys. Rev. B*, **78**, 195125 (2008).
35. A. Kitaev, *AIP Conf. Proc.*, **1134**, 22 (2009).
36. S. Ryu, A. P Schnyder, A. Furusaki, and A. W. W. Ludwig, *New J. Phys.*, **12**, 065010 (2010).
37. M. Z. Hasan and C. L. Kane, *Rev. Mod. Phys.*, **82**, 3045 (2010).
38. X.-L. Qi and S.-C. Zhang, *Rev. Mod. Phys.*, **83**, 1057 (2011).
39. Y. Tanaka, M. Sato, and N. Nagaosa, *J. Phys. Soc. Jpn.*, **81**, 011013 (2012).
40. Y. Ando, *J. Phys. Soc. Jpn.*, **82**, 102001 (2013).
41. T. Ando, T. Nakanishi, and R. Saito, *J. Phys. Soc. Jpn.*, **67**, 2857 (1998).

Chapter 3

# Electronic Structures of Graphene and Graphene Contacts

The electronic structure is a fundamental physical characteristic of graphene, and knowledge of this structure is necessary in order to understand the physics of graphene and to investigate its technological applications. Because graphene has a two-dimensional honeycomb lattice structure, its electronic structure, energy dispersion, and density of states can be calculated easily using the tight-binding (TB) model or the first-principles method. In this chapter, we present the results of electronic structure calculations for (i) single- and bilayer graphene sheets; (ii) graphene nanoribbons (GNRs) and those with hydrogen or vacancies; (iii) metal/graphene contacts, including metal/graphene/metal junctions; and (iv) graphene sheets with ripples.

## 3.1 Single- and Bilayer Graphene Sheets

### 3.1.1 *Single-Layer Graphene Sheet*

The electronic state of a carbon atom is composed of fully occupied $1s$- and $2s$-states, along with partially occupied $2p$-states with two

*Graphene in Spintronics: Fundamentals and Applications*
Jun-ichiro Inoue, Ai Yamakage, and Shuta Honda

ISBN 978-981-4669-56-6 (Hardcover), 978-981-4669-57-3 (eBook)
www.panstanford.com

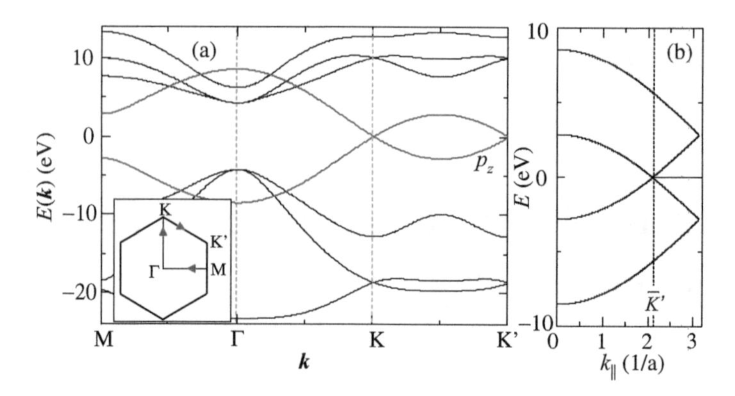


**Figure 3.1** (a) Energy dispersion of a graphene sheet along a high-symmetry wave vector and (b) energy dispersion along $k$ parallel to the zigzag-edge direction of a graphene nanoribbon.

electrons. Because the 1$s$-state has significantly lower energy than the other two states, the atomic orbitals of the 2$s$- and 2$p$-states are sufficient to describe the basic features of the electronic state of graphene. As the unit cell of a graphene sheet honeycomb lattice contains two carbon atoms, the electronic structure of a single-layer graphene sheet can be easily calculated using the tight-binding (TB) model and including 2$s$ and 2$p$ orbitals.

Figure 3.1a shows the energy dispersion relation of a single-layer graphene sheet obtained using the full orbital TB model in the Slater–Koster scheme. The overlap integrals between the different orbitals are neglected for simplicity. The energy levels of the 2$s$- and 2$p$-states and the hopping integrals between the nearest-neighbor (n.n.) sites are shown below. Note that these parameter values were determined according to Harrison's textbook [1]. The energy dispersion given in red consists of $p_z$ orbitals only, because it does not mix with the other orbitals; these are called $\pi$ bands. The dispersions represented by the black curves consist of $s$, $p_x$, and $p_y$ orbitals. The wave functions of these orbitals intermix with each other and form energy bands with a rather large band gap, which are called $\sigma$ bands. The Fermi energy is located at $E = 0$ eV in the figure (also see Table 3.1).

As can be seen from the results shown in Fig. 3.1, there are only two momentum states at the Fermi energy, $K$ and $K'$, and the energy

**Table 3.1** Values of band parameters used in the tight-binding calculation of the energy structure of graphene

| Energy levels (eV) | Hopping integrals (eV) |
|---|---|
| $\epsilon_s = -8.550$ | $ss\sigma = -4.926$ |
| $\epsilon_p = 0$ | $sp\sigma = 6.474$ |
| | $pp\sigma = 5.700$ |
| | $pp\pi = -2.850$ |

dispersion of the $\pi$ band is linear near these states. This means that $K$ and $K'$ are the Dirac points (DPs). Note that the energy dispersions near the DPs are called Dirac dispersions.

The orbital-decomposed density of states (DOS) of graphene is shown in Fig. 3.2. The calculation is performed for 20,000 $k$ points in the Brillouin zone with a small broadening $i\delta = 10^{-4}$ eV. These DOSs exhibit two-dimensional profiles, that is, a peaky shape and a discontinuous drop at the band edges, which is caused by van Hove singularities. The DOS value is infinitesimally small at the Fermi level, $E_{\mathrm{F}}$. Therefore, graphene is often called a zero-gap semiconductor. Because the energy bands consist of the $\pi$ band only

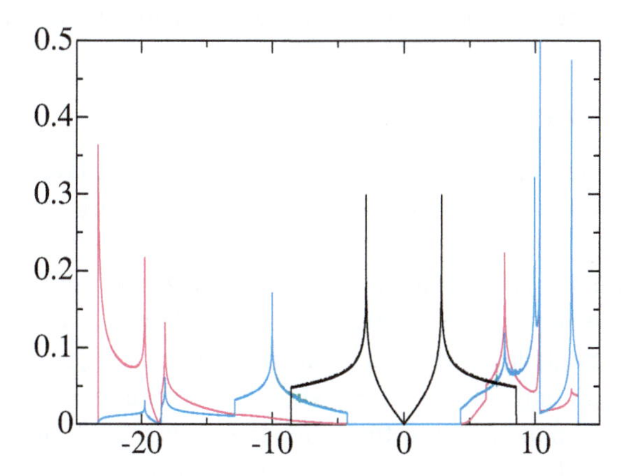


**Figure 3.2** Density of states projected to each orbital component. The $\pi$ bands are shown in black curves and the other curves belong to the $\sigma$ bands.

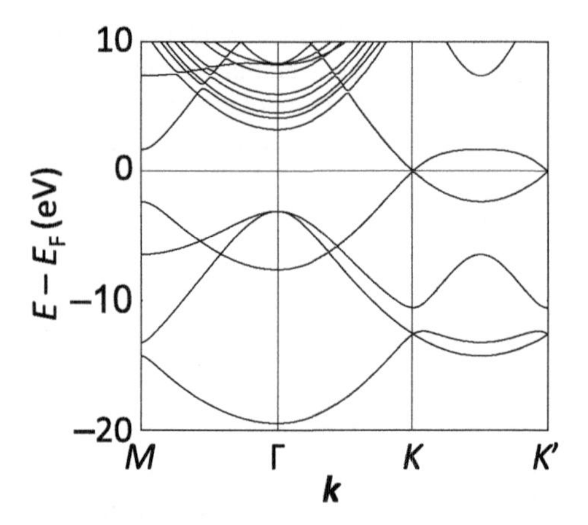


**Figure 3.3** Calculated results of electronic structure of graphene sheet by using the first-principles method.

near the Fermi energy, $\sigma$ bands are often neglected in discussions of the low-energy physics of graphene.

The results of graphene energy-state calculations performed using the first-principles method are shown in Fig. 3.3. The results of the electronic structure calculations near the Fermi energy are almost identical to those obtained using the TB model. The greatest difference is for the conduction bands above the Fermi energy, but this difference is irrelevant in terms of the properties of graphene.

### 3.1.2 *Bilayer Graphene*

Because graphite comprises multistacked graphene layers, it is quite natural for us to consider the electronic structures and physical properties of multilayered graphene lattices. Among them, bilayer graphene has been studied most extensively, because this material exhibits electronic structures that are qualitatively different from those of single-layer graphene. The most interesting point is that an energy gap may appear at $E_F$ when an external electric field $E$ is applied perpendicular to the graphene sheets. The opening of the energy band gap is due to a lack of inversion symmetry between the upper and lower graphene sheets. Note that the inversion symmetry is eliminated by the different charge distributions on the upper and

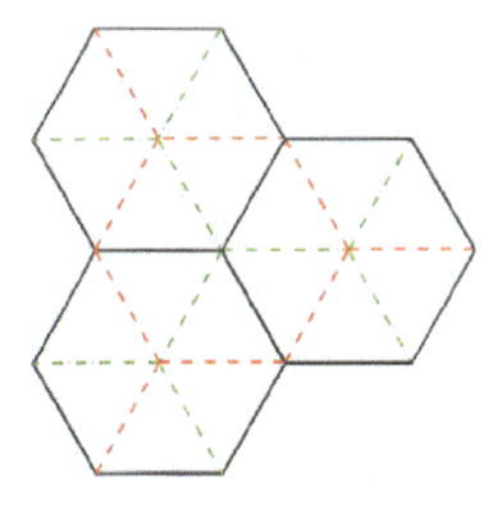

**Figure 3.4** Structures of the honeycomb lattice with three types of stacking. The honeycomb lattices A, B, and C in text are shown in solid black, broken red, and broken green lines, respectively.

lower graphene sheets. The appearance of the energy gap at $E_{\mathrm{F}}$ is crucial in technological applications of graphene where silicon is replaced by a graphene sheet.

The equilibrium interlayer distance $d_{\mathrm{eq}}$ of bilayer graphene is approximately 0.334 nm, which is significantly larger than the carbon–carbon distance of a single graphene sheet (approximately 0.142 nm). The honeycomb lattice of the overlayer graphene sheet has three inequivalent orientations, which are denoted as A, B, and C in Fig. 3.4 [2]. Stacking of the form ABAB ... (Bernal stacking) is the most common structure found in bulk graphite, although ABCABC ... (rhombohedral stacking) is also observed in this material. In multilayered graphene sheets, even a simple hexagonal stacking of form AA ... has been observed using high-resolution transmission electron microscopy (HRTEM) [3]. The same study has also shown that bilayer graphene edge openings can be found in limited areas, where the closed edge is partially broken.

The electronic structure of bilayer graphene can be calculated easily using the TB model or the first-principles calculation [2, 4–6]. Figure 3.5b shows the electronic structure near the DP for bilayer graphene, where a bias voltage $V$ is applied between the two layers [5]. The bilayer lattice structure is of AB type, as shown in Fig. 3.5a. The zero-gap structure of the unbiased bilayer, represented by the dotted curves, changes to the gapped structure indicated by the solid curves. We find that the band gap ($\Delta g$ in the figure) is related to the potential ($eV$ in the figure) induced by $V$. It should be noted that the Dirac (linear) dispersion of single-layer graphene changes to a dispersive form, as shown by the dotted curves in Fig. 3.5b.

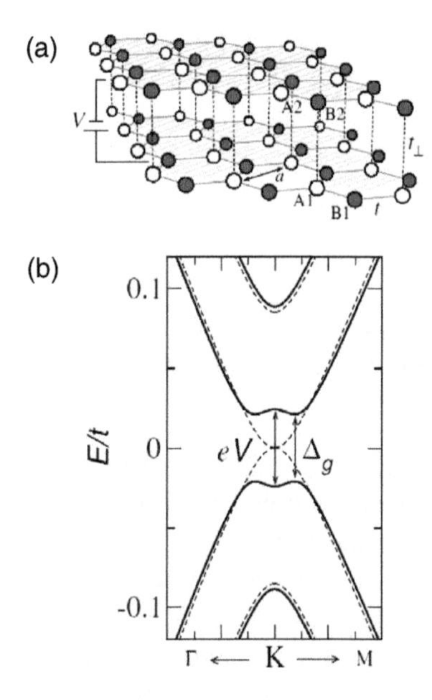


**Figure 3.5** (a) Lattice structure of bilayer graphene and (b) calculated energy state of bilayer graphene. Reprinted (figure) with permission from Ref. [5], Copyright (2007) by the American Physical Society.

Direct observation of the band gap in bilayer graphene has been performed in several experiments [7–10]. For example, Zhang et al. [8] performed an infrared microspectroscopy analysis of a dual-gate field-effect transistor (FET) composed of AB-stacked bilayer graphene; their experimental results are indicated by the square dots in Fig. 3.6. These researchers found that the band gap can be tuned up to 0.25 eV using a gate control. Their experimental results have been compared with theoretical values calculated using TB approximation [11] and the first-principles method [12], and agreement has been found between all three sets of results.

Detailed calculations have been performed using first-principles density functional theory (DFT) in order to examine the dependence of the energy band gap on the $V$ due to an $E$ induced by a gate voltage $V_g$ and also its dependence on the interlayer spacing $d$ [13]. The results are shown in Fig. 3.7 and indicate that the energy gap tends to increase with increased $E$ and saturates at higher values. The energy gap saturation values are larger for smaller $d$.

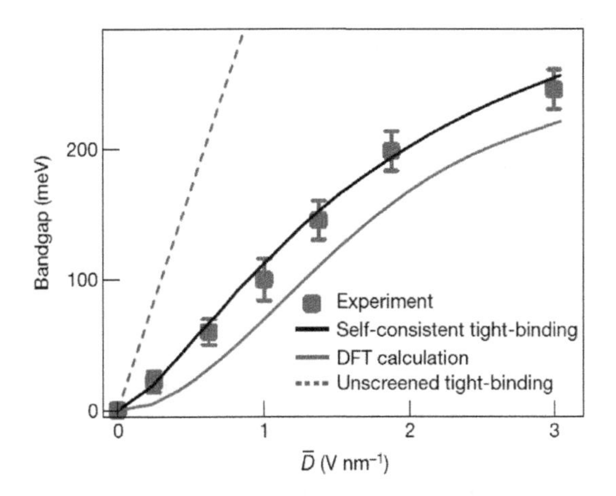


**Figure 3.6** Experimental and calculated results of a change in the energy band gap of bilayer graphene. Reprinted by permission from Macmillan Publishers Ltd: [*Nature*] (Ref. [8]), copyright (2009).

For the equilibrium spacing distance $d = 0.334$ calculated using the first-principles method, the saturated energy gap is almost 0.3 eV. Recently, the conductance $\Gamma$ of two types of suspended bilayer graphene was measured as a function of $V_g$, and it was suggested that one type of bilayer is insulating even at $V_g = 0$ [10].

## 3.2 Graphene Nanoribbons

A pseudo one-dimensional graphene structure is extremely important for technological applications in which electric currents are used for device functions. Such structures, which have nanometer-scale width $W$, are called graphene nanoribbons (GNRs). There are two basic types of GNRs, which have zigzag or armchair edges, as shown in Fig. 1.3 (Chapter 1). The electronic structures of these zigzag- and armchair-edge GNRs are easily calculated using the TB scheme. In fact, the electronic structures of GNRs had been studied intensively prior to the discovery of the graphene sheet [14, 15]. Representative results of GNR electronic structure calculations are shown in Fig. 3.8, in which the width $W$ is given in terms of $N =$ 4, 5, and 6, where $N$ indicates the number of zigzag chains [15]. Figure 3.8a shows the energy dispersion results for GNRs with

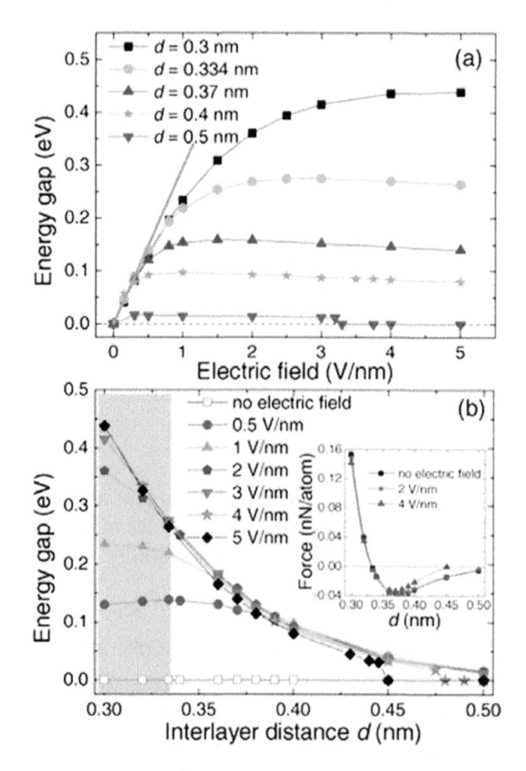


**Figure 3.7** (a) Energy band gap as a function of applied electric field and (b) calculated results of band gap as a function of interlayer lattice distance ($d$) in bilayer graphene. Reproduced with permission from Ref. [13]. Copyright 1992, AIP Publishing LLC.

armchair edges. It can be seen that an energy gap appears at the Fermi energy in armchair-edge GNRs, except for those with $N = 3M - 1$, where $M$ is an integer. On the other hand, in zigzag-edge GNRs, the conduction ($E > 0$) and valence ($E < 0$) states meet at the momentum $k = \pi$ and no energy gap appears, as shown in Fig. 3.8b. We find that a flat dispersion appears near $k = \pi$, which is identified as an edge-state (resonant-state, or Shockley-state) characteristic in zigzag-edge GNRs. Following discovery of the graphene sheet, similar calculations were performed using the Dirac Hamiltonian [16] and the first-principles method [17–19]. The mechanisms of these unusual GNR properties have already been debated, for example, in various review articles [2, 6].

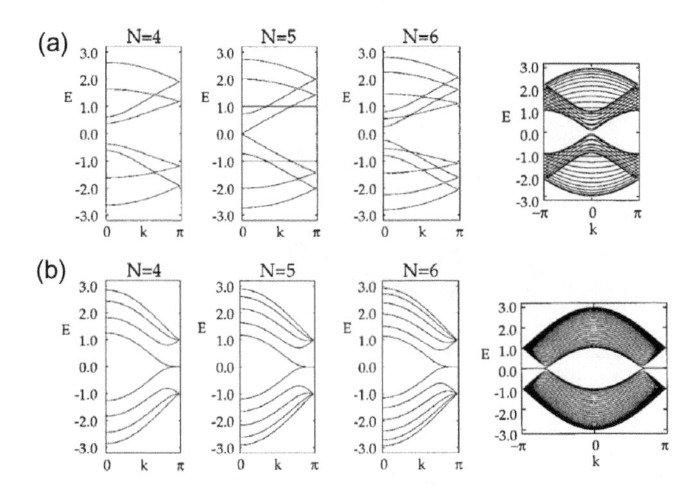


**Figure 3.8** (a) Energy states of armchair graphene nanoribbons and (b) those of zigzag graphene nanoribbons. Reprinted (figure) with permission from Ref. [15], Copyright (1996) by the American Physical Society.

## 3.2.1 *Energy Dispersion and Resonant States*

Here, we discuss further details of the electronic structures of single-layer GNRs. Let us first consider zigzag-edge GNRs with infinite length but finite $W$. In the following, $W$ is given in terms of the number of one-dimensional atomic chains, which are denoted as MCs. One zigzag chain composed of carbon atoms is counted as being two MCs in our notation, that is, $W = 2N$, where $N$ is defined as previously. The number of MCs in the GNRs shown in Fig. 3.9b is eight. Note that the present notation for the GNR $W$ differs from that used elsewhere in the literature. Because the GNR length $L$ is infinite and translational invariance holds in this direction, the momentum $k_{\parallel}$ in this direction is a good quantum number. The relationship between $k_{\parallel}$ and the Brillouin zone is also depicted in Fig. 3.9c.

Figure 3.9a shows the energy dispersions of a GNR with zigzag edges and $W = 100$ MCs. The band parameters are identical to those used in the two-dimensional graphene sheet calculations and the energy is given as a function of $k_{\parallel}$ parallel to the zigzag edge. The energy bands in black and blue are the $\pi$ and $\sigma$ bands, respectively. $K'$ is projected onto the $k_y$ ($k_{\parallel}$) axis and is referred to as $\bar{K}'$ hereafter, which appears at $k_{\parallel} = 2\pi/3$.

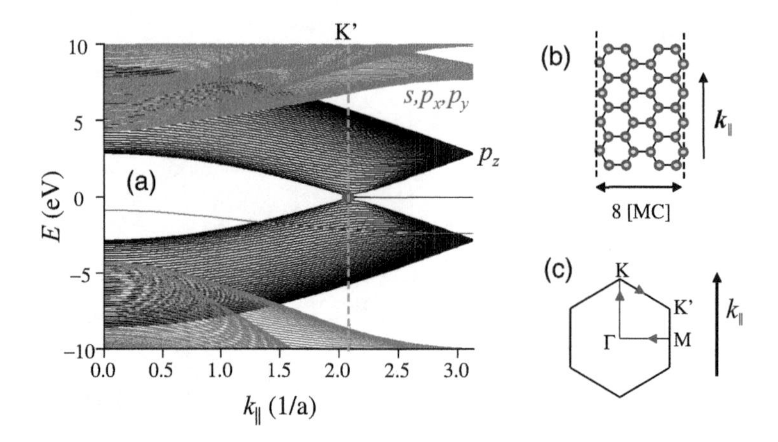


**Figure 3.9** (a) Calculated result of energy dispersion of graphene nanoribbon with zigzag edges with width $W = 100$ monochains (MCs). (b) Lattice structure of a zigzag-edge GNR with $W = 8$ MCs and (c) Brillouin zone of graphene and definition of momentum parallel to layer direction.

Resonant states appear in both the $\pi$ and $\sigma$ bands because of the presence of the edge. The $\pi$ band resonant state forms a flat band, which is represented by the black line in Fig. 3.9a. The red dot indicates the position of the $\bar{K}'$ point. Strictly speaking, the energy dispersion of the resonant state is not flat; rather, it depends on $W$. When $W \lesssim 20$, the DP is blurred and the lowest branch of the conduction band and the highest branch of the valence band merge as $k_{\|} \to \pi$. For wider ribbons, the resonant state tends to be dispersion-less. The amplitude of the resonant state of the $\pi$ bands in zigzag-edge GNRs is nonzero for only one of two sublattices, as shown in Fig. 3.11a. The amplitude is at a maximum at the zigzag edge and decays on sites further removed from the edge. The resonant states of the $\sigma$ bands, on the other hand, form a dispersive band, as shown by the red curve in Fig. 3.9a. The latter appears slightly below $E_{\text{F}}$. In contrast to the $\pi$ band case, the amplitude of the $\sigma$ band resonant state decays quickly as it extends into the inner region of the ribbon.

The energy band dispersions of armchair-edge GNRs are shown in Fig. 3.10a. Here, $k_{\|}$ corresponds to the momentum parallel to the armchair edges and $W = 30$, 31, and 32 MCs. $K$ is projected

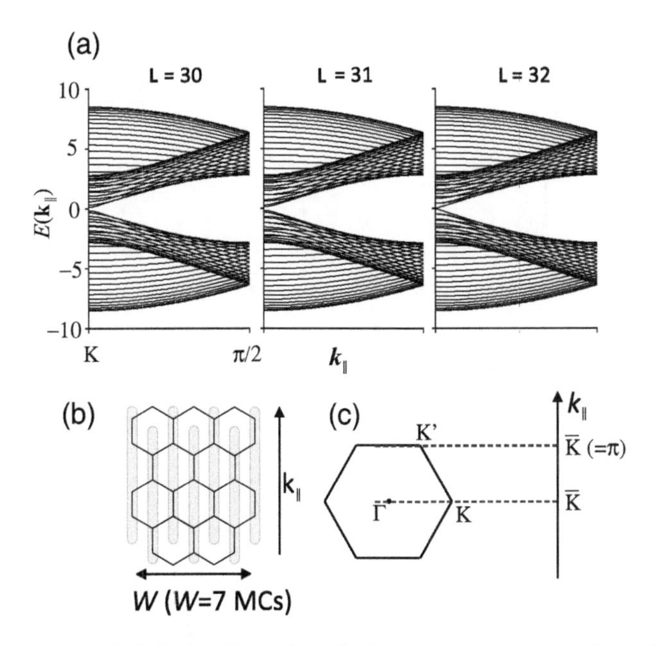


**Figure 3.10** (a) Calculated results of electronic structures of graphene nanoribbons with armchair edges. $L$ indicates the number of armchair chains. (b) Lattice structure of armchair graphene nanoribbon and (c) Brillouin zone.

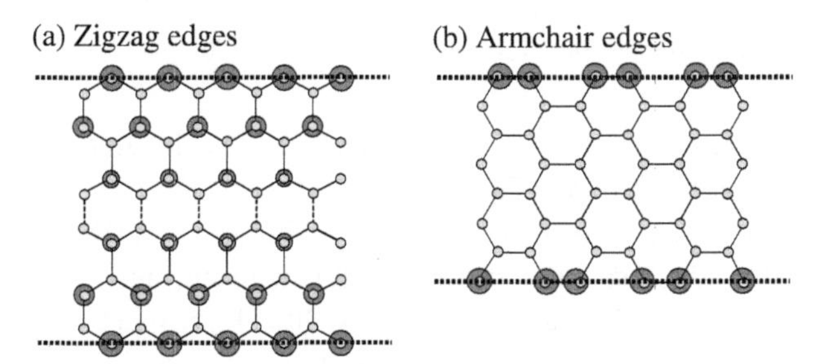


**Figure 3.11** (a) Calculated results of weight of resonant states near the zigzag edges and (b) those near the armchair edges.

onto the $\bar{\Gamma}$ point ($k_\parallel = (0, 0)$) and is denoted as $\bar{K}$. Note that there appears to be threefold periodicity in the electronic structure. The armchair ribbon is metallic when $W = 2(3n - 1)$, where $n$ is an integer. A resonance state also appears in the armchair edges, but

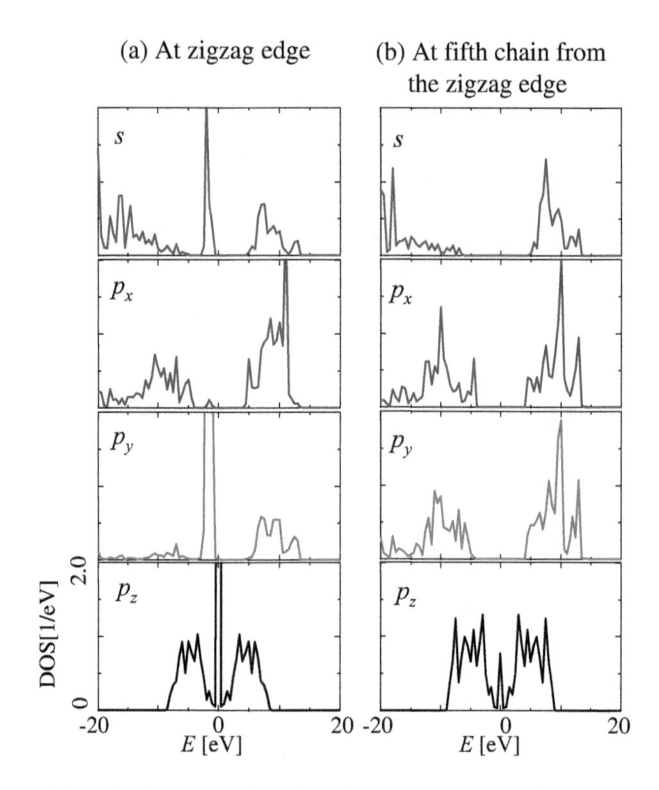


**Figure 3.12** Calculated results of local density of states decomposed into orbitals (a) at the zigzag edge and (b) at the fifth chain near the zigzag edge.

this originates from the $\sigma$ bands; therefore, it is located at the edge atoms only, as shown in Fig. 3.11b.

The local DOSs at the edge and at an inner chain (the fifth chain from the edge) calculated for a finite-sized zigzag-edge GNR are shown in Fig. 3.12. $W$ and $L$ are 20 and 300 MC, respectively. Overall, the calculated DOS exhibits peaky structures, which are attributed to the finite-sized GNRs used in the calculations. The DOS peak in the $p_z$ orbital at $E = 0$ shown in Fig. 3.12 is the resonant state of the $\pi$ band, and the peaks in the $s$ and $p_y$ orbitals immediately below $E = 0$ are the $\sigma$ band localized states at the edge of the GNR. In Fig. 3.12b, we find that the DOS peak of the $\pi$ band has finite amplitude, even at the fifth chain from the zigzag edge, while the peaks in the $\sigma$ bands disappear.

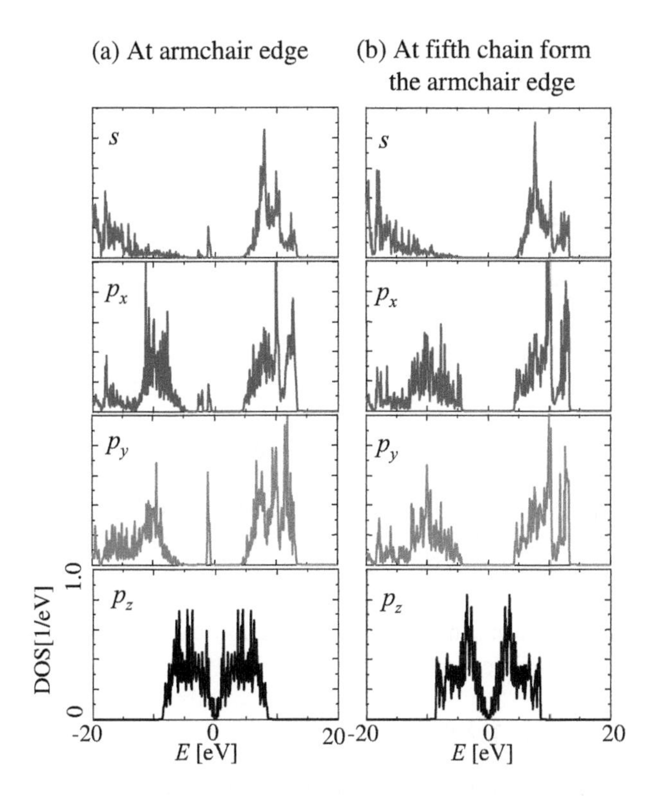


**Figure 3.13** Calculated results of local density of states decomposed into orbitals (a) at the armchair edge and (b) at the fifth chain near the armchair edge.

Figure 3.13 shows corresponding results for DOS calculations for an armchair-edge GNR. A sharp peak exists immediately below $E = 0$ in the DOS of the $p_y$ orbital, whereas no peak exists in that of the $p_z$ orbital. Thus, a clear difference between the DOSs at the zigzag and armchair edges can be seen in these figures. At the fifth chain from the armchair edge, the DOS peak of the $p_y$ orbital also disappears. Therefore, the resonant state at the armchair edge is caused by the $\sigma$ bands only.

Because the resonant state of the $\pi$ band at the zigzag edge appears at the Fermi energy, interesting physical properties (e.g., a current via the resonant state) may occur. Note that even the

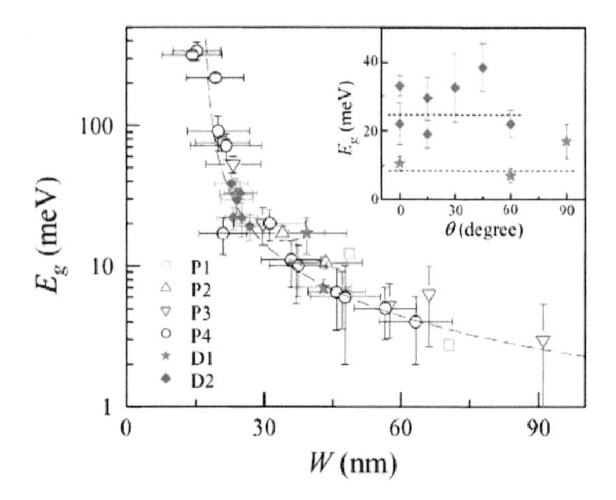


**Figure 3.14** Experimental results of energy band gap of graphene nanoribbon. Reprinted (figure) with permission from Ref. [20], Copyright (2007) by the American Physical Society.

resonant states of the $\sigma$ band may contribute under a finite $V$. These effects will be discussed in Chapter 4.

## 3.2.2 *Energy Band Gap*

The energy band gap of insulating armchair-edge GNRs increases with decreasing $W$. In contrast, no gap opening occurs for zigzag-edge graphene ribbons. Experimental observation of the energy band gap through measurement of temperature-dependent $\Gamma$ for lithographically patterned graphene ribbons has been reported [20], and the experimental results shown in Fig. 3.14 indicate that the band gap scales inversely with $W$, as predicted by theory. The energy gap at the Fermi energy becomes larger than 0.1 eV for $W < 10$ nm. Further, detailed calculations using the first-principles method have shown that GNR $W$ values should be 1–2 nm in order to obtain energy gaps with 1–3 eV [19, 21, 22].

## 3.2.3 *Hydrogen-Passivated Graphene Nanoribbons*

The results described before may be modified upon consideration of realistic scenarios for GNRs, that is, hydrogen passivation at the ribbon edges. The GNR edge atoms have a few n.n. atoms (one or two atoms for armchair- (zigzag-) edge atoms), and therefore, these atoms are unable to form complete covalent bonds with n.n. carbon atoms; this leads to interesting GNR structural features and physical properties. One such example is hydrogen passivation at GNR zigzag edges, and another is the appearance of edge states, as mentioned before.

Research on edge states was first performed for hydrogen-terminated graphite ribbons using the first-principles method [23]. Later, first-principles calculations were applied for hydrogen-passivated GNRs with zigzag and armchair edges [24, 25]. It has been shown that the band gap appears even in the case of armchair-edge nanoribbons with the same threefold periodicity. Further, the opening of the energy band gap has been attributed to the difference between the on-site potentials of the carbon atoms at the edge and those within the GNR, which is caused by hydrogen passivation. Note that the energy band gap also increases with decreased $W$ in the case of hydrogen-passivated GNRs.

Figure 3.15 shows results calculated for zigzag-edge GNRs without hydrogen passivation using the first-principles method [26]. The zigzag-edge GNR lattice structure is shown in Fig. 3.15g, whereas Figs. 3.15a and 3.15b show the energy dispersions of the up ($\uparrow$) and down ($\downarrow$) GNR spin states. The projected local DOSs at layers 1 and 16 (the numbering is defined in Fig. 3.15g) are shown in Figs. 3.15c and 3.15e, respectively, for up-spin states, and in Figs. 3.15d and 3.15f, respectively, for down-spin states. It should be noted that spin polarization appears in the hydrogen-unpassivated GNRs. The results indicate that the up-spin state corresponds to the majority and minority spin state at edges 1 and 16, respectively, indicating that the magnetic moments of the two edges align in antiparallel. The spin polarization appears near the zigzag edges only; however, a weak interaction exists between the spin polarizations of both edges. As a result of the interaction, the antiparallel alignment becomes stable. Note that the magnetism at the edges is discussed in detail in the next section.

Several characteristics should be noted: The edge states of the $\pi$ bands, (denoted by $\pi$ in the figures) of both majority and minority spin states are exchange-split and located near the Fermi energy. On the other hand, the edge states of the $\sigma$ bands of the majority

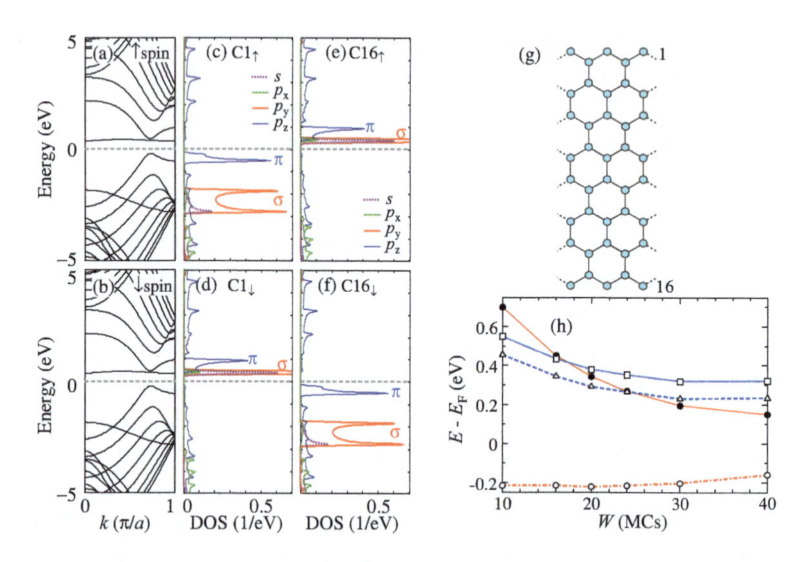

**Figure 3.15** Calculated results of hydrogen unpassivated graphene nanoribbon for (a) majority spin and (b) minority spin states. Local density of states of (c) up- and (d) down-spin states of layer 1 (C1↑ and C1↓), and those of (e) up- and (f) down-spin states at layer 16 (C16↑ and C16↓). (g) Size of the GNR. (h) Dependence of the bottom of the down-spin conduction band (solid red curve), top of the down-spin valence band (broken red curve), and position of the minority spin states of $\sigma$ edge state (region between two curves denoted by squares and triangles).

spin state are located far below $E_F$, while those of the minority spin state are located near $E_F$. The former result is very similar to that for the unpolarized states in GNRs shown in Fig. 3.9a; however, the result for the minority spin state is rather anomalous. In fact, the position of the edge state of the $\sigma$ band in the minority spin state case depends on $W$. Figure 3.15h shows the $W$ dependence of the $\sigma$ edge state. We find that the minority spin $\sigma$ edge state appears within the energy band for narrow GNRs only. This is because the energy band gap of the GNRs becomes large with decreasing $W$.

Figure 3.16 shows hydrogen-passivated GNR results calculated using the first-principles method. The hydrogen atoms are passivated at both edges and shown by yellow circles in Fig. 3.16g. Comparing the results shown in Fig. 3.16 with those in Fig. 3.15, we find that the $\sigma$ edge states disappeared but the edge states of the $\pi$

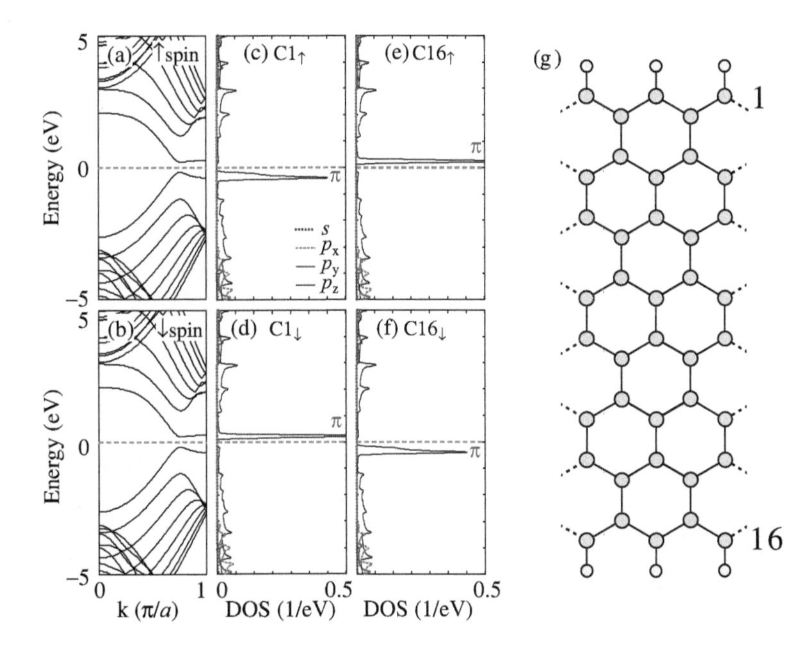


**Figure 3.16** Calculated results of hydrogen-passivated graphene nanoribbon, corresponding to the results shown in Fig. 3.15.

bands remain almost unchanged. These results may be attributed to a strong covalency between the graphene $\sigma$ band and the hydrogen 1$s$ state.

### 3.2.4 *Metal/Graphene Junctions*

When metals are in contact with GNRs with zigzag or armchair edges, the electronic structures near the contacts are modified. Here, we show some electronic structure results for calculations involving such metal/graphene junctions using the TB model. In these calculations, we prepare junctions in which a monoatomic metal layer is in contact with the zigzag or armchair edge of a GNR.

Figure 3.17a is a schematic of a metal/graphene junction structure. The GNR $W$ is 20 MCs, and a metallic monolayer with a square lattice is attached to both GNR edges. The metallic layer width is 50 MCs. Note that the junction width is infinite, while the length is finite. The graphene and metal are in contact at one atom

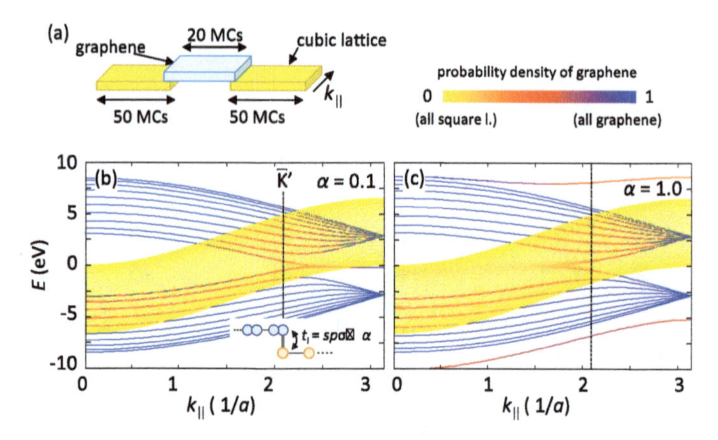


**Figure 3.17** (a) Junction structure of graphene with two cubic lattices. The widths of graphene and the cubic lattice are 20 and 50 MCs, respectively. (b) Calculated results of electronic structure with hopping integral factor 0.1 and (c) those with 1.0. Inset shows a definition of hopping integral between graphene and cubic lattices.

chain only, as shown in the inset of Fig. 3.17b. Because of the contact structure, the hopping integral $t_I$ is only nonzero between the atoms on both edges of the GNR and the square lattice. Here, $t_I = dd\sigma \times \alpha$, where $\alpha$ is a parameter that controls the magnitude of $t_I$ at the contact. Further, two values of $\alpha$ are adopted to facilitate comparison of the calculated results. The results for $\alpha = 0.1$ and 1.0 are shown in Figs. 3.17b and 3.17c, respectively.

Because of the band mixing between graphene and the metal, the obtained electronic states include both components of graphene and the metal. The relative weights of the components are expressed in color in Fig. 3.17: yellow (blue) indicates the weight attributed to the metal (graphene), whereas the intermediate color shows the degree of mixing of these weights. In the calculations, the atomic potential of the metal contact $s$ orbitals is taken to be zero, such that the $s$-band is located at the center of the overall band structure.

Although both graphene and metal bands maintain the original shapes of the isolated GNR and metallic ribbon, a striking change occurs in the electronic structure near $\bar{K}'$, as shown by the vertical dotted lines in the images. The electronic structure near the DP shown in Fig. 3.17b is similar to that shown in Fig. 3.8b; however,

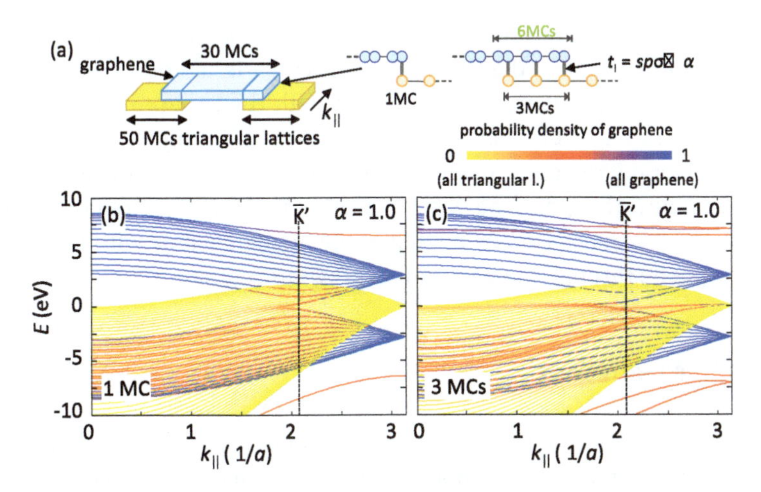


**Figure 3.18** (a) Junction structure of graphene with two triangular lattices. The widths of graphene and the triangular lattices are 20 and 50 MCs, respectively. (b) Calculated results of electronic structure with one MC overlapping and (c) those with four MCs overlapping. The value of $\alpha$ is 1.0 for both cases.

that shown in Fig. 3.17c differs from that shown in Fig. 3.8b. This difference may be attributed to the large interlayer $t_I = dd\sigma$. In Fig. 3.17c, split bands appear above and below the graphene band, which may correspond to bonding and antibonding states caused by the large value of $t_I$. The electronic structure near $\bar{\text{K}}'$ may be affected by these bonding and antibonding states, resulting in a different electronic structure from that calculated for $t_I = dd\sigma \times 0.1$.

Note that the square lattice of the metal may not be suitable for making good contact with the graphene sheet, because the latter has a honeycomb lattice. Metals with a triangular lattice are more suitable in this regard, as they can achieve wide overlapping contact with the graphene sheet. Figure 3.18 shows the results of graphene/metal junction calculations in which triangular-lattice metals were used. Figure 3.18a shows the schematics of the junction structure in which the $W$ of the metals and graphene are 50 and 20 MCs, respectively. The overlapping regions between the graphene and metal contacts are taken to be 1 and 4 MCs, the results of which are given in Figs. 3.18b and 3.18c, respectively. The value of $\alpha$ is 1.0 in both cases.

When the overlapping region at the contact is narrow (one MC, for example), the electronic structure is almost identical to that for the graphene/metal contact with the square-lattice metal shown in Fig. 3.17c. For junctions composed of contacts with wide overlapping regions, however, many additional states appear above and below the graphene band, as well as in the $E \approx$ eV region. Such states may be caused by mixing of the metallic bands with those of graphene through the wide contact regions.

## 3.3 Magnetism

We have shown that GNR zigzag edges produce edge states (resonant states called Shockley states) and have a $\delta$-function-type DOS at $E_F$. Because of the high DOS at $E_F$, the GNR edge states may satisfy the criterion for the appearance of ferromagnetism at zero temperature (Stoner condition). This condition is expressed as

$$U \times D(E_F) \geq 1,$$

where $U$ is the on-site Coulomb interaction between the up- and down-spin electrons and $D(E_F)$ is the DOS at the Fermi energy. Although the value of $U$ of the carbon atoms may be quite small, this condition may be satisfied only because of the infinitely large $D(E_F)$. As the magnetization occurs on one of two sublattices, a staggered sublattice potential appears on the carbon lattice, along with an energy band gap.

Prior to the discovery of graphene, the magnetism of nanographite ribbons was studied from a theoretical perspective [27], and experimental observation of ferromagnetism was reported for some graphite samples [28] and for proton-irradiated graphite [29]. Following the discovery of single-layer graphene, the ferromagnetism at GNR zigzag edges attracted considerable attention, and a number of studies were performed to confirm the spin polarization and to exploit this characteristic in novel devices [17, 18, 30–32]. Further, it was noted that defects in graphene and graphene nanomeshes also induce magnetism [33, 34].

For example, Son et al. [17] performed a first-principles calculation for zigzag-edge GNRs under an external in-plane (the

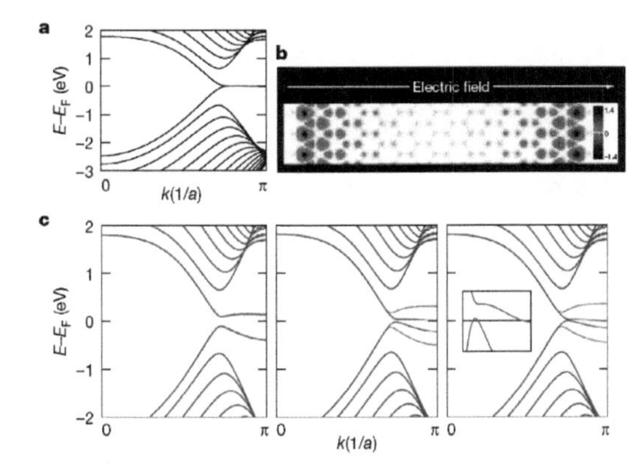


**Figure 3.19** Energy states of zigzag-edge graphene nanoribbons with and without an electric field. Reprinted by permission from Macmillan Publishers Ltd: [*Nature*] (Ref. [17]), copyright (2006).

transverse direction relative to the ribbon) electric field $E$. Figures 3.19a and 3.19c (left) show the electronic states of zigzag-edge GNRs with spin-unpolarized and spin-polarized states, respectively. Figure 3.19b shows the spin-resolved charge distribution in the GNR in the absence of the external $E$, where the red and blue colors indicate the up- and down-spins, respectively. The spin density at the edges was reported to be 0.43 $\mu_B$ per atom. Figures 3.19c (center) and (right) show the electronic structure under a 0.05- and 0.1-V Å$^{-1}$ $E$, respectively. The red and blue lines indicate the spin dependence, as previously.

Figure 3.20a,b shows schematics of the DOS without and with an external $E$, respectively. In the absence of the external $E$, the GNR exhibits antiparallel alignment of the spin polarization on the left (L) and right (R) regions of the ribbon. Here, $\alpha$ and $\beta$ indicate up- and down-spins, respectively. When the $E$ is applied, energy-state shifts occur, but the shift of the L-region energy is opposite to that of the R-region, as shown in Fig. 3.20b. As a result, only one spin state appears at the Fermi energy, which is a half-metallic state. Thus, Son et al. [17] reported that the half-metallic state can be controlled by an in-plane external $E$. Such a half-metallic state may be used for device applications, as mentioned next.

Many investigations of the appearance of ferromagnetism in GNRs have been conducted. Hence, rather large magnetic moments

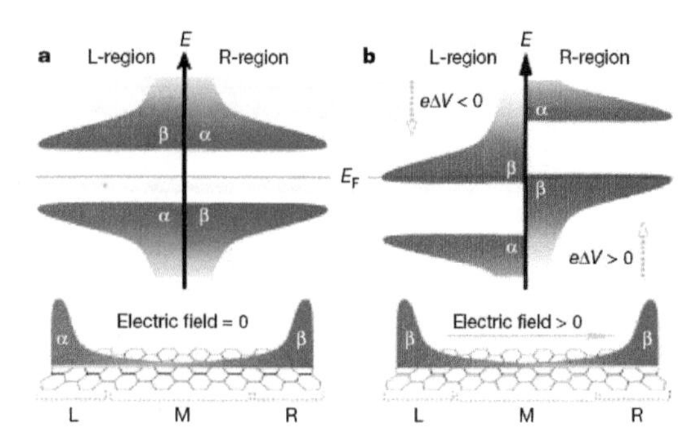


**Figure 3.20** Schematic of density of states of zigzag-edge graphene nanoribbons with and without an electric field. Reprinted by permission from Macmillan Publishers Ltd: [*Nature*] (Ref. [17]), copyright (2006).

obtained through implementation of first-principles calculations have been reported [32], and values of 1 and 1.12–1.53 $\mu_B$ per atom have been realized via hydrogen chemisorption and vacancy defects, respectively. The appearance of the long-range order has been claimed on the basis of a concept known as the Ruderman–Kittel–Kasuya–Yosida (RKKY) interaction between local moments [30]. The criteria for the appearance of ferromagnetism have also been discussed [35].

It should be noted, however, that the long-range order of the ferromagnetism of low-dimensional systems can be destroyed easily by spin fluctuations at elevated temperature and/or by lattice deformation (the Jahn–Teller effect). In fact, it has been noted that magnetism can seldom be realized in real graphene, because of the effects of edge reconstruction, edge passivation, and edge closure [37]. In a detailed study on the magnetization of graphene involving the direct ultrasonic cleavage of highly oriented pyrolytic graphite, no ferromagnetism was detected at any temperature down to 2 K [36]. Instead, the samples exhibited weak paramagnetism. Furthermore, Nair el al. [38] have shown that point defects in graphene carry magnetic moments with spin 1/2. However, these researchers did not detect any magnetic ordering down to 4.2 K.

Although ferromagnetism may not appear in graphene samples containing defects and edges, the magnetism of this material may still be interesting in that it may couple with the spin transport [39]. In fact, the spin relaxation of an electrical current through a graphene layer could be affected by the magnetism. Instead of the magnetism due to edges and defects in graphene, a formation of a local magnetism due to adatoms composed of materials such as iron, cobalt, and nickel has been reported in a first-principles study [40]. Investigations of changes in graphene spin transport due to the effects of such adatoms are of interest.

## 3.4 Metal and Insulator Contact with Graphene

Here, we review the electronic structures of various setups involving contact between a graphene sheet and two-dimensional (or quasi-two-dimensional) metallic (or insulating) lattices. Because these structures have translational invariance, we may perform the momentum space calculations using both the TB model and the first-principles method. As mentioned previously, satisfactory lattice matching is obtained for contact between a graphene sheet and a metallic (or insulating) sheet with a triangular lattice. Therefore, we first present the results of a calculation involving a model contact between graphene and a triangular-lattice material.

### 3.4.1 *Simple Model*

Figure 3.21 shows the electronic structure for a contact involving graphene and a two-dimensional material with a triangular lattice; this was calculated using a single $s$ orbital TB model for the triangular lattice. The $s$ and $p$ orbitals were incorporated into the graphene sheet calculation, and the contact structure is shown on the left-hand side of Fig. 3.21. Here, the position of the triangular lattice $s$-level is lowered by taking $\varepsilon_s = -10$ eV. The hopping integral is taken to be $t_s = -2$ eV and the band parameters of the graphene sheet are identical to those used previously. The hopping integral between the graphene and triangular lattice is taken to be $t' = \alpha \times t_g$ where $t_g$ is the hopping integral of the graphene $\pi$ band. Figure 3.21a–d shows the results obtained assuming $\alpha = 0$, 0.1, 0.5, and 1.0, respectively.

It is interesting to note that an energy gap appears at the DP when $t' \neq 0$. Even when $\alpha = 0.1$, a small energy gap appears. A

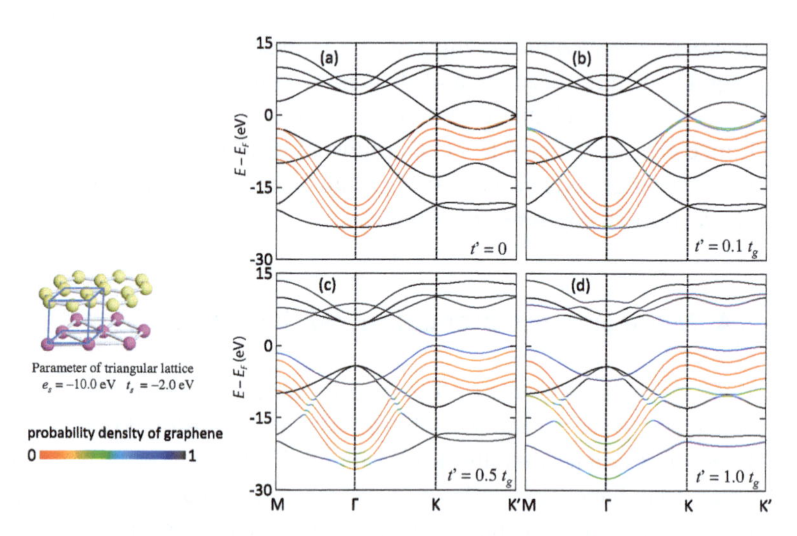


**Figure 3.21** Calculated results of electronic structure of graphene sheet on a triangular lattice with a hopping integral factor between graphene and triangular lattice of (a) 0.0, (b) 0.1, (c) 0.5, and (d) 1.0.

large energy gap appears for $\alpha = 0.5$ and 1.0, the appearance of which is attributed to bonding and antibonding effects caused by the orbital hybridization. This kind of effect may be used to control the graphene sheet energy gap by placing the graphene sheet in contact with suitable materials.

## 3.4.2 *Realistic Metal Contacts*

As described in Chapter 1, because of its high mobility, two-dimensionality, etc., graphene has potential applicability in electronics and/or spintronics devices. To realize graphene devices, graphene–metal contact is inevitable. Because the unique transport properties of graphene are attributed to the characteristic electronic structure of this material, only two states, that is, $K$ and $K'$ (which are DPs), and the linear dispersion near the DPs are responsible for the metallic behavior due to topological singularities. Therefore, changes in the electronic structure near the DP in the vicinity of the metal/graphene contact may strongly affect the graphene/metal junction transport properties. Therefore, a detailed study of the electronic structures of metal/graphene contacts is

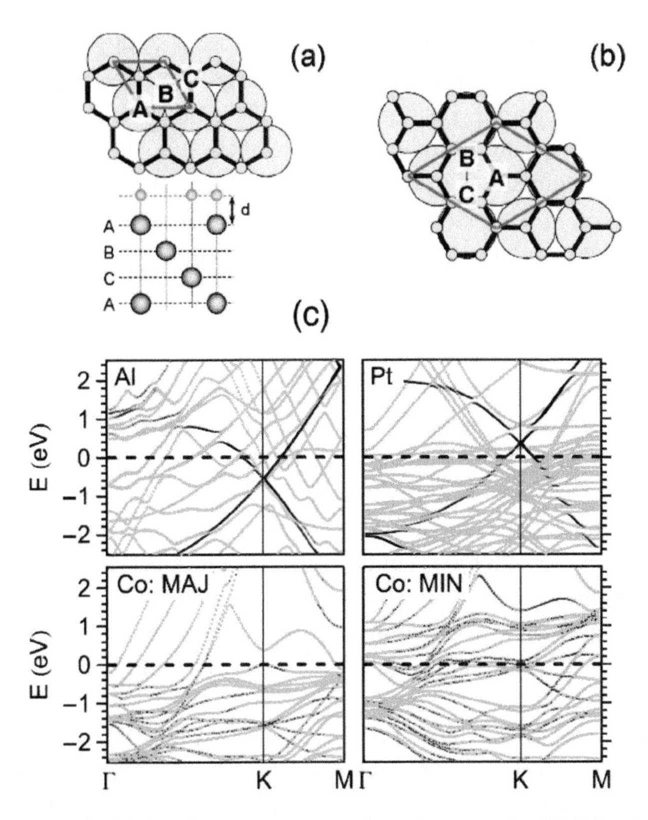


**Figure 3.22** (a, b) Overlayer structures of graphene on fcc (111) lattice and (c) calculated results of the electronic structures of grapehene on Al, Pt, Co (majority spin states) and Co (minority spin states). Reprinted (figure) with permission from Ref. [41], Copyright (2008) by the American Physical Society.

crucial for the successful realization of graphene devices. In this subsection, we provide a brief review of the electronic structures of graphene/metal contacts, depending on calculations using the first-principles method.

Giovannetti et al. [41, 42] have performed DFT calculations to examine graphene/metal contact electronic structures (for nickel, cobalt, palladium, aluminum, silver, copper, gold, and platinum). They adopted a contact comprising a single graphene sheet on a (111) surface of four metal layers, as shown in Fig. 3.22a for copper, nickel, and cobalt and in Fig. 3.22b for the other metals.

The graphene lattice constant was taken to be 2.445 Å and $d_{eq}$ was determined by minimizing the total energy of the contacts.

The calculated results for the electronic structures of the graphene/Al and graphene/Pt contacts and the majority and minority spin states of the graphene/cobalt contacts are shown in Fig. 3.22c. It is apparent that the trace of the graphene dispersion (the black lines in the images) remains rather strong for both the graphene/Al and graphene/Pt contacts. On the other hand, no trace of the graphene dispersion remains for the electronic structure of the graphene/cobalt contact.

Further, Giovannetti et al. [41, 42] showed that graphene/metal contacts can be classified into two types—physical and chemical contacts having weak and large binding energies, respectively. The $d_{eq}$ of the former is approximately 3.3 Å, while that of the latter is less than 2.3 Å. The mixing of the metal and graphene wave functions is large for the chemical contact case, and the electronic structure of the graphene, that is, the Dirac dispersion, is washed away. On the other hand, the characteristics of the graphene electronic structure remain visible in the case of the physical contact, as shown in Fig. 3.22c.

Because of the large band mixing between the graphene and metals, a Fermi energy shift $\Delta E_F$ occurs, as shown in Fig. 3.23; this results in a charge transfer from/to the graphene. Here, $\Delta E_F$

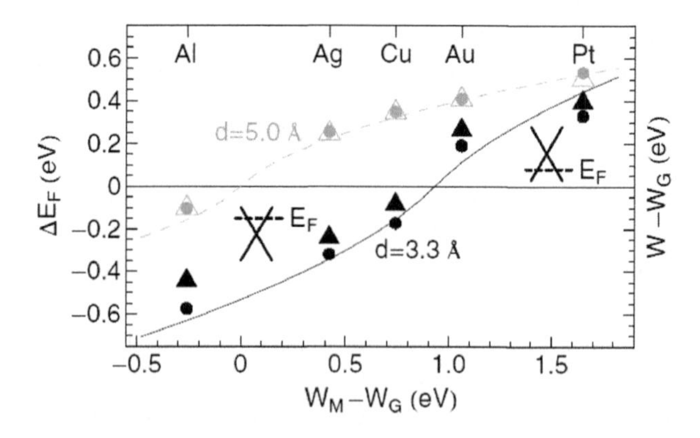


**Figure 3.23** Calculated results of shift in Fermi energy of graphene overlayer structures. Reprinted (figure) with permission from Ref. [41], Copyright (2008) by the American Physical Society.

**Table 3.2** Calculated results of work function $W$ in eV, equilibrium distance $d_{eq}^{(a)}$ in Å between graphene and metals, and shift of the Fermi energy $\Delta E$ in eV. (a), (b) are results obtained by Giovannetti et al. [41] and Ran et al. [43], and (c) is an experimental value

| Metal | $W_M^{(a)}$ | $W_M^{(b)}$ | $d_{eq}^{(a)}$ | $d_{eq}^{(b)}$ | $\Delta E^{(a)}$ | $\Delta E_F^{(a)}$ | $\Delta E_F^{(b)}$ |
|---|---|---|---|---|---|---|---|
| K | | 2.29 | | 2.93 | | | 0.640 |
| Ca | | 2.87 | | 2.24 | | | 0.228 |
| Li | | 2.93 | | 2.22 | | | 0.397 |
| Sc | | 3.5 | | 2.06 | | | 0.494 |
| Ti | | 4.33 | | 2.06 | | | 0.280 |
| Co | 5.44 | 5.0 | 2.05 | 2.08 | 0.160 | | −0.002 |
| Ni | 5.47 | 5.35 | 2.05 | 2.15 | 0.125 | | 0.009 |
| Al | 4.22 | 4.26 | 3.41 | 3.65 | 0.027 | 0.43 | 0.182 |
| Cu | 5.22 | 4.94 | 3.26 | 3.46 | 0.033 | −0.02 | −0.063 |
| Au | 5.54 | 5.31 | 3.31 | 3.63 | 0.030 | −0.31 | −0.257 |
| Pt | 6.13 | 5.67 | 3.30 | 3.59 | 0.038 | −0.48 | −0.260 |
| Pd | 5.67 | 5.6 | 2.30 | 2.79 | 0.084 | | −0.087 |
| Graphene | 4.48 | 4.6[c] | | | | | |

is defined as $\Delta E_F = E_F - E_{DP}$, where $E_{DP}$ indicates the energy at the DP. A positive (negative) value of $\Delta E_F$ indicates electron (hole) doping into the graphene. Giovannetti et al. showed that the values of $\Delta E_F$ can be basically understood as corresponding to the difference between the work functions of the graphene $W_G$ and the metals $W_M$. This will be explained in greater detail later. The calculated binding energy $\Delta E$, along with the calculated $\Delta E_F$ and $d_{eq}$ values, is presented in Table 3.2. The calculated and reported $W_G$ and $W_M$ are also presented.

Ran et al. [43] have extended the DFT calculations to more realistic junction structures, considering a finite-sized graphene ribbon in contact with metals at both ends. These researchers adopted a three-dimensional periodic boundary condition for unit cells composed of finite-sized clusters of graphene ribbon and metallic contacts. The widths and lengths of the clusters were $W =$ 3–5 and $L \approx 40$ zigzag chains, respectively, and the metallic contacts were composed of three layers. The examined metal species are listed in Table 3.2. Among them, the surface orientation is (100) for K and Li, (001) for Sc and Ti, and (111) for the others. The values of

$d_{eq}$ and $\Delta E_F$ were calculated at the contact region. The results are listed in Table 3.2, where they are compared with those obtained by Giovannetti et al.

As pointed out by Giovannetti et al. [41], Ran et al. [43] also classified the metallic contacts into chemical and physical contacts, apart from the potassium and Pd contacts. The $d_{eq}$ values of the former and latter are 2.1–2.2 Å and 3.5–3.6 Å, respectively. The origin of the difference between the chemical and physical contacts may be attributed to the difference between the band mixing of these contacts. Ran et al. [43] commented that this difference could be understood by noting the differences in the DOS at $E_F$. For example, the gold and titanium states at $E_F$ are $s$- and $d$-like, and are responsible for the physical and chemical contacts of the graphene/Au and graphene/Ti contacts, respectively. As for the Pd contact, $E_F$ is located at the top of the $d$ DOS and, therefore, the graphene/Pd contact exhibits an intermediate characteristic. Interestingly, these researchers showed that $d_{eq}$ varies at the graphene/Pd contact: it is short or long at the edge or internal region of the contact, respectively. They also showed that the contact resistance of graphene/Au is 2 orders of magnitude less than that of the graphene/Ti contact. Thus, the electronic structure at the contact is crucial as regard the electrical transport of graphene junctions. Further details of the contact resistance are given in the next section.

Note that $\Delta E_F$ may be interpreted by adopting a simple model incorporating $W_G$ and $W_M$, along with the effects of charge transfer between the graphene and metals. This model may be expressed as [41]

$$\Delta E_F = W_G - W_M - \frac{e^2}{\tilde{C}} N - V_{ci},$$

where $\tilde{C}$ is the effective capacitance at the contact given by the effective distance $d_{eff}$, $N$ is the number of electrons per unit cell transferred from the graphene to the metal (this is negative if the electrons are transferred from the metal to the graphene), and $V_{ci}$ indicates the chemical interaction at the contact. $N$ is expressed as $\pm D_0 \Delta E_F^2/2$, where $D_0$ is the slope of the linear DOS near the DP. By solving the equation for $\Delta E_F$ and fitting the expression with the calculated results of $\Delta E_F$ for various $d_{eq}$ values for the graphene/Cu contact, Giovannetti et al. determined the values of $d$–$d_{eff}$ and $V_{ci}$. Using these values, they explained the metal dependence of $\Delta E_F$ for physical contacts. The third term in the equation above indicates that the dipole field at the contact favors *p*-type doping.

The theoretically predicted tendency has been confirmed experimentally for Ti, Fe and Pt metals, with various coverage rates on the graphene sheet [44]. The bonding characteristic of graphene/metal contacts has also been reported based on theoretical investigation of the metal intercalation between graphene and a substrate [45]. Further, the effects of gold grains on graphene conduction and doping have been investigated by measuring the electrical resistance as a function of gate voltage [46]. That study showed that the gold grains cause *p*-doping and that the obtained $\Delta E_F$ is in disagreement with the results obtained via the first-principles calculations.

### 3.4.3 *Graphene on Silicon Carbide*

For technological applications of graphene, an easy fabrication process that yields large-area graphene sheets is desirable. To date, several graphene sheet fabrication methods have been proposed, including micromechanical cleaving of single-layer graphene, chemical vapor deposition of graphene on catalytic films, and graphene growth on silicon carbide (SiC) [22, 47–51]. Of these techniques, graphene formation on (SiC) seems to be the most promising. However, graphene that is grown on (SiC) may lose the characteristic features of the electronic states of freestanding graphene. As mentioned previously, first-principles calculations show that graphene/metal physical contacts may preserve Dirac dispersion. Therefore, in addition to experimental works on the structural properties of graphene fabricated epitaxially on Pt(111) or Au(111), theoretical studies on the electronic structure of graphene on (SiC) have been performed in order to confirm the occurrence of Dirac dispersion in the resultant graphene.

Hsu et al. [52] have performed first-principles calculations concerning the electronic structure of a graphene monolayer on SiC(0001) with metal intercalation. These researchers showed that 3/8 coverage of monolayer graphene by metal (Al, Ag, Au, Pt, and

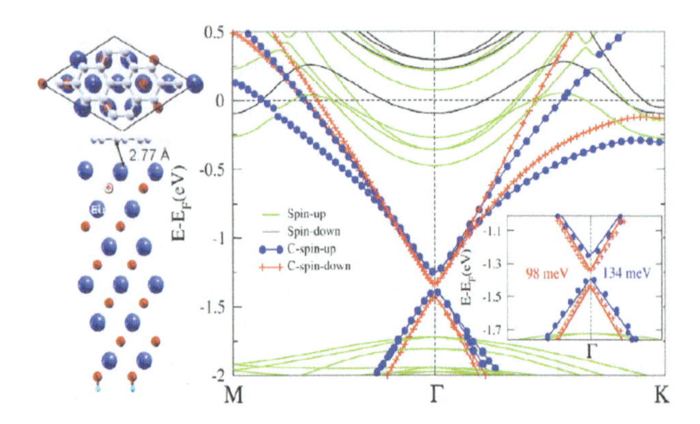


**Figure 3.24** Electronic structure graphene/EuO contacts. (Left) Side and top views of the lattice structure used in the calculation. (Right) Calculated results of the energy–momentum relation. Blue dots and red crosses are up- and down-spin states of graphene, respectively. Reprinted (figure) with permission from Ref. [54], Copyright (2013) by the American Physical Society.

Pd) may preserve the graphene Dirac dispersion. Furthermore, they showed that the metal intercalation gives rise to *n*-type doping of the graphene. Finally, the occurrence of Rashba-type energy–momentum dispersion has been examined in experiments on Ir(111) on a graphene sheet using photoemission spectroscopy [53].

## 3.4.4 *Graphene Contact with Insulator*

To apply graphene to spin-related devices, both an energy gap and spin splitting of the electronic states are desirable. Achieving such graphene contacts experimentally involves considerable difficulty; however, theoretical calculations can be performed in order to determine the conditions necessary for desirable graphene contacts. Figure 3.24 shows the electronic structure of a graphene/EuO contact calculated using the first-principles method [54, 55]. Here, EuO is known to be a ferromagnetic insulator, although the Curie temperature is below 20 K. The results show that the Dirac cone has an energy gap with spin splitting. The energy gaps of the up- and down-spin states are 134 and 98 meV, respectively. Because the graphene is electron doped, the EuO contact appears to be useless at present. Therefore, further study is desirable.

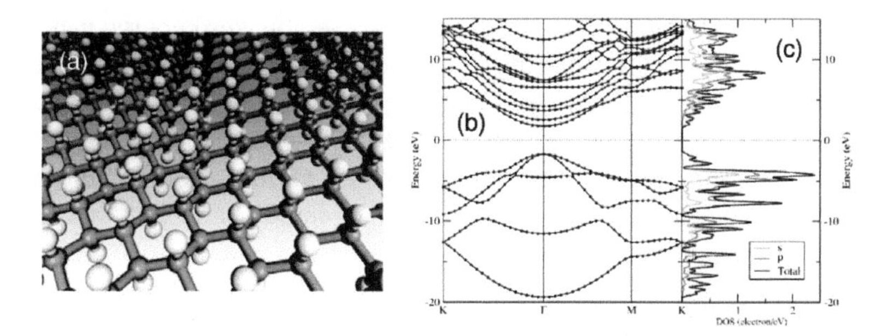


**Figure 3.25** (a) Lattice structure of graphene sheet coated by hydrogen atoms at both sides, (b) calculated results of energy band, and (c) density of states. Reprinted (figure) with permission from Ref. [56], Copyright (2007) by the American Physical Society.

### 3.4.5 *Hydrogen-Coated Graphene*

A large number of organic compounds made of carbon and hydrogen atoms exist. Hydrogen atoms also attach to graphene, yielding a material known as graphane. This material may have two types of lattice structure: One is in which the hydrogen atoms attach to both sides of the graphene sheet, called a chair conformer and shown in Fig. 3.25 [56], and the other is a structure called a boat conformer. Because of the bonding of the hydrogen atoms with the carbon atoms, the graphane electron configuration changes from the $sp^2$ of graphene to $sp^3$. An extra electron from a hydrogen atom enters the $p$–$\pi$ band. As a result, the $p$–$\pi$ band shifts to the lower energy region, leaving $p$–$\sigma$ bands near $E_F$.

The energy bands calculated for graphane are shown in Fig. 3.25b, and it is apparent that a large energy gap opens at $E_F$. Figure 3.25c shows the corresponding DOS. Sofo et al. [56] further showed that the graphane lattice is sufficiently stable. Further, control of graphane properties has been demonstrated by Eias et al. [57].

## 3.5 Ripples and Curved Graphene

The high mobility and weak spin–orbit coupling of graphene are excellent qualities, for example, as regard its use in spintronics-related technological applications. However, the zero energy gap is a disadvantage for replacing semiconductors with graphene. Above, we discussed the capacity of bilayer graphene and GNRs to produce energy band gaps at the Fermi energy with reasonable strength in certain situations. In this subsection, we review another type of graphene structure that may produce an energy band gap, that is , distorted graphene sheets.

### 3.5.1 *Ripples, Strain, and Spin–Orbit Interaction*

Graphene sheets are, in principle, flat because of the strong covalency between the $\sigma$ orbitals on the carbon atoms. Nevertheless, graphene sheets may wrinkle easily [58], and therefore, the features of graphene sheet ripples have been studied experimentally using scanning tunneling microscopy (STM) [50, 59, 60]. These measurements have shown that graphene on silicon dioxide has a curved structure, the lateral and vertical scales of which are 10 and 0.5 nm, respectively. This curved structure may be due to curvature of the substrate itself; however, it may be interesting to study the effect of this distorted structure on the electronic properties of the correspondingly distorted graphene.

The effects of lattice distortion or strain on graphene sheets have been studied numerically using both the TB scheme and first-principles methods [61–64]. The TB calculations using the $sp^3$ model have shown that an energy gap appears at the Fermi energy, because the strain induces band mixing between the $pp\sigma$ and $p_z$ orbitals [61, 62]. It has also been noted that homogeneous deformation does not produce a band gap. In fact, only inhomogeneous deformation gives rise to an energy gap, which is possibly less than 1 eV [63]. Further, strain-induced energy band gaps have also been studied for bilayer graphene samples using the TB model [64].

In addition, it is known that curved graphene may produce spin–orbit interaction (SOI) [65]. The appearance of SOI is attributed to a second n.n. hopping between the carbon atoms. Detailed analytical and numerical studies have been performed on this topic [66–69], and the SOI magnitude has been found to be rather small at approximately 4 meV, with the corresponding energy gap being $10^{-3}$ meV [68].

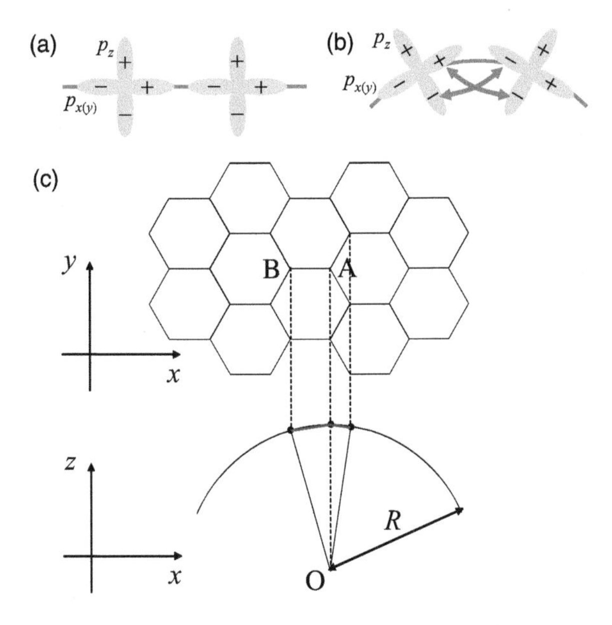


**Figure 3.26** Schematics of orbital overlap on graphene sheet (a) without curvature and (b) with curvature and (c) model used for curved graphene.

## 3.5.2 *Curved Graphene*

The effects of curvature on the graphene electronic structure are related to a change in the overlap of the wave functions on the n.n. carbon atoms. Figure 3.26a,b shows that the wave function overlaps of the $p_z$ orbitals on n.n. sites and those of the $p_x$ ($p_y$) orbitals are modified by a curved structure of graphene sheets.

Here, we show how the electronic structure near the DP is modified by the changes in the hopping matrix and the SOI using a TB model [70]. The change in the hopping matrix is derived by assuming a uniform curved structure around the $y$ axis (along zigzag chains) with a radius of curvature $R$, as shown in Fig. 3.26c. Three cases are considered: (1) $R = \infty$ (no curvature) and SOI $= 0$; (2) $R = 10a$ ($a$ is the intercarbon distance) and SOI $= 0$; and (3) $R = 10a$ and SOI $= 0.05 \times t_{pp\sigma}$.

The results of the calculation for the energy dispersion near the DP are shown in Fig. 3.27 for these three cases. In case 2 (dashed lines), the DP shifts in the $M$ direction with no change

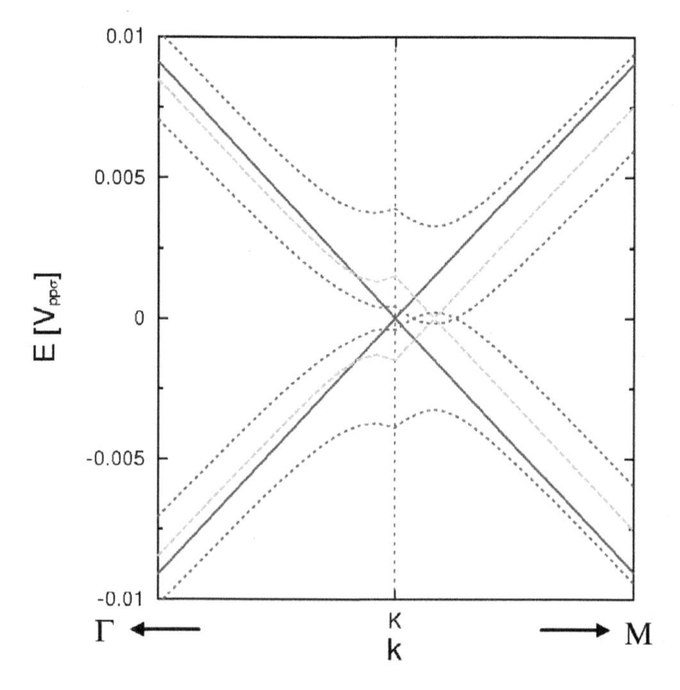


**Figure 3.27** Calculated results of electronic structure of curved graphene near Dirac point. Solid lines, dashed curves, and dotted curves correspond to cases (1), (2), and (3) defined in the text.

in the linear dispersion relation. Therefore, no change occurs in the DOS. In case 3, the energy dispersion is strongly modified, resulting in a change in the DOS. In this case, the linear relation of the energy dispersion disappears. The magnitude of the change in a realistic curved structure, however, may be significantly smaller than that shown in Fig. 3.27, because the magnitude of the SOI is very small.

The change in the electronic structure near the DP may be visualized as a shifting of the Dirac cone, as shown in Fig. 3.28. The shape and position of the Dirac cone for a flat graphene sheet are shown in Fig. 3.28a. In case 2, the Dirac cone shifts in the *M* direction, whereas it splits into double cones in case 3, as shown in Figs. 3.28b and 3.28c, respectively. Finally, both of the Dirac cones at the *K* and *K′* points shift, as shown on the right-hand side of Fig. 3.28c.

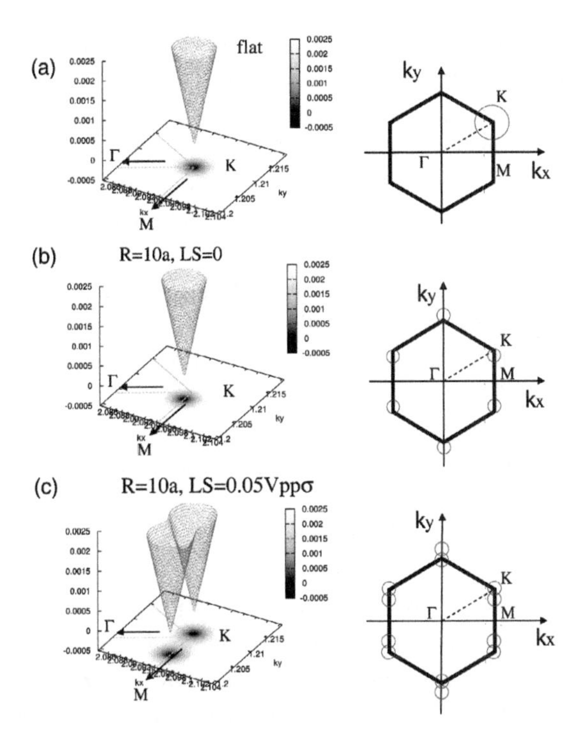


**Figure 3.28** Change in Dirac cone of (a) flat, (b) curved graphene without spin–orbit interaction, and (c) curved graphene with spin–orbit interaction.

## References


1. W. Harrison, *Electronic Structure and Properties of Solids*, W. H. Freeman, New York (1980).
2. A. H. Castro Neto, F. Guinea, K. S. Novoselov, and A. K. Geim, *Rev. Mod. Phys.*, **81**, 109 (2009).
3. Z. Liu, K. Suenaga, P. J. F. Harris, and S. Iijima, *Phys. Rev. Lett.*, **102**, 015501 (2009).
4. E. MacCann, *Phys. Rev. B*, **74**, 161403 (R) (2006).
5. E. V. Castro, K. S. Novoselov, S. V. Morozov, N. M. P. Peres, J. M. B. Lopes dos Santos, J. Nilsson, F. Guinea, A. K. Geim, and A. H. Castro Neto, *Phys. Rev. Lett.*, **99**, 216802 (2007).
6. S. Das Sarma, S. Adam, E. H. Hwang, and E Rossi, *Rev. Mod. Phys.*, **83**, 407 (2011).

7. J. B. Oostings, H. B. Heersche, X. Liu, A. F. Morpurgo, and L. M. K. Vandersypen, *Nat. Mater.*, **7**, 151 (2008).
8. Y. Zhang, T.-T. Tang, C. Girit, Z. Hao, M. C. Martin, A. Zettl, M. F. Crommie, Y. R. Shen, and F. Wang, *Nature*, **459**, 820 (2009).
9. X. Peng and R. Ahuja, *Phys. Rev. B*, **82**, 045425 (2010).
10. F. Freitag, J. Trbovic, M. Weiss, and C. Schönenberger, *Phys. Rev. Lett.*, **108**, 076602 (2012).
11. L. M. Zhang, Z. Q. Li, D. N. Basov, M. M. Fogler, Z. Hao, and M. C. Martin, *Phys. Rev. B*, **78**, 235408 (2008).
12. H. K. Min, B. Sahu, S. K. Banerjee, and A. H. MacDonald, *Phys. Rev. B*, **75**, 15515 (2007).
13. Y. Guo, W. Cuo, and C. Chen, *Appl. Phys. Lett.*, **92**, 243101 (2009).
14. M. Fujita, K. Wakabayashi, K. Nakada, and K. Kusakabe, *J. Phys. Soc. Jpn.*, **65**, 1920 (1996).
15. K. Nakada, M. Fujita, G. Dresselhaus, and M. S. Dresselhaus, *Phys. Rev. B*, **54**, 17954 (1996-II).
16. L. Brey and H. A. Fertig, *Phys. Rev. B*, **73**, 235411 (2006).
17. Y. W. Son, M. L. Cohen, and S. G. Louie, *Nature*, **444**, 347 (2006).
18. Y. W. Son, M. L. Cohen, and S. G. Louie, *Phys. Rev. Lett.*, **97**, 216803 (2006).
19. V. Barone, O. Hod, and G. E. Scuseria, *Nano Lett.*, **6**, 2748 (2006).
20. M. Y. Han, B. Özyilmaz, Y. Zhang, and P. Kim, *Phys. Rev. Lett.*, **98**, 206805 (2007).
21. L. Yang, C.-H. Park, Y.-W. Son, M. L. Cohen, and S. G. Louie, *Phys. Rev. Lett.*, **99**, 186801 (2007).
22. C. Berger, Z. M. Song, X. B. Li, X. S. Wu, N. Brown, C. Naud, D. Mayo, T. B. Li, J. Hass, A. N. Marchenkov, E. H. Conrad, P. N. First, and W. A. de Heer, *Science*, **312**, 1191 (2006).
23. Y. Miyamoto, K. Nakada, and M. Fujita, *Phys. Rev. B*, **59**, 9858 (1999).
24. S. Dutta, A. K. Manna, and S. K. Pati, *Phys. Rev. Lett.*, **102**, 096601 (2009).
25. S. Dutta and S. K. Pati, *J. Mater. Chem.*, **20**, 8207 (2010).
26. S. Honda, K. Inuzuka, T. Inoshita, N. Ota, and N. Sano, *J. Phys. D: Appl. Phys.*, **47**, 485004 (2014).
27. K. Wakabayashi, M. Fujita, H. Aoki, and M. Sigrist, *Phys. Rev. B*, **59**, 8271 (1999).
28. P. Esquinazi, A. Setzer, R. Hohne, C. Semmelhack, Y. Kopelevich, D. Spemann, T. Butz, B. Kohlstrunk, and M. Lösche, *Phys. Rev. B*, **66**, 024429 (2002).

29. P. Esquinazi, D. Spemann, R. Höhne, A. Setzer, K.-H. Han, and T. Buts, *Phys. Rev. Lett.*, **91**, 227201 (2003).
30. M. A. Vozmediano, M. P. Lopes-Sancho, T. Stauber, and F. Guinea, *Phys. Rev. B*, **72**, 155121 (2005).
31. H. Kumazaki and D. S. Hirashima, *J. Phys. Soc. Jpn.*, **76**, 064713 (2007).
32. O. V. Yazyev and L. Helm, *Phys. Rev. B*, **75**, 125408 (2007).
33. O. Y. Yazyev, *Rep. Prog. Phys.*, **73**, 056502 (2010).
34. H.-X. Yang, M. Chshiev, D. W. Boukhvalov, X. Waintal, and S. Roche, *Phys. Rev. B*, **84**, 214404 (2011).
35. P. Venezuela, P. B. Muniz, A. T. Costa, D. M. Edwards, S. R. Power, and M. S. Ferreira, *Phys. Rev. B*, **80**, 241413(R) (2009).
36. M. Sepioni, R. R. Nair, S. Rablen, J. Narayanan, F. Tuna, R. Winpenny, A. K. Geim, and I. V. Grigorieva, *Phys. Rev. Lett.*, **105**, 207205 (2010).
37. J. Kunstmann, C. Özdoğan, A. Quandt, and H. Fehske, *Phys. Rev. B*, **83**, 045414 (2011).
38. R. R. Nair, M. Sepioni, I.-L. Tsai, O. lehtinen, J. Keinonen, A. V. Krasheninnikov, T. Thomson, A. K. Geim, and I. V. Grigorieva, *Nat. Phys.*, **8**, 199 (2012).
39. K. M. MacCreary, A. G. Swartz, W. Han, J. Fabian, and R. K. Kawakami, *Phys. Rev. Lett.*, **109**, 186604 (2012).
40. D. W. Boukhvalov and M. I. Katsnelson, *Appl. Phys. Lett.*, **95**, 023109 (2009).
41. G. Giovannetti, P. A. Khomyakov, G. Brooks, V. M. Karpan, J. van den Bring, and P. J. Kelly, *Phys. Rev. Lett.*, **101**, 026803 (2008).
42. P. A. Kohmyakov, G. Giovannetti, P. C. Rusu, G. Brocks, J. van den Brink, and P. J. Kelly, *Phys. Rev. B*, **79**, 195425 (2009).
43. Q. Ran, M. Gao, X. Guan, Y. Wang, and Z. Yu, *Appl. Phys. Lett.*, **94**, 103511 (2009).
44. K. Pi, K. M. McCreary, W. Bao, W. Han, Y. F. Chiang, Y. Li, S.-W. Tsai, C. N. Lau, and R. K. Kawakami, *Phys. Rev. B*, **80**, 075406 (2009).
45. L. Adamska, Y. Lin, A. J. Ross, M. Batzill, and I. I. Oleynik, *Phys. Rev. B*, **85**, 195443 (2012).
46. C. E. Malec and D. Davidović, *Phys. Rev. B*, **84**, 033407 (2011).
47. K. S. Kim, Y. Zhao, H. Jang, S. Y. Lee, J. M. Kim, K. S. Kim, J. H. Ahn, P. Kim, J. Y. Choi, and B. H. Hong, *Nature*, **457**, 706 (2009).
48. A. Ouerghi, M. Marangolo, R. Belkhou, S. El Moussaoui, M. G. Silly, M. Eddrief, L. Largeau, M. Portail, B. Fain, and F. Sirotti, *Phys. Rev. B*, **82**, 125445 (2010).

49. A. Michon, S. Vézian, A. Ouerghi, M. Zielinski, T. Chassagne, and M. Portail, *Appl. Phys. Lett.*, **97**, 171909 (2010).
50. M. Gao, Y. Pan, L. Huang, H. Hu, L. Z. Zhang, H. M. Guo, S. X. Du, and H.-J. Gao, *Appl. Phys. Lett.*, **98**, 033101 (2011).
51. D. A. C. Brownson and C. E. Banks, *Phys. Chem. Chem. Phys.*, **14**, 8264 (2012).
52. C.-H. Hsu, W.-H. Lin, V. Ozolins, and F.-C. Chuang, *Appl. Phys. Lett.*, **100**, 063115 (2012).
53. A. Varykhalov, D. Marchenko, M. R. Scholz, E. D. L. Rienks, T. K. Kim, G. Bihlmayer, J. Sánches-Barriga, and O. Rader, *Phys. Rev. Lett.*, **108**, 066804 (2012).
54. H. X. Yang, A. Hallal, D. Terrada, X. Waintal, S. Roche, and M. Chshiev, *Phys. Rev. Lett.*, **110**, 046603 (2013).
55. S. Roche and S. O. Valenzuela, *J. Phys. D: Appl. Phys.*, **47**, 094011 (2014).
56. J. O. Sofo, A. S. Chaudhari, and G. D. Barber, *Phys. Rev. B*, **75**, 153401 (2007).
57. D. C. Elias, R. R. Nair, T. M. G. Mohiuddin, S. V. Morozov, P. Blake, M. P. Halsall, A. C. Ferrari, D. W. Boukhvalov, M. I. Katsnelson, A. K. Geim, and K. S. Novoselov, *Science*, **323**, 610 (2009).
58. J. C. Meyer, A. K. Geim, M. I. Katsnelson, K. S. Novselov, T. J. Booth, and S. Roth, *Nature*, **446**, 60 (2007).
59. E. Stolyarova, K. T. Rim, S. Ryu, J. Maultzsch, P. Kim, L. E. Brus, T. F. Heinz, M. S. Hybertsen, G. W. Flynn, *Proc. Nat. Acad. Sci. U S A*, **104**, 9209 (2007).
60. V. M. Pereira, A. H. Castro Neto, H. Y. Liang, and L. Mahadevan, *Phys. Rev. Lett.*, **105**, 156603 (2010).
61. V. M. Pereira, A. H. Castro Neto, and N. M. R. Peres, *Phys. Rev. B*, **80**, 045401 (2009).
62. G. Cocco, E. Cadelano, and L. Colombo, *Phys. Rev. B*, **81**, 241412(R) (2010).
63. I. I. Naumov and A. M. Bratkovsky, *Phys. Rev. B*, **84**, 245444 (2011).
64. B. Verberck, B. Partoens, F. M. Peeters, and B. Trauzettel, *Phys. Rev. B*, **85**, 125403 (2012).
65. T. Ando, *J. Phys. Soc. Jpn.*, **69**, 1757 (2000).
66. D. Huertas-Hernando, F. Guinea, and A. Brataas, *Phys. Rev. B*, **74**, 155426 (2006).
67. H. Min, J. E. Hill, N. A. Sinitsyn, B. R. Sahu, L. Kleinman, and A. H. MacDonald, *Phys. Rev. B*, **74**, 165310 (2006).

68. Y. Yao, F. Ye, X.-L. Qi, S.-C. Zhang, and Z. Fang, *Phys. Rev. B*, **75**, 041401 (R) (2007).
69. D. Gosálbez-Martínes, J. J. Palacios, and J. Fernández-Rossier, *Phys. Rev. B*, **83**, 115436 (2011).
70. T. Kato, S. Onari, and J. Inoue, *Physica E*, **42**, 729 (2010).

# Chapter 4

# Electrical and Spin Transport

In this chapter, we first discuss electrical transport in graphene sheets and graphene nanoribbons (GNRs), considering conductivity, mobility, and spin transport from both experimental and theoretical perspectives. The effects of disorder on the transport properties are also explained. Further, a brief report on the Hall effects is included. Subsequently, the transport properties of metal/graphene/metal junctions are reported. In the next section, we review experimental and theoretical studies on graphene field-effect transistors (FETs), including contact effects and the graphene spin valves. Finally, novel proposals for magnetoresistive devices using graphene are mentioned. The discussion of graphene/metal contacts given in this chapter is rather restricted compared to the extremely large volume of research data obtained to date, which have been summarized in excellent review articles [1–4]. The reader is advised to refer to these articles for topics omitted from this chapter.

*Graphene in Spintronics: Fundamentals and Applications*
Jun-ichiro Inoue, Ai Yamakage, and Shuta Honda

ISBN 978-981-4669-56-6 (Hardcover), 978-981-4669-57-3 (eBook)
www.panstanford.com

## 4.1 Electrical and Spin Transport in Graphene

### 4.1.1 *Conductivity and Mobility*

As mentioned in the previous chapter, the electronic structure of graphene exhibits a linear energy–momentum relation near the Fermi energy, and only two states, called the $K$ and $K'$ points, exist within the Brillouin zone. The electronic structure near the Fermi energy is called the Dirac cone, and graphene is often called a zero-gap semiconductor because of the zero density of states (DOS) at the Fermi energy. The linear energy–momentum relation (Dirac dispersion) causes the electrons to behave as massless relativistic particles. Thus, one expects that the electrons in graphene can move significantly faster than those in ordinary metals and semiconductors. On the other hand, because there is no DOS at the Fermi energy and, therefore, no carrier density $n$ at this energy, it might be naively concluded that the conductivity tends to be vanish. These contradictory conclusions suggest that the electrical transport in graphene may be quite anomalous.

Theoretical investigation of electrical transport had already been initiated before the discovery of the graphene sheet [5]. The primary focus was the effects of the two-dimensionality and the characteristic electronic structures of graphene on the electrical transport. Ando et al. [6] predicted that the linear dispersion relations lead to the absence of conduction electron backscattering. Note that backscattering is also prohibited in carbon nanotubes, but in that case, the effect is significantly larger than in graphene because scattering can only occur in the forward and backward directions in a nanotube. Further, Suzuura et al. argued for weak localization corrections to the conductivity for disordered electrons in a two-dimensional honeycomb lattice [7]. The latter work shows that the universality class of electrical conduction in graphene changes from sympletic to orthogonal with increasing temperature. Note that the universality class in graphene nanoribbons (GNRs) has also been said to be dependent on short- or long-range impurity scatterings in graphene [8]. This crossover of the universality class may be related to the shape of the graphene Fermi surface. In addition, strong suppression of the weak localization in graphene has been reported on the basis of measurements of the low-field magnetoresistance (MR) [9]. However, the authors of the latter study argued that mesoscopic corrugations of the graphene sheet induce a dephasing effect. Finally, suppression of the backscattering caused by hydrogen passivation was also reported on the basis of measurements of the edge magnetotransport in disordered GNRs [10].

Conductance ($\Gamma$) quantization has been predicted for zigzag-edge GNRs, with $\Gamma$ increasing stepwise with $(2n+1)e^2/h$ [11]. The effects of disorder scattering on the conductance quantization seem to be controversial. For example, Mucciolo et al. [12] have reported that the conductance-step disappears as a result of edge and bulk roughness, depending on a numerical study, while Ihnatsenko and Kirczenow [13] have reported that the conductance steps survive in the presence of disorder in GNRs. Experimentally, Tombros et al. [14] have reported a quantized $\Gamma$ of $2e^2/h \times n$ in GNRs and further showed that a pronounced plateau of $0.6 \times 2e^2/h$ exists when an external magnetic field $B$ of 0.2 T is applied. These researchers have argued that this result may be related to the 0.7 anomaly in the quantized $\Gamma$ of point contacts, which is caused by electron–electron interactions. It has also been reported that conductance quantization could be affected by the transmission probability at the contacts in field-effect transistors (FETs) composed of graphene [15].

Although the effective mass $m_{\text{eff}}$ of electrons in graphene may be extremely small, $n$ tends to zero when the Fermi energy approaches the Dirac point. Therefore, the value of the Drude conductivity $\sigma = ne^2\tau/m_{\text{eff}}$ at the limit of low $n$ is nontrivial, where $\tau$ is the transport lifetime. The observed conductivity of graphene is approximately $\sigma = 4e^2/h$ and may not be smaller than this value [16]. On the basis of a theoretical study, Katsnelson has reported that the minimum $\sigma$ values of single- and bilayer graphene are $\sigma_{\text{min}} = 4e^2/\pi h$ and $4e^2/2h$, respectively [17]. He has further argued that the intrinsic disorder caused by the relativistic effect (Zitterbewegung) yields a Fano factor of $1 - 2/\pi \approx 1/3$. The Fano factor for $\sigma$ has also been studied numerically [18]. The observed value of the maximum resistivity, on the other hand, is $\rho_{\text{max}} = 6.5$ k$\Omega$, which is close to the 6.45 k$\Omega$ value for $h/4e^2$. The numerical factor of 4 may

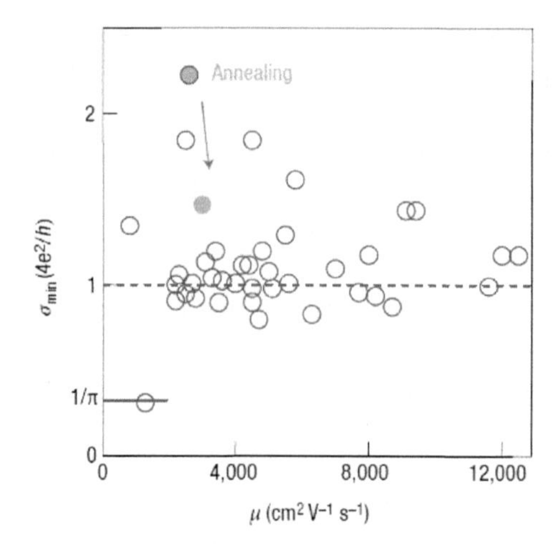


**Figure 4.1** Experimental results of minimum conductivity as a function of mobility density in graphene. The solid line indicates the theoretical results at a low-density limit. Reprinted by permission from Macmillan Publishers Ltd: [*Nature Materiels*] (Ref. [1]), copyright (2007).

indicate the valley degrees of freedom ($K$ and $K'$ points) and the spin degrees of freedom, and it appears that $\sigma$ is quantized, rather than Γ [1]. Figure 4.1 shows experimental values for the minimum $\sigma$ [1]. Although the values are scattered, the average seems to be close to $\sigma = 4e^2/h$. Several theories have been proposed as regard the possible origins of the $\pi$ factor, for example, the effects of contact-induced states [19], electron–electron interaction, charged impurities [20], and the difference between the effects of intraband and interband transitions [21]. Note also that the numerical results for $\sigma$ obtained using the first-principles method and tight-binding (TB) formalism give $\sigma = 4e^2/h$, as will be discussed later. The origin of the difference between the experimental and theoretical results has not yet been clarified.

Because the observed electron velocity $v$ in graphene is almost 400 times smaller than that of light in vacuum, but is still higher than that of electrons in semiconductors, the electrical $\sigma$ at finite carrier density is expected to be extremely high. The observed results, however, suggest that $\sigma$ saturates with increasing electric field $E$. Hwang et al. [22] have studied carrier transport in graphene layers using Boltzmann theory and have shown that this theory gives reasonable results for $n > 10^{12}/\text{cm}^2$, in agreement with the experimental results. Furthermore, these researchers have shown that $\sigma$ saturation depends on the charged impurities for $n < 10^{12}/\text{cm}^2$. The $\sigma$ saturation mechanism has been studied further. For example, Da Silva et al. [23] have presented theoretical results obtained by including all the scattering mechanisms in the Boltzmann formalism and compared the results with their experimental data. Hence, they suggested that the substrate optical phonons are crucial to the $\sigma$ saturation. Li et al. [24] have combined a first-principles calculation and Monte Carlo simulation and demonstrated that the energy dispersion $E(\boldsymbol{k})$ varies under high $E$, resulting in an increase in the electron scattering.

Rather than the $\sigma$ defined as a proportionality constant of the current $J$ for the given $E$, that is, $J = \sigma E$, another quantity, namely the carrier mobility $\mu$, is often used to represent the transport properties of materials. $\mu$ is defined as a proportionality constant in the relation between the average electron velocity $\langle v \rangle$ and $E$, such that $\langle v \rangle = \mu E$. As $E$ is given as the gradient of the electric potential $V$, the unit of $\mu$ is $\text{cm}^2/\text{Vs}$. Because the number of carrier electrons is not well defined in graphene as a result of the electronic structure of the graphene zero-gap semiconductor, the experimental-measurement-based $\mu$ is frequently reported. As explained in Chapter 1, it is expected that the $\mu$ of graphene may be extremely large because no electrical scattering occurs in ideal or nearly ideal graphene sheets.

The $\mu$ of the graphene sheet was first reported to be $\sim 10^4$ $\text{cm}^2/\text{Vs}$ at room temperature (RT) [5]. On the basis of measurements of the linear dependence of $\sigma$ on a gate voltage $V_g$ [25], it was reported that the $\mu$ value is approximately $1.5 \times 10^4$ $\text{cm}^2/\text{Vs}$ for both electrons and holes, which are independent of temperature between 10 and 100 K. Subsequently, $\mu$ was measured for samples fabricated using several methods, that is, for epitaxial graphene FETs on silicon carbide [26, 27], for graphene on copper produced by chemical vapor deposition (CVD) [28], and for mechanically exfoliated graphene [29], The measured $\mu$ values are summarized in Table 4.1. The $\mu$ observed for flat suspended graphene can be quite high; however, the values for graphene on substrates seem to be lower, as shown in the table. These values of the mobility should be

**Table 4.1** Observed data of carrier mobility $\mu$ with carrier density $n$

| $\mu$ (cm$^2$/Vs) | $n$ (1/cm$^2$) | Sample | Reference |
|---|---|---|---|
| 10,000 | | FET | [5] |
| 15,000 | $10^{13}$ | | [25] |
| 3,000 – 27,000 | | QHE | [34] |
| 2,000 | $3.6\times10^{12}$ | | [35] |
| 40,000 | | on h-BN | [36] |
| 125,000 – 275,000 | | on h-BN | [33] |
| 10,000 | | RT | [37] |
| 23,600 | | mechanical exfoliated | [29] |
| 500 – 700 | | Au top gate | [31] |
| 6,000 | | FET on SiC | [26] |
| 4,050 | | CVD on Cu | [28] |
| 27,000 | | | [38] |
| <600,000 | $5\times10^9$ | stably suspended, 77K | [32] |
| 2-3$\times10^4$ | $4\times10^{12}$ | theory | [39] |

compared with those in semiconductors, for example, the electron mobility $\mu_e = 490$ cm$^2$/Vs and hole mobility $\mu_h = 95$ cm$^2$/Vs of doped silicon, and $\mu = 1 \times 10^4$ cm$^2$/Vs of high-electron-mobility transistors (HEMTs).

The $\mu$ values reported for some samples are lower than expectations. These results could be attributed to electron scattering due to extrinsic sources caused by sample structures established on certain kinds of substrates. The extent to which $\mu$ is increased following elimination of such extrinsic scattering sources may be an interesting topic of research. Neugebauer et al. [30] have performed cyclotron resonance response studies for well-defined graphene flakes on the surface of bulk graphite, with the flakes being well decoupled from the surface; they found that $\mu$ may be as high as $10^7$ cm$^2$/Vs. This high $\mu$ indicates a long scattering time, which has been reported to be 20 ps. Although the mean free path $l$ depends on the Fermi velocity or $n$, it could be as long as 0.3–0.4 μm [1, 31]. Thus, considering the current status of modern Si technology, the fabrication of devices using ballistic transport could become possible.

Tombros et al. [32] have noted that the metals available for use as contacts in graphene FETs are limited (Au, Pd, Pt, Cr, and Nb) because of a problem affecting the reactive acid etching method used in the device fabrication process. These researchers have proposed a new fabrication technique to produce high $\mu$ values of up to 600,000 $cm^2/Vs$ at $n_e = 5 \times 10^9/cm^2$ at 77 K. This technique may yield mechanically stable, suspended, high-$\mu$ graphene devices. Further, the same group [33] has reported a technique to transfer high-$\mu$ graphene sheets (single- and bilayer graphene) onto commercially available hexagonal boron nitride (BN) substrates.

### 4.1.2 *Effects of Disorder*

As seen in Table 4.1, the $\mu$ values depend on the samples, indicating that there are extrinsic effects determining electrical transport in graphene, that is, scattering due to impurities, adatoms, substrate roughness, edge effects, contact with electrodes and/or gates, etc. There may be several primary factors determining the electrical transport in graphene: First, a fundamental issue regarding the extent to which the graphene resistivity $\rho$ can be reduced; second, a practical issue, that is, how unavoidable disorder affects the $\rho$ in graphene sheets fabricated using ordinary methods; and finally, the effects of contacts in graphene junctions as regard technological applications. In this subsection, we briefly discuss the former two issues.

Prior to the discovery of the single-layer graphene sheet, electrical transport was examined theoretically for carbon nanotubes and two-dimensional honeycomb lattices, and weak-localization corrections to $\sigma$ and the universality class had been discussed. In the Drude picture of electrical conductivity, electron scattering affects $\tau$. However, considering the electronic structure of graphene and the low $n$, it seems necessary to consider $\sigma$ from a quantum mechanical perspective and to examine the applicability of the Boltzmann approach.

Using the Dirac Hamiltonian, Nomura and MacDonald [40] studied quantum transport for massless Dirac fermions and showed that the effects of the short-range potential on the electron scattering differ from those of the Coulomb potentials. This difference may

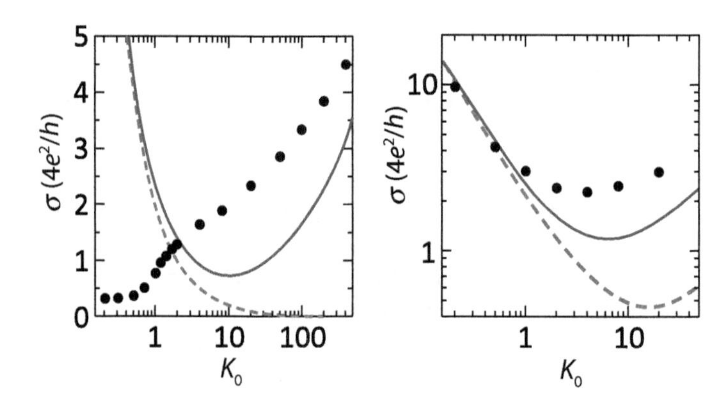


**Figure 4.2** Theoretical results of conductivity of graphene sheet indicating a crossover from quantum to Boltzman transport. Reprinted (figure) with permission from Ref. [41], Copyright (2009) by the American Physical Society.

be attributed to the range of charge screening in graphene. For doped graphene, Hwang et al. [22] adopted a Boltzmann transport theory to model electron scattering using random charged impurity centers as well as short-range scattering, and obtained quantitative agreement between calculated and experimental results. These researchers thus suggested that the dominant carrier-scattering mechanism is Coulomb scattering by random charged impurities near graphene and The substrate. They further suggested that $\mu$ could be as high as $2 \times 10^6$ cm$^2$/Vs for clean graphene, even at RT. Later, Adam et al. [41] performed a comparison of a quantum-mechanical numerical calculation using Boltzmann-type semiclassical theory for electron transport in graphene specimens with various disorder strengths. Figure 4.2 shows the theoretical results. These researchers showed that, for doped graphene, both results agree independently of the disorder strength but, for graphene with the Fermi energy at the Dirac point, the Boltzmann results do not agree with quantum simulation for weak disorder. Thus, the numerical results show a crossover between quantum and semiclassical behavior with changes in $n$.

The Boltzmann approach has been further examined using the Landauer approach for both short- and long-range scattering in the TB model [42]. The calculated $\sigma$ results are proportional to $n$

and agree with the results obtained using the Boltzmann approach beyond the Born approximation. Because defects and impurities in a graphene sheet induce resonant states that are different to the impurity states in metals and semiconductors, first-principles calculations were performed for realistic resonant impurity states produced by various organic molecules on graphene sheets. The $\sigma$ calculated quantum mechanically following TB mapping of the resonant states is consistent with experimental results [43, 44]. Further, a similar approach has been adopted for graphene sheets with structural defects [45], with the results indicating that the minimal $\sigma$ of $4e^2/\pi h$ is realized for $n > 0.5 \times 10^{14}/\text{cm}^2$ and that Anderson localization may be possible for defects when the concentration is less than 1%.

Quantum-theory-based studies of GNR electrical conductivity have been performed by several groups using the TB model [46–48]. For example, Martin and Blanter [47] have studied transport in disordered armchair-edge GNRs. Because the armchair nanoribbon can be metallic or insulating depending on the ribbon width $W$, opening or closing of the energy gap varies from position to position for armchair nanoribbons with fluctuating $W$. As a result, the transport in such armchair GNRs can be modeled as a one-dimensional variable range hopping. In fact, these researchers showed that low-temperature $\sigma$ is expressed in an effective one-dimensional hopping between nearby segments. When the disorder strength is increased, the effects of Coulomb potential interaction appear and the Coulomb potential blockade begins to play a role. Further, Schubert and Fehske [48] have numerically confirmed that a metal–insulator transition is observed in hydrogen-terminated GNRs and have also shown that electron–hole puddles induce charge inhomogeneity, resulting in suppression of the Anderson localization. Using the first-principles method, spin-dependent conductance has been studied for hydrogen-terminated short GNRs by Sahin and Senger [49]. Pieper et al. [50] have studied the effects of disorder on transport through GNRs using exact diagonalization of finite but large-sized graphene ribbons. These researchers have introduced structural fluctuations at the edges as well as mesh structures within the GNRs and calculated the optical conductivity using the Landauer-Büttiker formalism.

There are some noteworthy aspects of the treatment of disorder scattering in graphene, for example, Coulomb scattering due to the weak screening of extra charges caused by adatoms on the graphene sheet, along with so-called vertex corrections in the Green's function formalism (Kubo formalism) for $\sigma$. We have already mentioned the former point. Li et al. [51] have reported electrical transport for correlated disorder due to the long-range potential. Further, Wang et al. [39] have developed density functional theory (DFT) incorporating the coherent potential approximation (CPA), including nonequilibrium vertex corrections, so as to manage the disorder caused by nitrogen or boron impurities. These researchers have shown that $\mu$ can be $2\text{–}3\times 10^4$ cm$^2$/Vs for an $n$ of $4\times 10^{12}$/cm$^2$. Such vertex corrections are not essential for inclusion in the numerical calculation of finite-sized systems; however, they should be appropriately considered when one uses the Kubo formalism analytically. Finally, a completely different method, that is, a semiclassical Monte Carlo approach, has been adopted by Ferry [54] to manage short-range potential scattering on graphene on boron nitride, silicon carbide, and silicon dioxide substrates.

Although a large number of experimental studies exist, we mention only a few results here. Pi et al. [52] have studied the effects of transition metal clusters, which induce a $1/r$ Coulomb potential in a graphene sheet, on electron scattering and doping. Han et al. [53] have performed experiments on GNRs with back gates, for $20 < W < 120$ nm and $0.5 < L < 2$ μm, where $L$ is the GNR length. The transport measurements suggest the existence of localized in-gap states that result in thermally activated hopping and variable range hopping at high and low temperature, respectively.

**Note on impurity states in graphene:** From a theoretical perspective, it may be worth noting that the localized impurity states in graphene are distinct from those in other metals and semiconductors. Let the unperturbed Hamiltonian and impurity potential be $H_0$ and $V$, respectively. Then, the impurity state on site $i$ may be given as

$$G_{ii}(E+i0) = g_{ii}(E+i0)\left[1 - V_i g_{ii}(E+i0)\right]^{-1},$$

using the unperturbed Green's function

$$g(E+i0) = \left[E + i0 - H_0\right]^{-1}.$$

The energy dependence of the on-site Green's function $g_{ii}(E + i0)$ (imaginary and real parts) calculated for graphene sheet has been compared with those of metals and semiconductors [43]. We find that the characteristic features of the graphene electronic structure are also reflected in the energy dependence of the Green's function.

**Note on vertex corrections:** The Kubo–Greenwood formula is often used for calculating the $\sigma$ of disordered materials, where $\sigma$ is given by a product of two one-particle Green's functions. Therefore, in the $\sigma$ calculation, one should take the average of the product of two Green's functions. Because the average of the product of two Green's functions is not, in general, equal to the product of two averaged Green's functions, the so-called vertex correction should be properly evaluated. The vertex correction does not vanish for materials with a linear dispersion relation, which is the case for graphene, and careful analysis is therefore necessary. Such analysis for graphene has been performed by, for example, Wang et al. [39]. As regard the vertex corrections, see also Inoue et al. [55–57].

### 4.1.3 *Spin Transport*

Because graphene is composed of a light element, that is, carbon, the spin–orbit interaction (SOI) is weak and, therefore, one can expect that the spin diffusion length $\lambda_{sp}$, which is defined as the length for which the memory of spin of an electron is preserved, may be quite long. $\lambda_{sp}$ was first reported to be $\sim$1.5–2 μm at RT on the basis of a nonlocal measurement for graphene junctions with cobalt electrodes [35]. The same researchers did not observe any significant $\lambda_{sp}$ dependence on temperature. In the experimental setup used in that study, the graphene sheet is fabricated on silicon dioxide, and a thin aluminum oxide insulator is inserted between the cobalt electrode and graphene sheet. The experimental schematics of the nonlocal measurement are shown in Fig. 4.3. Electrons are injected from terminal 3 and a voltage difference is detected between terminals 1 and 2. Because of the spin polarization $P$ of the cobalt electrodes, the injected electrons produce spin accumulation near terminal 3. Thus, the spin accumulation decays with distance and becomes small at voltage terminals 1 and 2; however, it

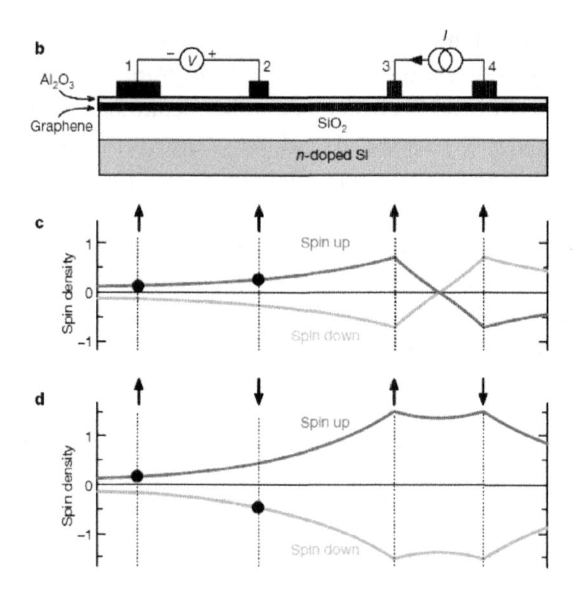


**Figure 4.3** Schematics of experimental setup for nonlocal measurements of spin diffusion length. Reprinted by permission from Macmillan Publishers Ltd: [*Nature*] (Ref. [35]), copyright (2007).

induces a voltage difference between these terminals. Because this voltage difference depends on the magnetization direction of the Co electrode, as shown in Fig. 4.3, one can estimate the length of the spin accumulation decay, that is, $\lambda_{sp}$. Goto et al. [58] also performed nonlocal measurements, but for multilayer graphene samples with Co and Cr/Au electrodes in direct contact with the multilayer graphene sheet. The reported value of $\lambda_{sp}$ was longer than 8 μm at 4 K. Increased $\lambda_{sp}$ in response to an increasing number of layers in a multilayer graphene sample was later confirmed by Tombros et al. [59].

**Note on nonlocal measurements:** The lower two panels in Fig. 4.3 are schematics of the spin density (up- and down-spins) for antiparallel (AP) and parallel (P) magnetization alignments of ferromagnetic metals (FMs) on electrodes 3 and 4. The magnetization direction is indicated by the up or down arrow in the figure. For the P alignment, because of the positive $P$ (up-spin electrons are dominant in cobalt ferromagnetic electrodes), more up- than down-spin electrons are injected from terminal 3 into the graphene.

Therefore, the latter is reduced in the region near terminal 3 so as to satisfy the charge neutrality condition in graphene. At terminal 4, more up- than down-spin electrons are extracted, and therefore, the spin density near terminal 4 is reversed from that near terminal 3. In the regions near terminals 1 and 2, the spin density decays with distance from terminal 3, as indicated by the red and green lines for the up- and down-spin electrons, respectively. Because no voltage is supplied to the region left of the terminal 3, the up- and down-spin densities are essentially identical to the chemical potentials of the up- and down-spin electrons (spin accumulation). The difference between the chemical potential at terminals 1 and 2 is measured by a voltmeter connected between these terminals. Because the up-spin electrons are dominant in ferromagnetic electrodes 1 and 2, the chemical potential difference between the two values denoted by the black dots is observed.

In AP alignment, where the magnetization at terminals 2 and 4 is reversed, the charge density is modified as shown in the lowest panel in Fig. 4.3. In this case, more down-spin electrons are extracted at terminal 4, and the spin density near that terminal is similar to that near terminal 3. Because the magnetization at terminal 2 is reversed, the down-spin electrons are dominant for that terminal. Therefore, the chemical potential difference at the sites shown by the black dots can be measured. Note that the sign of the chemical potential difference is reversed. This is the most important indication of the spin injection from an FM to graphene, and $\lambda_{sp}$ is estimated by changing the spacing between ferromagnetic electrodes 2 and 3.

One may use nonmagnetic metals (NMs) for terminals 1 and 4. In this case, the up- and down-spin densities are identical at both of these terminals.

The measured values of $\lambda_{sp}$ are significantly shorter than those predicted by theory, incorporating the negligibly small SOI of graphene. This means that the observed spin relaxation time $\tau_{sp}$ is shorter than the theoretical value by an order of magnitude.

The relationship between $\tau_{sp}$ and $\mu$ may clarify the spin relaxation mechanism as follows. When $\tau_{sp} \propto \mu$, the Elliot–Yafet (EY) mechanism dominates the spin relaxation, whereas when $\tau_{sp} \propto$

$1/\mu$, the D'yakonov–Perel (DP) mechanism is dominant. Therefore, various experimental results have been reported.

The van Wees group has performed extensive experiments on spin transport in graphene on the basis of measurements of the MR and Hanle spin precessions in nonlocal spin devices [35, 36, 59–63]. Further, Tombros et al. [61] have studied the effects of the directions of spins injected from ferromagnetic electrodes to the nonlocal geometry of a graphene sample by applying a high $B$ perpendicular to the graphene plane. Hence, the $\tau_{sp}$ for the injected spins in the absence of $B$ was shown to be 160 ps, and these researchers demonstrated that $\tau_{sp}$ decreases by 20% when a high $B$ is applied. For that measurement, it was assumed that the longitudinal spin relaxation time was equal to the transverse relaxation time, $T_1 = T_2$, respectively. Subsequently, the same group [62] showed that this assumption does, in fact, hold and that $\tau_{sp}$ is of the order of 50–200 ps for graphene flakes of 0.1–0.2 μm in width. Furthermore, they showed that the relaxation mechanism is of EY type. They also demonstrated that the spin accumulation $n_s$ is well described by the drift diffusion model

$$D\nabla^2 n_s + \mu E \nabla n_s - \frac{n_s}{\tau_{sp}} = 0,$$

where $D$ is a diffusion constant and $E$ is an applied electric field. These researchers have suggested that the increase in $\lambda_{sp}$ in multilayer graphene may be attributed to a decrease in the charged impurity scattering, which is caused by an increase in the charge-screening effect in multilayer graphene [59].

In spite of these intensive experiments, the large difference between the experimental and theoretical values of $\tau_{sp}$ may suggest extrinsic sources for spin dephasing, for example, roughness at the graphene/substrate interface caused by impurities and adatoms. For example, edge effects and ripples may produce a finite SOI. Contact between graphene and electrodes may also produce such effects. Therefore, further studies have been performed on the spin transport in order to clarify the spin diffusion mechanism. For example, Pi et al. [64] have performed nonlocal spin valve measurements of a graphene sheet with a chemically doped gold atom surface, and showed that $\tau_{sp}$ increases with increasing gold coverage. These researchers suggested that the increase in $\tau_{sp}$

may be attributed to certain unknown mechanisms; however, an increase in the charge screening effect due to the gold coverage remains possible. Further, Han et al. [65] have studied spin injection into single-layer graphene by inserting a titanium dioxide insulating layer into cobalt/graphene contacts. They have observed a long $\tau_{sp} = 450$–$500$ ps, and suggested that the tunnel barrier reduces the contact-induced spin relaxation. In addition, in nonlocal measurements for single- and bilayer graphene, Volmer et al. [66] have shown that $\tau_{sp}$ increases with increasing area resistance $R_c A$ at the contact, indicating the importance of the contact quality.

As for the scattering mechanism, Han and Kawakami [67] have performed nonlocal spin valve measurements of spin relaxation for single- and bilayer graphene and showed that the EY and DP mechanisms are dominant at low temperature for single- and bilayer graphene, respectively. Yang et al. [68] have also shown that the spin relaxation mechanism in bilayer graphene is of DP type both at low temperature and RT. These researchers obtained a long $\tau_{sp} = 2$ ns at RT. Zomer et al. [36] have observed spin transport for the case of a high-$\mu$ graphene sample on hexagonal boron nitride (BN), in which $\mu \lesssim 40,000$ cm$^2$/Vs and $\lambda_{sp} \lesssim 4.5$ μm, and determined that $\tau_{sp} \approx 200$ ps. These researchers then suggested that the scattering mechanism could be of both EY and DP type.

Thus, the spin-scattering mechanism remains controversial, and the puzzle concerning the large difference between the experimental and theoretical values of $\tau_{sp}$ remains unsolved [69]. In a theoretical study, Ochoa et al. [70] have shown that the $\tau_{sp}/\tau_p$ (transport lifetime) ratio depends strongly on $n$ but is independent of defect type, suggesting that unknown mechanisms different from the EY type could be responsible for the spin relaxation, even though the $\tau_{sp} \propto \tau_p$ relation holds. Because contact with electrodes is inevitable in graphene junctions, the $\tau_{sp}$ value may depend on the contact type and quality. Further studies on the contact quality and the spin relaxation mechanism are desirable from both experimental and theoretical perspectives.

Finally, other spin transport results have been reported; for example, Han et al. [71] performed nonlocal spin valve measurements at RT for single-layer graphene and reported electron–hole asymmetry in the nonlocal MR. The nonlocal MR is roughly independent of the electron bias but varies significantly with hole bias. Further, spin transport in graphene attached to a ferromagnetic insulator has been analyzed using the Dirac Hamiltonian [72].

**Note on Hanle measurement:** There are two electron spin relaxation types, spin polarization relaxation (relaxation of the longitudinal components of the spins) and relaxation of the spin phase coherence (relaxation of the spin transverse components). The former and the latter are characterized by relaxation times $T_1$ and $T_2$, respectively. The decay in the spin density in the nonlocal spin injection explained before corresponds to the longitudinal spin relaxation, while the transverse spin relaxation can be measured through application of an external $B$. Because the magnetization of ferromagnetic electrodes is in plane, spins parallel to the graphene plane are injected. The injected spins are rotated by an external $B$ perpendicular to the graphene plane and lose phase coherence. Then, the voltage signal measured between terminals 1 and 2 reduces. By fitting the voltage dependence on $B$, one may estimate $T_2$.

**Note on EY and DP mechanisms for spin relaxation:** There are two primary spin relaxation mechanisms in metallic systems, namely the EY and DP mechanisms. In the EY mechanism, electrons are scattered by impurities, such as magnetic or nonmagnetic impurities with SOI. The electron spin is reversed in the scattering process and the spins of the conduction electrons are thereby diffused. In this mechanism, the spin lifetime varies linearly with the momentum scattering time as the carrier concentration is varied, that is, $\tau_{\mathrm{sp}} \propto \mu \propto \tau_p$. In the DP mechanism, the existence of a SOI caused by structural asymmetry is assumed. An example is the Rashba-type SOI, which depends on the momenta of the conduction electrons. Because of this dependence, the direction of the conduction electron spins is changed by scattering from the initial to the final momentum state. In this case, $\tau_{\mathrm{sp}}$ is inversely proportional to $\mu$, that is, $\tau_{\mathrm{sp}} \propto 1/\mu \propto 1/\tau_p$.

### 4.1.4 *Hall Effects*

The transverse transport properties (Hall effects) also appear anomalously and in an interesting manner in graphene. Electrons drift in a perpendicular direction relative to $\boldsymbol{E}$ when an out-of-plane magnetic field is applied, that is, $J_{\perp} \propto (\boldsymbol{E} \times \boldsymbol{H})_{\perp}$. This phenomenon can be observed in any metal and is called the ordinary Hall effect, because this phenomenon was discovered by E. H. Hall in 1879. In the quantum mechanical picture, quantization of the energy levels due to $\boldsymbol{H}$ occurs, which is known as Landau level formation. As $n$ is affected by $V_g$, the Fermi energy crosses the Landau levels one by one, and the conductance increases stepwise by $e^2/h$. This type of Hall effect is called the quantum Hall effect ($q$-Hall effect).

In graphene, the $q$-Hall effect behavior is anomalous, that is, the transverse conductance increases as

$$\sigma_{xy} = \pm\frac{4e^2}{h}\left(N + \frac{1}{2}\right),$$

where the + (−) indicates that the carrier type is electron (hole) and $N$ is the Landau level index. The factor of 4 comes from the degrees of spin and pseudo spin caused by the nonequivalent sublattices. As $n$ is increased by $V_g$, occupation of the Landau levels (the number of which corresponds to the $N$ in the above expression) increases discontinuously as a function of $n$. Therefore, $\sigma_{xy}$ increases in stepwise fashion with $V_g$. The lowest value of $\sigma_{xy}$ is $(4e^2/h) \times 1/2$, which is distinct from the zero value in the conventional $q$-Hall effect, and $\sigma_{xy}$ increases by $4e^2/h$ with a further increase in $V_g$. This anomalous $q$-Hall effect was first reported by Novoselov et al. and Zhang et al. [25, 34, 73]. The observed results for $\sigma_{xy}$ are shown in Fig. 4.4. The above expression was also confirmed theoretically [74]. The difference between the $q$-Hall effect observed in normal metals and that in graphene is shown in Fig. 4.4.

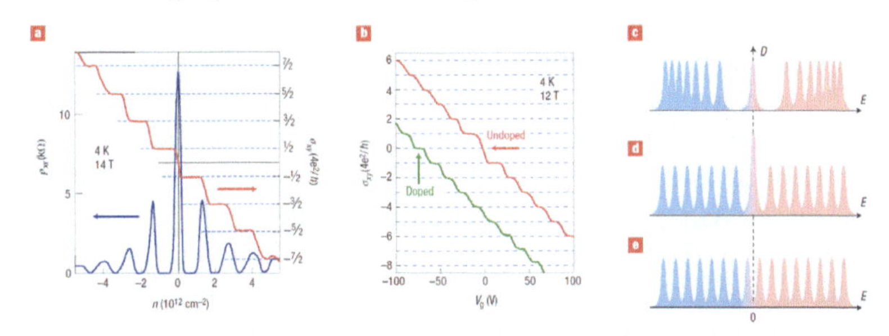


**Figure 4.4** Results for quantum Hall effects. Reprinted by permission from Macmillan Publishers Ltd: [*Nature Materiels*] (Ref. [5]), copyright (2007).

The anomalous behavior of the $q$-Hall effect in graphene has been attributed to the pseudo spin caused by the existence of two different sublattices. This result can also be explained in terms of the Berry's phase produced by the pseudo spin. Thus, the $q$-Hall effect is often called the chiral $q$-Hall effect.

The properties of the $q$-Hall effect in graphene have been studied for various scenarios. For example, the effects of edge and surface states on the $q$-Hall effect have been examined, along with electron–electron interaction [75]. In addition, the effects of disorder on the $q$-Hall effect in graphene $p$-$n$ junctions have been investigated, along with the electrically tunable $q$-Hall effect in a graphene sample decorated with 5d transition metal adatoms [76–79]. It has been noted that a large SOI is induced in graphene when heavy elements are attached to the graphene sheet. In addition, studies on fractional $q$-Hall states [80, 81] and on the $q$-Hall effect in twisted bilayer graphene [82] have been performed.

The introduction of the SOI in graphene induces both a spin-Hall effect and a quantum spin-Hall effect [83, 84], although the SOI induced in the graphene sheet is extremely weak [85]. Several groups have studied the spin Hall effects in this material [86–88], and the study of electrical transport in graphene has led researchers to the field of topological insulators. Further, the $q$-Hall effect introduced by valley isospins has also been discussed [89].

### 4.1.5 *Other Transport Properties*

Graphene exhibits anomalous behavior not only as regard electrical transport but also for thermoelectrical transport. In a theoretical study, Zhang et al. [90] examined the topological nature of the thermoelectric power in graphene for gapped single- and bilayer sheets, and Zhu et al. [91] performed a numerical study of the thermoelectrical transport of Dirac fermions under a strong $B$ and disorder. As regard experimental studies, Wei et al. [92] reported that the Seebeck coefficient diverges with $1/\sqrt{n}$ and the Nernst signal become very large at the Dirac point. This observation is quite interesting, not only as regard fundamental physics, but also for technological applications, because effective heat flow in nanoscale devices is one of the important components of technical applications. One of the important quantities is the figure of merit, which is defined as $T\Gamma S^2/\kappa$, where $T$ is temperature, $S$ is the thermopower, and $\kappa$ is the thermal conductance. Theoretical studies have been performed by some groups, and it has been shown that the figure of merit can be larger than one [93, 94].

## 4.2 Graphene Junctions and Field-Effect Transistors

As described before, the characteristic features of graphene's electronic states have attracted intensive study on the intrinsic properties of graphene transport phenomena. Nevertheless, detailed study of the transport phenomena of graphene junctions with electrodes is essential for application of graphene to electronics and spintronics. Because of the two-dimensionality and high $\mu$ of graphene, the most interesting graphene sheet application is in FETs. Because the translational symmetry along the current direction is broken in the junction structure, simple models were first adopted using the TB approximation and realistic models were then examined using the first-principles method. The application of graphene in FETs has already been discussed in the first report on graphene by Novoselov et al. [5], and a large number of experimental studies have been performed. In this section, we first report briefly on the results obtained via model calculations intended to clarify the characteristics of the graphene junctions with NM electrodes and then review various experimental and theoretical results. Because the contact properties have a significant effect on graphene junctions, including those in FETs, some details on the contact resistance $R_c$ will be given.

### 4.2.1 *Nonmagnetic Metal/Graphene/Nonmagnetic Metal Junctions*

Schomerus [95], Robinson and Schomerus [96], and Blanter and Martin [97] have performed analytical and numerical calculations on the conductance of NM/graphene/NM junctions using a simple TB model, in which graphene $\pi$ electrons and NM $s$ electrons are treated in a single-orbital model. Square-lattice $s$ orbital electrodes

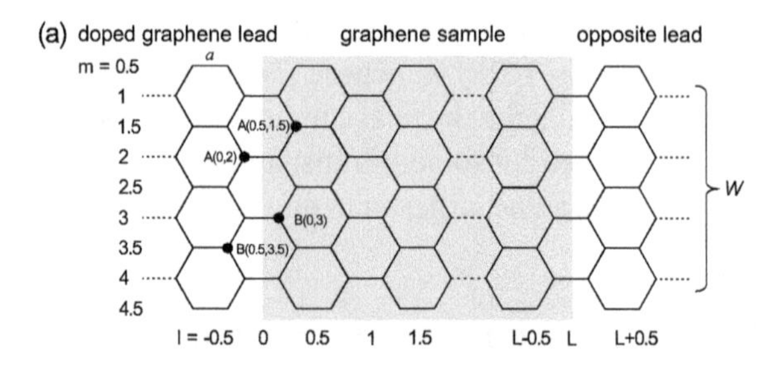


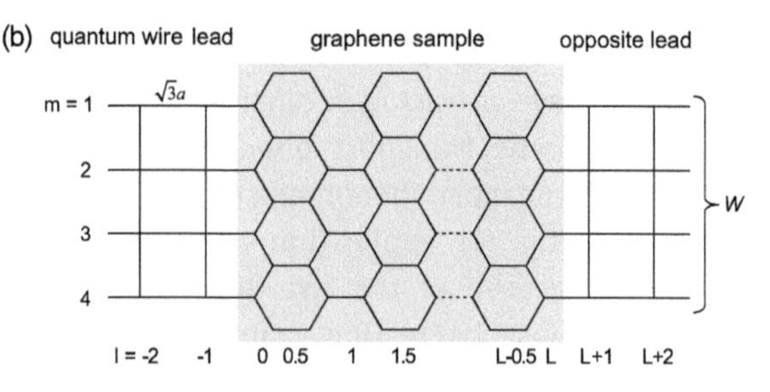


**Figure 4.5** Lattice structures used for model calculation of conductance of graphene sheet with (a) graphene electrodes and (b) square-lattice electrodes. Reprinted (figure) with permission from Ref. [95], Copyright (2007) by the American Physical Society.

are connected to the graphene sheet at the zigzag edges. For large-sized graphene sheets with sufficiently large $W$ and $L$, one may use the Dirac dispersion for the graphene electronic state, and analytic expressions of the conductance for arbitrary doping of electrons or holes into the graphene can be derived. See also Chapter 1.

Schomerus [95] has derived analytical expressions of $\Gamma$ for graphene junctions with graphene electrodes and for those with square-lattice $s$ orbital electrodes. Their structures are shown in Fig. 4.5a,b. For the graphene junctions with graphene electrodes

$$\Gamma = \frac{4e^2}{\pi h}\frac{W}{L}\frac{\sqrt{4\gamma^2 - V^2}}{V}\arcsin\frac{V}{2\gamma}, \tag{4.1}$$

where $h$ is the Planck constant, $\gamma$ is the hopping integral between the graphene and electrodes, which is taken to be identical to that in the graphene sheet in this model, and $V$ is the electrode potential. When $V \rightarrow 0$, the expression tends to $\Gamma = (4e^2/\pi h)(W/L)$, as derived from the Dirac equation for the ideal graphene sheet. For junctions with square-lattice electrodes

$$\Gamma = \frac{4e^2}{\pi h}\frac{W}{L}\frac{\sqrt{4\gamma^2 - (V+\gamma)^2}}{V+\gamma}\arcsin\frac{V+\gamma}{2\gamma}. \qquad (4.2)$$

In this case, when $V \to 0$, $\Gamma = (2e^2/\sqrt{3}h)(W/L)$. In typical diffusive metals, the relation $\Gamma = \sigma W/L$ holds between $\Gamma$ and $\sigma$ when no $R_c$ exists. Equation 4.1 is interpreted as indicating that $\sigma$ is determined by a $\Gamma$ that exhibits linear dependence on the sample length. On the other hand, Eq. 4.2, obtained for graphene junctions, indicates how the $\sigma$ of the junction is modified by the existence of the junction interface. The results show interesting graphene transport properties, with the Dirac dispersion being assumed for the electronic states of the $\pi$ electrons.

Blanter and Martin have shown the $\Gamma$ of junctions comprising graphene doped with some dopant $n$. Two contributions are made to the conductance, from the propagating and evanescent modes. The former is given by the product of the quantum conductance and the effective number of the transport channels, and vanishes in the zero-carrier-doping limit, for which the Fermi level $E_F$ is located precisely at the Dirac point. The results are almost independent of the electrode properties. The evanescent conduction may be interpreted as follows. Undoped graphene has only two momentum states in the Brillouin zone, $K$ and $K'$, and the other states removed from these two points are gapped. These gapped states form an insulating barrier in the graphene junction. When the Fermi energy is located at the Dirac point, the energy gap can be infinitesimally small, which results in $\Gamma$ with a $1/L$ dependence. Further, the surface (edge) state at the zigzag edges also contributes to the evanescent modes; however, this contribution is only dominant for small $L$.

The $1/L$ dependence of the evanescent modes can be simply understood: The tunneling probability at the $k_{\parallel}$ wave vector shown in Fig. 4.6 is expressed as

$$T\left(k_{\parallel}, L\right) \propto \exp\left[-c(k_{\parallel} - K')L\right],$$

with a constant $c$. $\Gamma$ is obtained by integrating $T\left(k_{\parallel}, L\right)$ over $k_{\parallel}$, resulting in

$$\begin{aligned}\Gamma(L) &\propto \int_0^{2\pi} e^{-cL|k_{\parallel}-K'|}dk_{\parallel} \\ &\propto 1/L.\end{aligned}$$

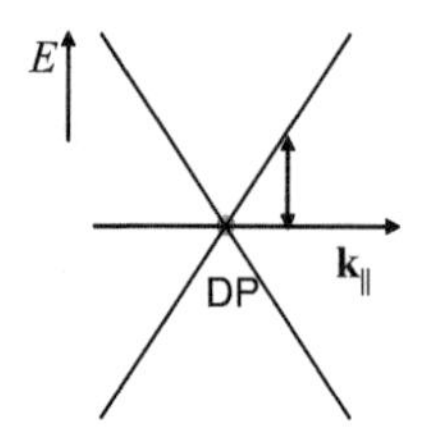


**Figure 4.6** An intuitive understanding of $1/L$ dependence of conductance of graphene junctions.

For junctions composed of doped graphene, the propagating modes contribute to $\Gamma$ in proportion to the $\Delta E_F$ or $V_g$ applied to the graphene sheet. In that case, the contribution of the evanescent modes (tunneling modes) decays in proportion to $L^{-4}$. Therefore, it is negligibly small compared to the contribution from the propagating modes. On the other hand, the propagating modes exhibit an oscillatory dependence on $L$, as shown in Fig. 6.5 in Section 6 [98], which may be interpreted as an effect of Friedel oscillation.

### 4.2.2 *Field-Effect Transistors*

Control of electric currents via tuning of electric fields is the core principle behind silicon electronics, and the FET is one of the most important devices in that technological field. Further development of modern electronics, especially in relation to information technology, is reliant on the downsizing of silicon devices. However, a serious difficulty limits silicon electronic device sizes, and therefore, novel materials with superior performance to silicon are desirable. Because of their high $\mu$, low $\rho$, and two-dimensionality, graphene sheets have high potential applicability to FETs or to any field-effect devices (FEDs). For technological applications of graphene, the following features are required: band-gap opening, sufficient productivity of high-quality graphene, and improved processability [99]. At present, however, the basic features of FETs should be examined in order to realize sufficiently high $\mu$ in FET structures and a large on/off ratio for the FET current by gate control. Both of these parameters are related to the characteristics of FET contacts.

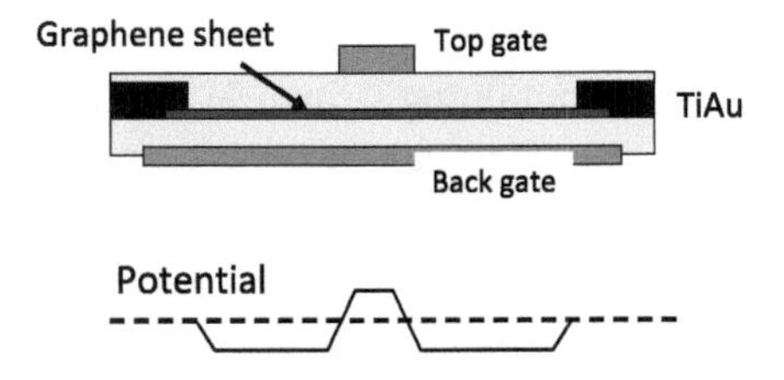


**Figure 4.7** Schematics of *p-n* junctions using a graphene sheet with a gate contact.

The electric field effects on graphene transport were first reported by Novoselov et al. [5], and a high $\mu$ of $1 \times 10^4$ $cm^2/Vs$ was obtained through application of a $V_g$. The sample used in that experiment was not a single graphene sheet, but rather a few-layer graphene specimen. Soon after the discovery of single-layer graphene, several experiments on the transport properties of FETs composed of single- or bilayer graphene or GNRs were performed [31, 100–103]. Huard et al. [100], for example, measured the resistance $R$ across the *p-n-p* and *n-p-n* bipolar junction barrier produced by a $V_g$ as a function of the barrier height and $n$. A schematic of the examined sample, which had top and back gate, is shown in Fig. 4.7. These researchers showed that the barrier is sufficiently strong to form bipolar junctions within the graphene sheet and that the $R$ across the barrier increases sharply. As for the $\mu$ of graphene FETs, although the values are reduced in comparison with those of suspended graphene, they are appropriately large for effective application of these devices, as indicated in Table 4.1. In addition, the effects of disorder scattering on carrier transport in *p-n* junctions have been studied theoretically, and it has been shown that the conditions for ballistic transport in the *p-n* junction are only marginally satisfied [104].

Despite the large number of studies on the characteristics of FETs composed of graphene, these devices have a significant disadvantage, because graphene is a zero-gap semiconductor. The on/off ratio of the current under a $V_g$ is quite small in graphene FETs. Therefore, fabrication of a graphene structure with a finite energy gap is highly desirable. Several methods to open the energy gap in graphene have been proposed, as discussed in Chapter 3.

For example, the fabrication of nanoribbons with $W$ of less than 5 nm and nanomeshes containing structural holes has been proposed [105]. In addition, the band gap in bilayer graphene opens under application of an $E$ perpendicular to the graphene sheet. The details of this setup have been studied experimentally by Castro et al. [106]. Another method is to allow the graphene sheet to react with atomic hydrogen [107]. The hydrogen reaction transforms the metallic graphene into an insulator. This structure is called graphane, and its energy gap has been reported to be 3.5 eV [108]. Some of the related experimental results have already been reviewed in Chapter 3.

Wang et al. [102] have used sub-10-nm graphene ribbons to fabricate graphene FETs. The typical $W$ of the sub-10-nm ribbon is $2 \pm 0.5$ nm, which may produce an energy gap that is sufficiently large for FETs, as mentioned in Chapter 3. These researchers used Pd contacts for the source and drain in order to reduce the Schottky barrier height and observed a large on/off ratio in the $I$–$V$ relation that could reach $10^6$.

Further, Echtermeyer et al. [101] have prepared a double-gated graphene FED with rather large-sized graphene. They placed the graphene sheet on silicon dioxide and positioned $SiO_x$ above the graphene sheet. A tungsten top gate was further grown on the $SiO_x$ layer. Echtermeyer et al. then measured the current change under variation of the top-gate voltage $V_{tg}$, and observed only a small current change for $-4$ V $< V_{tg} <$ 4 V; however, they found that an extremely large current change corresponding to a $10^6$ on/off ratio occurs when the $V_{tg}$ range is extended to $\pm 5$ V. These researchers attributed this extremely large on/off ratio to a chemical modification of the graphene, such as to graphane and/or graphane oxide, that is, graphene sheets with a large amount of hydroxyl (-OH) or similar chemical groups.

The electrical transport in dual-gated bilayer graphene has been measured by Taychatanapat and Jarillo-Ferrero [103], who reported that the energy gap caused by an applied $E$ could be 250 meV. However, the $\Gamma$ does not decrease significantly, indicating that the effective transport gap may be only a few millielectron volts. These researchers attributed this result to disorder scattering in the graphene sheets.

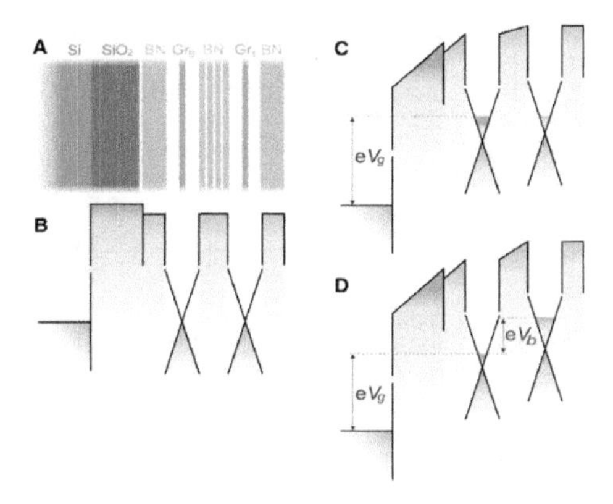


**Figure 4.8** Schematics of junction structure presented by Britnell et al. and change in the Fermi energies of two graphene sheets after applying gate voltage. From Ref. [109]. Reprinted with permission from AAAS.

Thus, achieving a sufficiently high on/off ratio, that is, larger than $10^6$, seems to be quite difficult in graphene FETs. Britnell et al. [109] have proposed a structure that is completely different from that of a typical FET. These researchers prepared a sample with a silicon/boron nitride/graphene/insulator/graphene/boron nitride structure, as shown in Fig. 4.8, in which the current is injected into one graphene layer tunnel through the barrier and exits via the second graphene layer. A $V_g$ is applied through the silicon layer, resulting in a charge imbalance between two graphene layers. Because of the low DOS near the graphene Fermi energy, the charge imbalance produces a large change in the height of the DOS at the Fermi energy in both of the graphene layers. The change in the DOS results in a significant change in the tunnel rate and produces a large on/off ratio. The performance of this kind of novel FET structure could be improved through the selection of suitable barrier materials.

### 4.2.3 *Contact Resistance*

In contrast to the high-$\mu$ case, the electronic states that contribute to the electrical transport through graphene are those near the Dirac points, and therefore, $n$ cannot be sufficiently large. The metal/graphene contact region may, in general, produce a contact

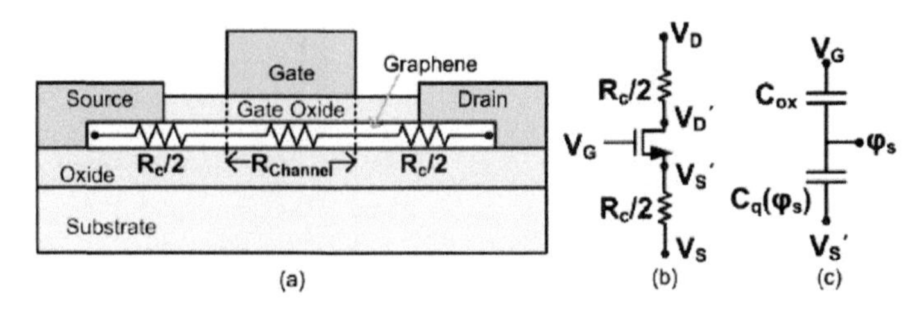


**Figure 4.9** Contact resistance in a graphene FET. Reproduced with permission from Ref. [111]. Copyright 2011, AIP Publishing LLC.

resistance $R_C$, and the resistance at the contact is usually large. In graphene FETs, the $R_C$ should be sufficiently lower than the $R$ in the graphene channel in order to realize gate controllability of the junction Γ, that is, to realize high on/off ratios.

In metal/graphene/metal junctions with $L \approx 1$ μm, the electronic states of the contact region and those of the central region of the graphene sheet may differ from each other. At the contact region, the graphene electronic state may be highly modified due to band mixing between the graphene and metals, as described in Chapter 3. In the central region of the graphene, on the other hand, the Dirac dispersion, which is an intrinsic feature of the electronic state, may be reserved without large modification. When the $R_C$ is larger than the $R$ of the graphene channel itself, the junction does not function as an FET. Therefore, the most important issue in the case of graphene FETs is the extent to which the $R_C$ is reduced. Although the $R_C$ at the source and drain dominates the total junction resistance, the contacts (including the gate contacts) should also be considered, as shown in Fig. 4.9 [110, 111].

We have already described the properties of metal/graphene contacts in Chapter 3, where changes in the graphene electronic structure at the contact, electron or hole doping into graphene, and the difference between chemical and physical contacts were discussed. Supplementary to the above, we here review several reports on the following topics: controlling the electrical transport using contacts on the basis of the selection of suitable metals [112], DFT calculations for graphene on a nickel(111) layer [113], ripple formation in graphene on a ruthenium(0001) layer [114], a graphene/ferromagnetic insulator contact [115], and the difference

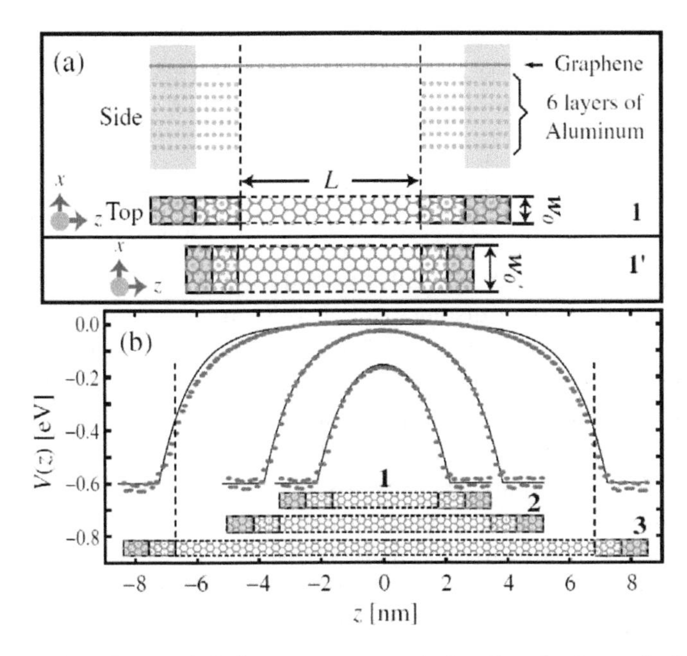


**Figure 4.10** Potential shift near contacts calculated in the first-principles band calculations. Reprinted (figure) with permission from Ref. [120], Copyright (2010) by the American Physical Society.

between graphene contacts with titanium and titanium dioxide [116].

Model and realistic calculations of electron transport through metal/graphene contacts have been performed by several groups [50, 117–121]. For example, Barraza-Lopez et al. [120] studied the above problems, adopting the first-principles method, as well as using a $\pi$-electron TB model. They applied the first-principles calculations to Al/graphene/Al junctions with $L = 3.40, 6.80$, and 13.60 nm with six-layer Al electrodes, as shown in Fig. 4.10a. Note that the contact comprising Al electrodes with graphene incorporates an overlayer structure. The Al/graphene contact was classified as a physical contact with a weak interaction between the graphene and Al. Therefore, the characteristic features of the graphene electronic state survive at the contact region. The Fermi-level shift at the contact was shown to be $\Delta E_{\rm F} = E_{\rm F} - E_{\rm D} = 0.6$ eV (electron doping), which is consistent with the result obtained by

Giovannetti et al. The position dependence of $\Delta E_{\rm F}$, or the potential profile $V(z)$, where $z$ indicates the position measured from the center of the graphene sheet, is shown in Fig. 4.10b. The profile is well fit by the function

$$V(z) = \begin{cases} \Delta \cosh(z/\lambda)/\cosh(L_{\rm eff}/2\lambda) & \text{for } |z| < L_{\rm eff}/2 \\ \Delta & \text{otherwise} \end{cases},$$

where $\Delta = -0.6$ eV, $\lambda = 1.05$ nm, and $L_{\rm eff} = L + 5a_0$, with a carbon–carbon interatomic distance $a_0 = 0.142$ nm.

The plot of the calculated $\Gamma$ as a function of energy exhibits two minima for long graphene junctions, at $E = -0.6$ and 0 eV, which are attributed to the Dirac points at the contact and central regions of the graphene sheet, respectively. For short graphene junctions, the $\Gamma$ minimum at $E = 0$ eV disappears. Because the interaction between electrodes composed of aluminum and a graphene sheet is weak, the Dirac dispersion characteristic remains and, therefore, the $\Gamma$ minimum appears even in the case of short graphene junctions.

Many experimental studies on $R_{\rm c}$ have been performed by a large number of groups [26, 27, 110, 122–128]. For example, Malec and Davidović [125] have measured the $R_{\rm c}$ for gold grains on a single-layer graphene sheet exfoliated mechanically from natural graphite. From four-probe $J$-$V$ measurements, they predicted $\Delta E_{\rm F} = E_{\rm F} - E_{\rm D}$ of $-0.19$ eV for an interlayer distance of 3.3 Å calculated using the first-principles technique. (Note that the definition of the sign of $\Delta E_{\rm F}$ differs from those in Malec and Davidović.) The value of $|\Delta E_{\rm F}|$ is smaller than the calculated value of 0.35 eV; however, for an the interlayer distance of ~4 Å, $|\Delta E_{\rm F}|$ is 0.35 eV, which is in agreement with the theoretical value. The contact resistance has been reported to be $R_{\rm c} = 915\ \Omega$, which results in an effective contact resistance $R_{\rm c}W \equiv \rho_{\rm c}^{W} = 128\ \Omega\ \mu$m for an average grain diameter of $W = 140$ nm. Because the average area of the contact grains is $A = 0.016\ \mu\text{m}^2$, the effective contact resistivity $\rho_{\rm c}^{A} = R_{\rm c}A = 14.6 \times 10^{-8}\ \Omega\text{cm}^2$.

The $\rho_{\rm c}^{A}$ value of metal/graphene contacts has been measured and reported to be approximately $5 \times 10^{-6}\ \Omega\text{cm}^2$, according to several groups [26, 27, 110, 124]. These measurements provide evidence that a work function difference $\Delta W$ at the graphene/metal contacts may reduce $\rho_{\rm c}^{A}$; this can be realized via doping into the graphene. Huang et al. [126] have reported that the effective contact resistance ($\rho_c^W$) is 7500 Ω μm for a Ti/Au contact, 2100 Ω μm for a Ni/Au contact, and 750 Ω μm for a Ti/Pd/Au contact using top-gated graphene FETs. Further, they reported $\rho_c^A = 2 \times 10^{-6}$ Ωcm$^2$ for the Ti/Pd/Au contact. Moon et al. [27] have reported that $\rho_c^A$ is of the order of $10^{-7}$ Ωcm$^2$ for Ti/Pd/Au contacts.

Nagashio and Toriumi [127] have investigated $\rho_c^W$ for several metal/graphene contacts using data obtained via four-probe measurements. They reported $\rho_c^W$ = 500–2000 Ω μm for Ni contact, $7 \times 10^3$–$5 \times 10^4$ Ω μm for Ti/Au, contacts, and $2 \times 10^3$–$4 \times 10^5$ Ω μm for Cr/Au contacts. They observed that the important quantity is $\rho_c^A$, rather than $\rho_c^W$, and deduced $\rho_c^A \approx 5 \times 10^{-6}$ Ωcm$^2$ for Ni/graphene contacts.

Note that $R_c$ should be sufficiently smaller than the graphene channel resistance $R_{ch}$, which is given as $R_{ch}^S(L/W)$, where $R_{ch}^S$ is the graphene sheet resistance (250 Ω) and $L$ is the length of the graphene part of the junction. If we assume $R_c/R_{ch} = 0.1$, then

$$\rho_c^A = R_c Wd = 0.1 R_{ch} Wd = 0.1 R_{ch}^S (L/W) Wd = 0.1 R_{ch}^S Ld.$$

This result indicates that, for a given $L$, $\rho_c^A$ increases with increasing overlapping graphene-length $d$ at the contact. However, Nagashio and Toriumi [127] have noted that the value of $\rho_c^A$ should saturate to be constant for $d$ larger than a certain value of $d_T$. This is because the current transfer from the metal to graphene should occur near the edge of the metal at the contact. For $\mu$ = 5000 cm$^2$V$^{-1}$s$^{-1}$ and a graphene carrier number of $n = 5 \times 10^{12}$ cm$^{-2}$, these researchers estimated $\rho_c^A$ to be $10^{-8}$ Ωcm$^2$ for junctions with $L = 1$ μm, using the equation above. The value of $\rho_c^A$ is one or two orders of magnitude lower than the experimental value of $10^{-6}$–$10^{-7}$ Ωcm$^2$.

Robinson et al. [128] have shown that $R_c$ can be largely reduced via process optimization for a combination of oxygen plasma etching and subsequent heat treatment. They observed a $\rho_c^A \approx 1 \times 10^{-7}$ Ωcm$^2$ for a Ti/Au contact, and showed that the value of $\rho_c^A$ is rather independent of the $\Delta W$ between graphene and metal by examining the $R_c$ for Ti, Cu, Pd, Ni and Pt metal contacts. Thus, we find that $R_c$ tends to be decreased; however, further reduction seems to be realized for graphene FETs.

Finally, Nouchi and Tanigaki [123] have commented that distortion of the $J$ - $V_g$ relation reduces the performance of graphene FETs, and showed that this distortion can be attributed to contacts with or without charge density pinning effects, rather than to impurity effects as is widely believed.

## 4.3 Spin Valves and Magnetoresistance

Because graphene is comprised of carbon, which is a light element, the SOI should be extremely weak and, therefore, spin-conserving transport through the graphene sheet may be highly likely. Measurements of $\lambda_{sp}$ have indicated that it is sufficiently long to support this behavior, as mentioned previously in the discussion on magnetoresistive devices. The issues to be overcome in order to realize such devices, however, differ from those for graphene FETs.

The important criteria for the realization of graphene FETs are high $\mu$ and a large on/off ratio, which are governed by the junction $R_c$. In graphene magnetoresistive junctions, on the other hand, the important components may be the $\lambda_{sp}$ (or $\tau_{sp}$) of the conduction electrons in the junction and the $R$ at the contact in order to realize sufficient spin injections into the graphene. As mentioned above, $\lambda_{sp}$ is usually measured in the nonlocal geometry in four-terminal devices, and $\tau_{sp}$ is measured using the Hanle effect, also in the nonlocal geometry. In these nonlocal devices, a spin accumulation is realized through injection of high-density currents at an FM and graphene contact, and the decay rate as a function of distance or time under an applied $B$ is measured. The observed values of $\lambda_{sp}$ ($\tau_{sp}$) are much shorter than those obtained theoretically [69]. Therefore, some extrinsic mechanisms strongly affect $\lambda_{sp}$. Note also that, in two-terminal devices, the contacts (which are inevitable in such a junction) may influence $\lambda_{sp}$.

The spin injection, spin transport, and spin detection for FM/semiconductor/FM junctions have been studied both through experiment and theory. However, it has been noted that spin injection from ferromagnets to semiconductors is not easy, because of the large difference between the resistivities of these materials. As this difference is of the order of $10^3$–$10^4$, spin injection is only possible for junctions using 100% spin-polarized half-metals as the FM. In other words, the spin transport of the entire junction is governed by semiconductors with high electrical resistivity, but without spin-dependent conductivity, as compared with the FM. This effect is called conductivity (resistivity) mismatch between ferromagnets and semiconductors [129]. However, this obstacle to spin injection into the semiconductor has been shown to be eliminated via the insertion of insulating materials between the ferromagnets and semiconductors or by using a Schottky barrier. Because the conductivity mismatch may also appear in junctions with FMs and doped graphene, insertion of a tunneling barrier between the FM and graphene is desirable; however, the spin relaxation at the tunneling barrier should not be enhanced. Thus, distinct problems occur as regard realization of magnetoresistive graphene junctions.

### 4.3.1 *FM/Graphene/FM Junctions*

We begin with a model study of the MR of graphene junctions with ferromagnetic electrodes, as reported by Brey and Fertig [130]. In their study, these researchers adopt model junctions in which square lattice electrodes are in contact with the zigzag edges of the graphene. The electronic structure of the square lattice is given by a TB $s$ orbital. A simple modification to add d-like states is also performed. Using both analytical and numerical calculations, Brey and Fertig show that the MR effect of such junctions is, in general, small, except for cases where one of the spin bands is close to or within the graphene energy gap. The theoretical results are consistent with the experimental values, in which the observed MR ratio is, at most, 10%.

The theoretical results seem to be reasonable when we consider the fact that the spin polarization of the single orbital $s$-band may not be large and that the graphene itself is nonmagnetic. More precisely, when the momentum states of the Dirac points reside within the up- and down-spin bands of the exchange-split electrode $s$-bands, the difference between the junction up- and down-spin conductance values, represented by $\Gamma_{\uparrow}$ and $\Gamma_{\downarrow}$, respectively, cannot be large.

Therefore,

$$P = \frac{\Gamma_\uparrow - \Gamma_\downarrow}{\Gamma_\uparrow + \Gamma_\downarrow},$$

may be small, resulting in a small MR.

The exceptional cases in which large MR ratios are obtained can be interpreted as follows. Let us consider a zigzag-edge junction in which the $\bar{K}'$ point is located at $k_\parallel = 2\pi/3$, while $\bar{K}$ is located at the band edge. In this case, the Dirac point ($\bar{K}'$ point) may contribute to the electrical conduction; however, it does not always reside within both the up or down electrode spin bands. When it resides within up- or down-spin band, either $\Gamma_\uparrow$ and $\Gamma_\downarrow$ can be zero, resulting in $P = 1$. This yields an extremely large MR ratio. The mechanism of this large MR may be attributed to matching/mismatching of the $\bar{K}'$ point with propagating modes in the electrodes specified by $k_\parallel$ [95–97]. The simple electronic states of the $s$ orbital square-lattice electrodes, however, cannot realize such exceptional cases under reasonable $s$ orbital band splitting, and only a small MR ratio is obtained, as mentioned before. In graphene junctions with realistic ferromagnetic electrodes, the situation is altered, as is explained in detail in Chapter 6.

### 4.3.2 *Novel Magnetoresistance Theory*

Various theories regarding the MR effects in GNRs have been presented [131–135]. As mentioned in the previous chapter, on the basis of first-principles calculation, it has been predicted that the edge states of zigzag GNRs exhibit half-metallic ferromagnetism, as shown in Fig. 4.11 [136, 137]. Kim and Kim [131] have predicted a very large MR for such GNRs in contact with two FMs, as shown in Fig. 4.12. When the magnetization is in parallel, an edge current appears along one of the two zigzag edges. When the magnetization is in antiparallel, the $P$ of each edge at the magnetic contacts is reversed and no edge current flows (see Fig. 4.11). Therefore, a large MR effect is expected. It has, however, been commented that such edge magnetism may not persist at elevated temperatures. Spin transport and MR have also been studied for GNRs passivated by hydrogen and oxygen [132] and for iron-terminated zigzag GNRs

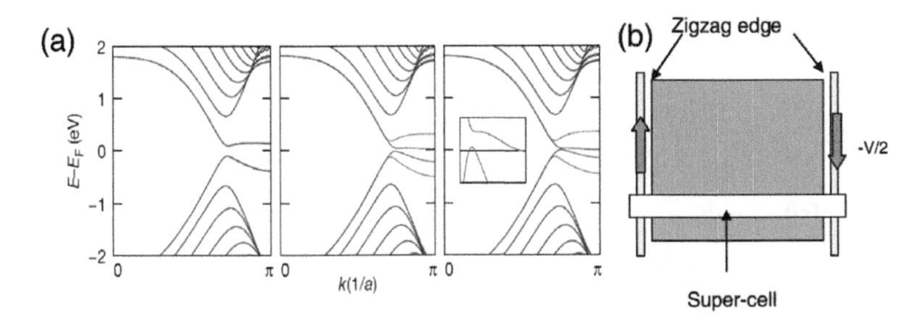


**Figure 4.11** Electronic structure of edge states. Reprinted by permission from Macmillan Publishers Ltd: [*Nature*] (Ref. [136]), copyright (2006).

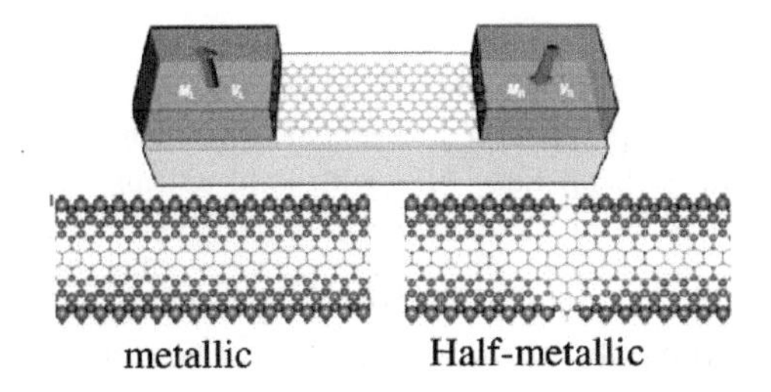


**Figure 4.12** Schematic figure for large MR using half-metallic states at edges. Reprinted by permission from Macmillan Publishers Ltd: [*Nature Nanotechnology*] (Ref. [131]), copyright (2008).

[134]. Further, Dubois et al. [135] have studied spin filtering and the MR effect at a graphene and hexagonal boron nitride interface.

Spin filtering and ballistic MR have been studied using the Dirac Hamiltonian [138], and MR in multilayer junctions has also been examined [139, 140]. In addition, alternative sample structures may merit investigation, some theoretical studies have been performed along this direction. For example, Tian et al. [141] have proposed possible spin injection from ferromagnetic graphene resulting in a magnetic proximity effect. In this case, no contact-induced influences arise.

## 4.3.3 *Nonlocal Spin Valve*

Nonlocal MR measurements using four-terminal devices were first adopted by the van Wees group [35, 60]. Because of the effect of the conductivity mismatch, Tombros et al. [35] used samples in which aluminum oxide was inserted between graphene and cobalt layers.

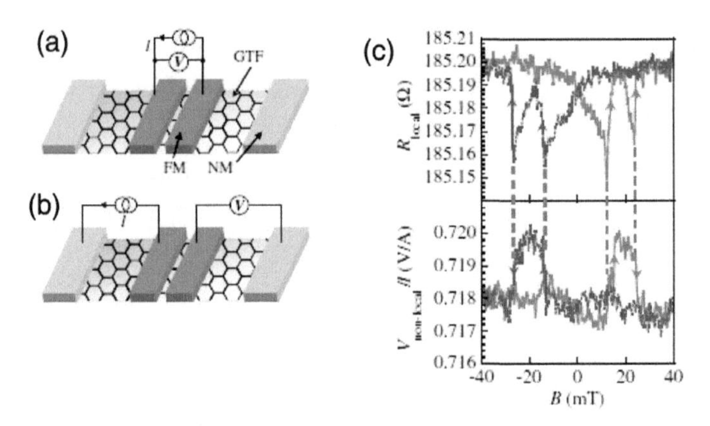


**Figure 4.13** (a and b) Schematics of experimental setup for nonlocal measurements and (c) observed results of resistance change for graphene sheet. Reproduced from Ref. [142] with kind permission from the Japan Society of Applied Physics.

These researchers coated the entire graphene sheet with aluminum oxides using the natural oxidation of aluminum metal on graphene. The observed value of $\lambda_{sp} = 2\ \mu m$ is relatively long; however, it is significantly shorter than that estimated theoretically for suspended graphene. The discrepancy of the $\lambda_{sp}$ (or spin relaxation time) values could be attributed to the contact quality, as mentioned in the previous section. Nevertheless, spin-dependent transport was realized in the four-terminal devices. In fact, the spin injection was confirmed using nonlocal measurements for samples composed of cobalt electrodes on graphene multilayers with approximately 10 nm thickness by Ohishi et al. [142]. The change in resistivity as a function of an external $B$ is shown in Fig. 4.13. However, this change is rather small and was observed in the nonlocal measurements only. Further, it was attributed to an $R_c$ that was 20 times higher than the graphene $R$.

Because the insulating layer at the contact is crucial, insulating materials for metal/graphene contacts have also been examined experimentally. By comparing contacts composed of thin aluminum oxide and magnesium oxide layers to single- and bilayer graphene samples for two-terminal devices, Dlubak et al. [143] have reported that the use of aluminum oxide for bilayer graphene junctions seems to induce improved spin injection performance. On the other

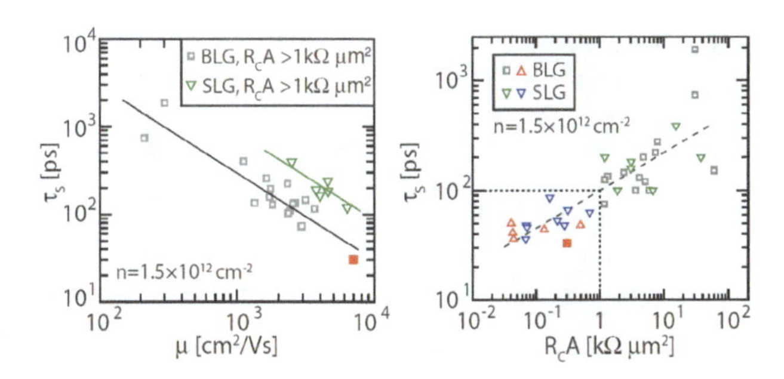


**Figure 4.14** Observed results of dependence of spin relaxation time on (a) mobility and (b) area resistance. Reprinted (figure) with permission from Ref. [66], Copyright (2013) by the American Physical Society.

hand, Han et al. [65] have performed nonlocal measurement of the spin injection from cobalt electrodes separated from graphene by magnesium oxide barriers, obtaining a large change in resistivity at RT caused by application of an external $B$. Their result indicates that a magnesium oxide barrier is efficient for spin injection into graphene. Further, their result for the nonlocal MR (a change in resistance $\Delta R$ of approximately 130 Ω at RT) is the largest yet observed in any material. It has also been shown that tunnel barriers reduce the contact-induced spin relaxation. Volmer et al. [66] have studied the role of magnesium oxide barriers in spin and charge transport, and showed that the spin relaxation time is inversely proportional to the $\mu$ in devices with large contact-resistance area $R_cA > 1$ kΩ μm$^2$. It was also demonstrated that $\tau_{sp}$ increases with $R_cA$, as shown in Fig. 4.14. Thus, a discrepancy may exist between the experimental results as regard the roles of aluminum oxide and magnesium oxide tunnel barriers.

## 4.3.4 *Two-Terminal Spin Valve*

Two-terminal measurements of MR for FM/graphene/FM junctions have also yielded finite values for the MR effects. Hill et al. [144] first reported on the spin valve effect in two-terminal graphene ferromagnetic junctions. They used single-, two-, and three-layer graphene sheets attached to nickel-iron electrodes, as shown in Fig. 4.15a. By applying an external $B$, the magnetization alignment of the two nickel-iron electrodes became antiparallel at a certain

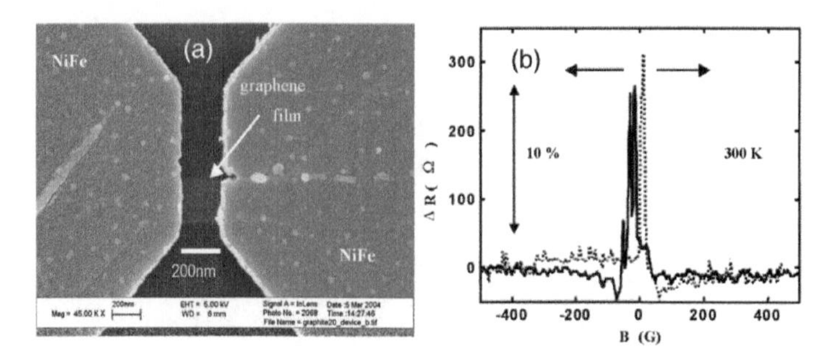


**Figure 4.15** (a) Sample image of NiFe/graphene/NiFe junction and (b) observed results of resistance change caused by a change in magnetization alignment. © [2006] IEEE. Reprinted, with permission, from Ref. [144].

$B$ and the $R$ increased (the spin valve effect). $\Delta R$ (MR) was approximately 10% in these measurements (Fig. 4.15b), which is rather small compared with those observed in metallic multilayers and in magnetic tunnel junctions.

Nishioka and Goldman [145] reported a 0.39% $\Delta R$ for Co/multilayer graphene/Co junctions at 2K. A ferromagnetic junction using graphite was also reported. Figure 4.16c shows the MR

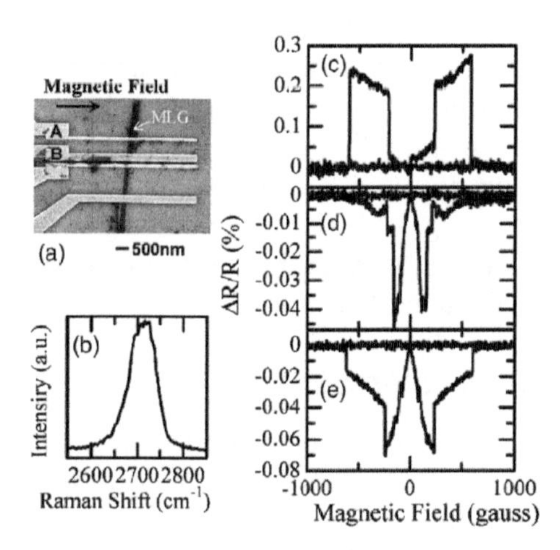


**Figure 4.16** MR measurements. (a) Sample configuration measured by scanning electron microscopy, (b) Raman spectrum of the multilayer graphene device, (c) results of magnetoresistance, and (d and e) AMR of the 300 and 100 nm wide Co electrodes at 10 K, respectively.

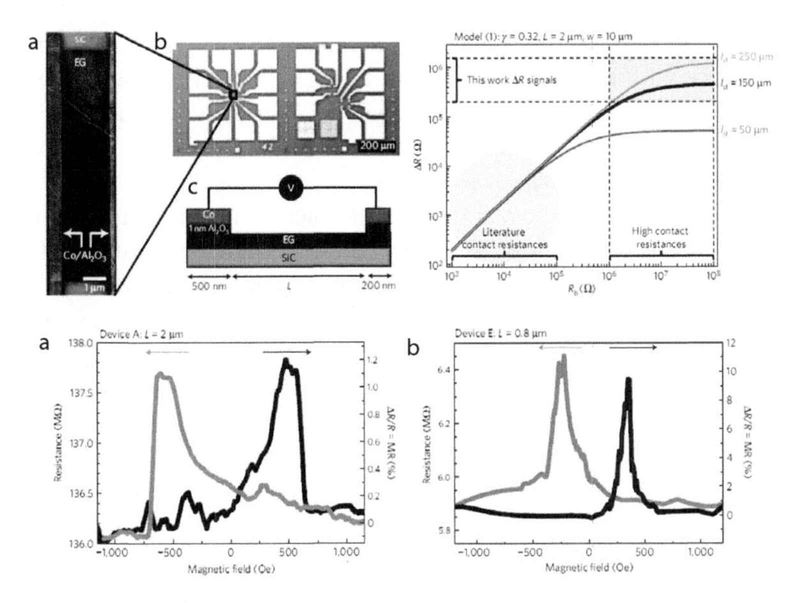


**Figure 4.17** Experimental setup and observed results for two-terminal measurements of magnetoresistance. Reprinted by permission from Macmillan Publishers Ltd: [*Nature Physics*] (Ref. [148]), copyright (2012).

experimental results for multilayered graphene measured at 2 K. Wang et al. [146] used a Co/graphite/Co junction in which an MgO layer was inserted between Co electrodes and a graphite layer. The MR was 12% at 7 K. Graphene ferromagnetic junctions using magnetite, which is believed to be half-metallic, were also fabricated and MR was observed [147] However, this MR is small compared to that observed in the other junctions.

Dlubak et al. [148] performed a detailed study on MR in two-terminal devices; their experimental setup is shown in Fig. 4.17a. They used epitaxial graphene on SiC with Co electrodes and $Al_2O_3$ tunnel barriers between the Co and graphene. The distance between the two cobalt electrodes was $L = 1$–$2$ μm, and the graphene $W$ was approximately 10 μm, with the contact width being 500 or 200 nm. For $L = 2$ μm, $\Delta R$ was approximately 1% (Fig. 4.17, bottom right), and for $L = 0.8$ μm, it was approximately 10% (Fig. 4.17, bottom left). The $R$ values themselves for the former and latter were approximately 137 and 6 MΩ, respectively. The $\lambda_{sp}$ was determined to be 100–200 μm and μ was 17,000 cm$^2$/Vs at

$n = 10^{12}/\text{cm}^2$. Thus, Dlubak et al. demonstrated that a two-terminal device with graphene on SiC and an $Al_2O_3$ tunnel barrier exhibits MR effects. It should be noted that $\Delta R$ increases with increasing barrier resistance $R_b$ and tends to saturate for $R_b > 10^7\,\Omega$, as shown in Fig. 4.17 (top left).

Using a permalloy/graphene interface, spin pumping was investigated by Singh et al. [149]. These researchers showed that single-layer graphene absorbs the spin angular momentum more effectively than bilayer graphene. They attributed their result to the SOI of adatom copper. Finally, vertical tunneling through two graphene sheets with an inserted insulating layer has been studied from a theoretical perspective [150], and an experiment on vertical tunneling through a single-layer graphene sheet has been performed [151]. The spin valve structure used in the latter experiment consisted of Co/Graphene/Ni-Fe, and the observed MR value was approximately 4% at 4.2 K. The use of this spin valve structure may be interesting, especially when we note the recently proposed graphene FET using tunneling between two graphene sheets.

## 4.4 Novel Graphene Devices

Numerous proposals for novel devices exploiting the characteristics of graphene have been presented. The most interesting concept utilizes the pseudo spin degree of freedom caused by the valley structure. Rycers et al. [152, 153] and Xiao et al. [154] have proposed utilizing a spin valve effect in nanoconstricted GNRs, the origin of which is completely different from the conventional spin valve effect. Because the valley degree of freedom plays a role similar to the spin degree of freedom, an effect similar to a spin valve appears in zigzag-edge GNRs with constriction (Fig. 4.18). The concept of the valley degree of freedom has led to the development of a novel field of electronics called "valleytronics" (see also [155]). The use of a valley filter effect has also been proposed by Fujita et al. [156] for a strain-engineered graphene sheet and by Li et al. [157] for bilayer graphene.

As a more realistic example of graphene applications, a graphene magnetic field sensor (or Hall sensor) has been proposed and

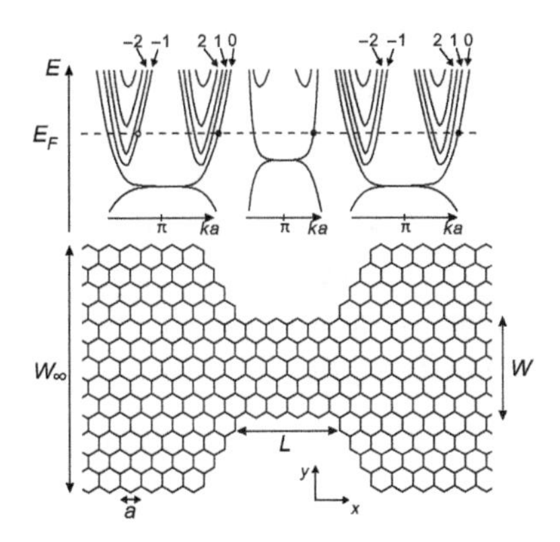


**Figure 4.18** Schematic figure to explain the valley filter. Reprinted by permission from Macmillan Publishers Ltd: [*Nature Physics*] (Ref. [152]), copyright (2007).

developed. Pisana et al. [158, 159] fabricated a quasi two-dimensional sensor and detected nanoscale-sized magnetic domains using the Hall effect and enhanced geometric MR effects. This result is attributed to the fact that a graphene sheet is a mono-atomic layer with high $\mu$, and therefore, the spatial resolution can be sufficiently high for this application. The sensitivity of graphene Hall sensors has been also examined by another group [160, 161].

Transistor operation at very high frequencies should be realized in future electronics. Lin et al. [162, 163] have presented an FET fabricated on a 2-inch graphene wafer with a cutoff frequency as high as 100 GHz. The same group has also demonstrated a wafer-scale graphene circuit in which all the circuit components are monolithically integrated on a single silicon carbide wafer. A graphene barristor, which is a triode device with a gate-controlled Schottky barrier, has also been demonstrated by Yang et al. [164]. A negative differential resistance, which is an important concept and component in semiconductor technology, has been theoretically predicted on the basis of band-gap engineering for nanoribbons, nanomeshes, and Bernal stacking of graphene on hexagonal boron nitride [165].

Transparent conductive graphene electrodes for dye-sensitized solar cells have been proposed for optoelectronics, as they can realize high stability, transparency, and conductivity [166]. The application of graphene in individual gas molecule detectors has also been proposed [167] because molecules adsorbed on graphene produce a change in the local $n$, which results in a step-like $\Delta R$. Graphene is sufficiently sensitive to detect such fine molecules electronically, because of the exceptionally low-noise transport. Applications of graphene to optical equipment have also been proposed [168–171]. Finally, applications of graphene to flash memory devices and a review of the chemical functionalization of GNRs can be found in Refs. [99, 172].

## References

1. A. K. Geim and K. S. Novoselov, *Nat. Mater.*, **6**, 183 (2007).
2. A. H. Castro Neto, F. Guinea, K. S. Novoselov, and A. K. Geim, *Rev. Mod. Phys.*, **81**, 109 (2009).
3. K. Wakabayashi, Y. Takane, M. Yamamoto, and M. Sigrist, *New J. Phys.*, **11**, 095016 (2009).
4. S. Das Sarma, S. Adam, E. H. Hwang, E. Rossi, *Rev. Mod. Phys.*, **83**, 407 (2011).
5. K. S. Novoselov, A. K. Geim, S. V. Morozov, D. Jiang, Y. Zhang, S. V. Dubonos, I. V. Grigorieva, and A. A. Firsov, *Science*, **306**, 666 (2004).
6. T. Ando, T. Nakanishi, and R. Saito, *J. Phys. Soc. Jpn.*, **67**, 2857 (1998).
7. H. Suzuura and T. Ando, *Phys. Rev. Lett.*, **89**, 266603 (2002).
8. K. Wakabayashi, Y. Takane, and M. Sigrist, *Phys. Rev. Lett.*, **99**, 036601 (2007).
9. S. V. Morozov, K. S. Novoselov, M. I. Katsnelson, F. Schedin, L. A. Ponomarenko, D. Jiang, and A. K. Geim, *Phys. Rev. Lett.*, **97**, 016801 (2006).
10. J.-M. Poumirol, A. Cresti, S. Roche, W. Escoffier, M. Goiran, X. Wang, X. Li, H. Dai, and B. Raquet, *Phys. Rev. B*, **82**, 041413 (R) (2010).
11. F. Muñoz-Rojas, D. Jacobs, J. Fernández-Rossier, and J. J. Palacios, *Phys. Rev. B*, **74**, 195417 (2006).
12. E. R. Mucciolo, A. H. Castro-Neto, and C. H. Lewenkopf, *Phys. Rev. B*, **79**, 075407 (2009).
13. S. Ihnatsenko and G. Kirczenow, *Phys. Rev. B*, **80**, 201407 (R) (2009).
14. N. Tombros, A. Veligura, J. Junesch, M. H. D. Guimarães, I. J. Vera-Marun, H. T. Jonkman, and B. J. van Wees, *Nat. Phys.*, **5**, 697 (2009).

15. C. Lian, K. Tohy, T. Fang, G. Li, H. G. Xing, and D. Jena, *Appl. Phys. Lett.*, **96**, 103109 (2010).
16. K. S. Novoselov, E. McCann, S. V. Morozov, V. I. Falko, M. I. Katsnelson, U. Zeitler, D. Jiang, F. Schedin, and A. K. Geim, *Nat. Phys.*, **2**, 177 (2006).
17. M. I. Katsnelson, *Eur. Phys. J.*, **52**, 151 (2006).
18. C. H. Lewenkopf, E. R. Mucciolo, and A. H. Castro Neto, *Phys. Rev. B*, **77**, 081410(R) (2008).
19. R. Golizadeh-Majarad and S. Datta, *Phys. Rev. B*, **79**, 085410 (2009).
20. J. J. Paracios, *Phys. Rev. B*, **82**, 165439 (2010).
21. M. Lewkowicz, B. Rosenstein, and D. Nghiem, *Phys. Rev. B*, **84**, 115419 (2011).
22. E. H. Hwang, S. Adam, and S. Das Sarma, *Phys. Rev. Lett.*, **98**, 186806 (2007).
23. A. M. Da Silva, K. Zou, J. K. Jain, and J. Zhu, *Phys. Rev. Lett.*, **104**, 236601 (2010).
24. X. Li, K. M. Borysenko, M. Buongiorno Nordelli, and K. W. Kim, *Phys. Rev. B*, **84**, 195453 (2011).
25. K. S. Novoselov, A. K. Geim, S. V. Morozov, D. Jiang, M. I. Katsnelson, I. V. Grigorieva, S. V. Dubonos, and A. A. Firsov, *Nature*, **438**, 197 (2005).
26. J. S. Moon, D. Curtis, S. Buis, T. Marshall, D. Wheeler, I. Valles, S. Kim, E. Wang, X. Weng, and M. Fanton, *IEEE Electron Device Lett.*, **31**, 1193 (2010).
27. J. S. Moon et al., *IEEE Electron Device Lett.*, **31**, 260 (2010).
28. X. Li, W. Cai, J. An, S. Kim, J. Nah, D. Yang, R. Piner, A. Welamakanni, I. Jung, E. Tutuc, S. K. Banerjee, L. Colombo, and R. S. Ruoff, *Science*, **324**, 1312 (2009).
29. L. Liao, Y.-C. Lin, M. Bao, R. Cheng, J. Bai, Y. Liu, Y. Qu, K. Wang, Y. Huang, and X. Duan, *Nature*, **467**, 305 (2010).
30. P. Neugebauer, M. Orlitra, C. Faugeras, A.-L. Barra, and M. Potemski, *Phys. Rev. Lett.*, **103**, 136403 (2009).
31. M. C. Lemme, T. J. Echtermeyer, M. Baus, and H. Kurz, *IEEE Electron Device Lett.*, **28**, 282 (2007).
32. N. Tombros, A. Veligura, J. Junesch, J. Jasper van den Berg, P. J. Zomer, M. Wojtaszek, I. J. Vera Marun, H. T. Jonkman, and B. J. van Wees, *J. Appl. Phys.*, **109**, 093702 (2011).
33. P. J. Zomer, S. P. Dash, N. Tombros, and B. J. van Wees, *Appl. Phys. Lett.*, 232104 (2011).
34. Y. Zhang, Y.-M. Tan, H. L. Stormer, and P. Kim, *Nature*, **438**, 201 (2005).

35. N. Tombros, C. Jozsa, M. Popinciuc, H. T. Jonkman, and Bart J. van Wees, *Nature*, **448**, 571 (2007).
36. P. J. Zomer, M. H. D. Guimarães, N. Tombros, and B. J. van Wees, *Phys. Rev. B*, **86**, 161416(R) (2012).
37. K. Bolotin, K. Sikes, Z. Jiang, M. Klima, G. Fundenberg, J. Hone, P. Kima, and H. Stormer, *Solid State Commun.*, **146**, 351 (2008).
38. C. Berger, Z. Song, X. Li, X. Wu, N. Brown, C. Naud, D. Mayou, T. Li, J. Hass, A. N. Marchenkov, E. H. Conrad. P. N. First, and W. A. de Heer, *Science*, **312**, 1191 (2006).
39. Z. Wang, H. Guo, and K. H. Bevan, *J. Comput. Electron.*, **12**, 104 (2013).
40. K. Nomura and A. H. MacDonald, *Phys. Rev. Lett.*, **98**, 076602 (2007).
41. S. Adam, P. W. Brouwer, and S. Das Sarma, *Phys. Rev. B*, **79**, 201404(R) (2009).
42. J. W. Klos and I. V. Zozoulenko, *Phys. Rev. B*, **82**, 081414 (R) (2010).
43. T. O. Wehling, M. I. Katsnelson, and A. I. Lichtenstein, *Chem. Phys. Lett.*, **476**, 125 (2009).
44. T. O. Wehling, S. Yuan, A. I. Lichtenstein, A. K. Geim, and M. I. Katsnelson, *Phys. Rev. Lett.*, **105**, 056802 (2010).
45. A. Lherbier, S. M.-M. Dubois, X. Declerck, S. Roche, Y.-M. Niquet, and J. C. Charlier, *Phys. Rev. Lett.*, **106**, 046803 (2011).
46. E. Louis, J. A. Verges, F. Guinea, and G. Chiappe, *Phys. Rev. B*, **75**, 085440 (2007).
47. I. Martin and Ya. M. Blanter, *Phys. Rev. B*, **79**, 235132 (2009).
48. G. Schubert and H. Fehske, *Phys. Rev. Lett.*, **108**, 066402 (2012).
49. H. Şahin and R. T. Senger, *Phys. Rev. B*, **78**, 205423 (2008).
50. A. Pieper, G. Schubert, G. Wellein, and H. Fehske, *Phys. Rev. B*, **88**, 195409 (2013).
51. Q. Li, E. H. Hwang, E. Rossi, and S. Das Sarma, *Phys. Rev. Lett.*, **107**, 156601 (2011).
52. K. Pi, K. M. McCreary, W. Bao, W. Han, Y. F. Chiang, Y. Li, S.-W. Tsai, C. N. Lau, and R. K. Kawakami, *Phys. Rev. B*, **80**, 075406 (2009).
53. M. Y. Han, J. C. Brant, and P. Kim, *Phys. Rev. Lett.*, **104**, 056801 (2010).
54. D. K. Ferry, *J. Comput. Electron.*, **12**, 76 (2013).
55. J. Inoue, G. E. W. Bauer, and L. W. Molenkamp, *Phys. Rev. B*, **67**, 033104 (2003).
56. J. Inoue, G. E. W. Bauer, and L. W. Molenkamp, *Phys. Rev. B*, **70**, 041303(R) (2004).

57. H. Itoh and J. Inoue, *J. Magn. Soc. Jpn.*, **30**, 1 (2006).
58. H. Goto, A. Kanda, T. Sato, S. Tanaka, Y. Ootuka, S. Okada, H. Miyazaki, K. Tsukagoshi, and Y. Aoyagi, *Appl Phys. Lett.*, **92**, 212110 (2008).
59. T. Maassen, F. K. Dejena, M. H. Guimarães, C. Józsa, and B. J. van Wees, *Phys. Rev. B*, **83**, 115410 (2011).
60. C. Józsa, M. Popinciuc, N. Tombros, H. T. Jonkman, and B. J. van Wees, *Phys. Rev. Lett.*, **100**, 236603 (2008).
61. N. Tombros, S. Tanabe, A. Veligura, C. Jozsa, M. Popinciuc, H. T. Jonkman, and B. J. van Wees, *Phys. Rev. Lett.*, **101**, 046601 (2008).
62. M. Popinciuc, C. Józsa, P. J. Zomer, N. Tombros, A. Veligura, H. T. Jonkman, and B. J. van Wees, *Phys. Rev. B*, **80**, 214427 (2009).
63. C. Józsa, T. Maassen, M. Popinciuc, P. J. Zomer, A. Veligura, H. T. Jonkman, and B. J. van Wees, *Phys. Rev. B*, **80**, 241403(R) (2009).
64. K. Pi, W. Han, K. M. McCreary, A. G. Swartz, Y. Li, and R. K. Kawakami, *Phys. Rev. Lett.*, **104**, 187201 (2010).
65. W. Han, K. Pi, K. M. McCreary, Y. Li, J. J. I. Wong, A. G. Swartz, and R. K. Kawakami, *Phys. Rev. Lett.*, **105**, 167202 (2010).
66. F. Volmer, M. Drögeler, E. Maynicke, N. von den Driesch, M. L. Boschen, G. Guntherrodt, B. Beschoten, *Phys. Rev. B*, **88**, 161405(R) (2013).
67. W. Han and R. K. Kawakami, *Phys. Rev. Lett.*, **107**, 047207 (2011).
68. T.-Y. Yang, J. Balakrishnan, F. Volmer, A. Avsar, M. Jaiswal, J. Samm, S. A. Ali, A. Pachoud, M. Zeng, M. Popinciuc, G. Güntherrodt, B. Beschoton, and B. Ozyilmaz, *Phy. Rev. Lett.*, **107**, 047206 (2011).
69. S. Roche and S. O. Valenzuela, *J. Phys. D: Appl. Phys.*, **47**, 094011 (2014).
70. H. Ochoa, A. H. Castro Neto, and F. Guinea, *Phys. Rev. Lett.*, **108**, 206808 (2012).
71. Z. Han, W. H. Wang, K. Pi, K. M. McCreary, W. Bao, Y. Li, F. Miao, C. N. Lau, and R. K. Kawakami, *Phys. Rev. Lett.*, **102**, 137205 (2009).
72. H. Haugen, D. Huertas-Hernand, and A. Brataas, *Phys. Rev. B*, **77**, 115406 (2008).
73. C. Kane, *Nature*, **438**, 168 (2005).
74. V. P. Gusynin and S. G. Sharapov, *Phys. Rev. Lett.*, **95**, 146801 (2005).
75. A. H. Castro Neto, F. Guinea, and N. M. R. Peres, *Phys. Rev. B*, **73**, 205408 (2006).
76. J. R. Williams, L. DiCarlo, and C. M. Marcus, *Science*, **317**, 638 (2007).
77. D. A. Abanin and L. S. Levitov, *Science*, **317**, 641 (2007).
78. J. Li and S.-Q. Shen, *Phys. Rev. B*, **78**, 205308 (2008).

79. H. Zhang, C. Lazo, S. Blügel, S. Heinze, and Y. Mokrousov, *Phys. Rev. Lett.*, **108**, 056802 (2012).
80. W. Bao, Z. Zhao, H. Zhang, G. Liu, P. Kratz, L. Jing, J. Velasco,Jr., D. Smirnov, and C. N. Lau, *Phys. Rev. Lett.*, **105**, 246601 (2010).
81. F. Ghahari, Y. Zhao, P. Caddeu-Zimansky, K. Bolotin, and P. Kim, *Phys. Rev. Lett.*, **106**, 046801 (2011).
82. J. D. Sanchez-Yamagishi, T. Taychatanapat, K. Watanabe, T. Taniguchi, A. Yacoby, and P. Jarillo-Herrero, *Phys. Rev. Lett.*, **108**, 076601 (2012).
83. C. L. Kane and E. J. Male, *Phys. Rev. Lett.*, **95**, 226801 (2005).
84. M. Z. Hasan and C. L. Kane, *Rev. Mod. Phys.*, **82**, 3045 (2010).
85. H. Min, J. E. Hill, N. A. Sinitsyn, B. R. Sahu, L. Kleinman, and A. H. MacDonald, *Phys. Rev. B*, **74**, 165310 (2006).
86. T. Kato, S. Onari, and J. Inoue, *Physica E*, **42**, 729 (2010).
87. D. A. Abanin, R. V. Gorbachev, K. S. Novoselov, A. K. Geim, and L. S. Levitov, *Phys. Rev. Lett.*, **107**, 096601 (2011).
88. A. Varykhalov, D. Marchenko, M. R. Scholz, E. D. L. Rienks, T. K. Kim, G. Bihlmayers, *Phys. Rev. Lett.*, **108**, 066804 (2012).
89. J. Tworzydlo, I. Snyman, A. R. Akhmerov, and C. W. J. Beenaker, *Phys. Rev. B*, **76**, 035411 (2007).
90. C. Zhang, S. Tewari, and S. Das Sarma, *Phys. Rev. B*, **79**, 245424 (2009).
91. L. Zhu, R. Ma, L. Sheng, M. Liu, and D.-N. Sheng, *Phys. Rev. Lett.*, **104**, 076804 (2010).
92. P. Wei, W. Bao, Y. Pu, C. N. Lau, and J. Shi, *Phys. Rev. Lett.*, **102**, 166808 (2009).
93. Y. Ouyang and J. Guo, *Appl. Phys. Lett.*, **94**, 263107 (2009).
94. H. Kageshima, H. Hibino, M. Nagase, Y. Sekine, and H. Yamaguchi, *Jpn. J. Appl. Phys.*, **50**, 070115 (2011).
95. H. Schomerus, *Phys. Rev. B*, **76**, 045433 (2007).
96. J. P. Robinson and H. Schomerus, *Phys. Rev. B*, **76**, 115430 (2007).
97. Ya. M. Blanter, I. Martin, *Phys. Rev. B*, **76**, 155433 (2007).
98. A. Yamamura, S. Honda, J. Inoue, and H. Itoh, *J. Magn. Soc. Jpn.*, **34**, 34 (2010).
99. B. Genorio and A. Znidarsic, *J. Phys. D: Appl. Phys.*, **47**, 094012 (2014).
100. B. Huard, J. A. Sulpizio, N. Stander, K. Todd, B. Yang, and D. G-Gordon, *Phys. Rev. Lett.*, **98**, 236803 (2007).
101. T. J. Echtermeyer, M. C. Lemme, M. Baus, B. N. Szafranek, A. K. Geim, and H. Kurz, *IEEE Electron Device Lett.*, **29**, 952 (2008).

102. X. Wang, Y. Ouyang, X. Li, H. Wang, J. Guo, and H. Dai, *Phys. Rev. Lett.*, **100**, 206803 (2008).

103. T. Taychatanapat and P. Jarillo-Herrero, *Phys. Rev. Lett.*, **105**, 166601 (2010).

104. M. N. Fogler, D. S. Novikov, L. I. Glazman, and B. I. Shklovskii, *Phys. Rev. B*, **77**, 075420 (2008).

105. V. H. Nguyen, J. Saint-Martin, D. Querlioz, F. Mazzamuto, A. Bournel, Y.-M. Niquet, and P. Dollfus, *J. Comput. Electron.*, **12**, 85 (2013).

106. E. V. Castro, K. S. Novoselov, S. V. Morozov, N. M. R. Peres, J. M. B. Lopes des Santos, J. Nilsson, F. Guinea, A. K. Geim, and A. H. Castro Neto, *Phys. Rev. Lett.*, **99**, 216802 (2007).

107. D. C. Elias, R. R. Nair, T. M. G. Mohiuddin, S. V. Morozov, P. Blake, M. P. Halsall, A. C. Ferrari, D. W. Doukhvalov, M. I. Katsnelson, A. K. Geim, and K. S. Novoselov, *Science*, **323**, 610 (2009).

108. J. O. Sofo, A. S. Chaudhari, and G. D. Barber, *Phys. Rev. B*, **75**, 153401 (2007).

109. L. Britnell, R. V. Gorbachev, R. Jalil, B. D. Belle, F. Schedin, A. Mishchenko, T. Georgiu, M. I. Katsnelson, L. Eaves, S. V. Morozov, N. M. R. Peres, J. Leist, A. K. Geim, K. S. Novoselov, and L. A. Ponamarenko, *Science*, **335**, 947 (2012).

110. A. Venugapal, L. Colombo, and E. M. Vogel, *Appl. Phys. Lett.*, **96**, 013512 (2010).

111. K. N. Parrish and D. Akinwande, *Appl. Phys. Lett.*, **98**, 183505 (2011).

112. Y. Ren, S. Chen, W. Cai, Y. Zhu, C. Zhu, and R. S. Ruoff, *Appl. Phys. Lett.*, **97**, 053107 (2010).

113. F. Mittendorfer, A. Garhofer, J. Redinger, J. Klimeš, J. Harl, and G. Kresse, *Phys. Rev. B*, **84**, 201401 (R) (2011).

114. S. J. Altenburg, J. Kröger, B. Wang, M.-L. Bocquet, N. Lorente, and R. Berndt, *Phys. Rev. Lett.*, **105**, 236101 (2010).

115. P. Michetti and P. Recher, *Phys. Rev. B*, **84**, 125438 (2011).

116. K. M. MacCreary, K. Pi, and R. K. Kawakami, *Appl. Phys. Lett.*, **98**, 192101 (2011).

117. N. Nemec, D. Tománek, and G. Cuniberti, *Phys. Rev. B*, **77**, 125420 (2008).

118. Q. Ran, M. Gao, X. Guan, Y. Wang, and Z. Yu, *Appl. Phys. Lett.*, **94**, 103511 (2009).

119. I. Deretzis, G. Fiori, G. Iannaccone, and A. La Magna, *Phys. Rev. B*, **82**, 161413(R) (2010).

120. S. Barraza-Lopez, M. Vanević, M. Kindermann, and M. Y. Chou, *Phys. Rev. Lett.*, **104**, 076807 (2010).
121. S. Barraza-Lopez, *J. Comput. Electron.*, **12**, 145 (2013).
122. Y.-B. Zhou, B.-H. Han, Z.-M. Liao, Q. Zhao, J. Xu, and D.-P. Yu, *J. Chem. Phys.*, **132**, 024706 (2010).
123. R. Nouchi and K. Tanigaki, *Appl. Phys. Lett.*, **96**, 253503 (2010).
124. S. Russo, M. F. Craciun, M. Yamamoto, A. F. Morprgo, and S. Tarucha, *Physica E*, **42**, 677 (2010).
125. C. E. Malec and D. Davidović, *Phys. Rev. B*, **84**, 033407 (2011).
126. B.-C. Huang, M. Zhang, Y. Wang, and J. Woo, *Appl. Phys. Lett.*, **99**, 032107 (2011).
127. K. Nagashio and A. Toriumi, *Jpn. J. Appl. Phys.*, **50**, 070108 (2011).
128. J. A. Robinson, M. Labella, M. Zhu, M. Hollander, R. Kasarda, Z. Hughes, K. Trumbull, R. Cavalero, and D. Snyder, *Appl. Phys. Lett.*, **98**, 053101 (2011).
129. G. Schmidt, D. Ferrand, L. W. Molenkamp, A. T. Filip, and B. J. van Wees, *Phys. Rev. B*, **62**, 4290(R) (2000).
130. L. Brey and H. A. Fertig, *Phys. Rev. B*, **76**, 205435 (2007).
131. W. Y. Kim and K. S. Kim, *Nat. Nanotech.*, **3**, 408 (2008).
132. M. Zeng, L. Shen, M. Yang, C. Zhang, and Y. Feng, *Appl. Phys. Lett.*, **98**, 053103 (2011).
133. Z. F. Wang and F. Liu, *Appl. Phys. Lett.*, **99**, 042110 (2011).
134. C. Cao, Y. Wang, H.-P. Cheng, and J.-Z. Jiang, *Appl. Phys. Lett.*, **99**, 073110 (2011).
135. S. M.-M. Dubois, X. Declerck, J.-C. Charlier, and M. C. Payne, *ACS Nano*, **7**, 4578 (2013).
136. Y. W. Son, M. L. Cohen, and S. G. Louie, *Nature*, **444**, 347 (2006).
137. Y.-W. Son, M. L. Cohen, and S. G. Louie, *Phys. Rev. Lett.*, **97**, 216803 (2006).
138. V. V. Cheianov and V. I. Fal'ko, *Phys. Rev. B*, **74**, 041403 (R) (2006).
139. K. H. Ding, Z.-G. Zhu, Z.-H. Zhang, and J. Berakolar, *Phys. Rev. B*, **82**, 155143 (2010).
140. S. B. Kumar and J. Guo, *J. Comput. Electron.*, **12**, 165 (2013).
141. H. Y. Tian, K. S. Chan, and J. Wang, *Phys. Rev. B*, **86**, 245413 (2012).
142. M. Ohishi, M. Shiraishi, R. Nouchi, T. Nozaki, T. Shinjo, and Y. Suzuki, *Jpn. J. Appl. Phys.*, **46**, L605 (2007).

143. B. Dlubak, P. Seneor, A. Anane, C. Barraud, C. Deranlot, D. Deneuve, B. Servet, R. Mattana, F. Petroff, and A. Fert, *Appl. Phys. Lett.*, **97**, 092502 (2010).

144. E. W. Hill, A. K. Geim, K. Novoseloc, F. Schedin, and P. Blake, *IEEE Trans. Magn.*, **42**, 2694 (2006).

145. M. Nishioka and A. M. Goldman, *Appl. Phys. Lett.*, **90**, 252505 (2007).

146. W. H. Wang, K. Pi, Y. Li, Y. F. Chiang, P. Wei, J. Shi, and R. K. Kawakami, *Phys. Rev. B*, **77**, 020402(R) (2008).

147. Z.-M. Liao, H.-C. Wu, J.-J. Wang, G. L. W. Cross, S. Kumar, I. V. Shvets, and G. S. Duesberg, *Appl. Phys. Lett.*, **98**, 052511 (2011).

148. B. Dlubak, M.-B. Martin, C. Deranlot, B. Servet, S. Xavier, R. Mattana, M. Spinkle, C. Berger, W. D. De Herr, F. Petroff, A. Anane, P. Seneor, and A. Fert, *Nat. Phys.*, **8**, 557 (2012).

149. S. Singh, A. K. Patra, B. Barin, E. del Barco, and B. Özyilmaz, *IEEE Trans. Magn.*, **49**, 3147 (2013).

150. R. M. Feenstra, D. Jena, and G. Gu, *J. Appl. Phys.*, **111**, 043711 (2012).

151. E. Cobas, A. L. Friedman, O. M. J. van't Erve, J. T. Robinson, and B. T. Jonker, *IEEE Trans. Magn.*, **49**, 4343 (2013).

152. A. Rycerz J. Tworzydlo, and C. W. J. Beenakker, *Nat. Phys.*, **3**, 172 (2007).

153. A. R. Akhmerov, J. H. Bardarson, A. Rycerz, and C. W. Beenakker, *Phys. Rev. B*, **77**, 205416 (2008).

154. D. Xiao, W. Yao, and Q. Niu, *Phys. Rev. Lett.*, **99**, 236809 (2007).

155. M. Ezawa, *Phys. Rev. B*, **88**, 161406(R) (2013).

156. T. Fujita, M. B. A. Jalil, and S. G. Tan, *Appl. Phys. Lett.*, **97**, 043508 (2010).

157. X. Li, Z. Zhang, and D. Xiao, *Phys. Rev. B*, **81**, 195402 (2010).

158. S. Pisana, P. M. Braganca, E. E. Marinero, and B. A. Gurney, *Nano Lett.*, **10**, 341 (2010).

159. S. Pisana, P. M. Braganca, E. E. Marinero, and B. A. Gurney, *IEEE Trans. Magn.*, **46**, 1910 (2010).

160. V. Panchal, Ó. Iglesis-Freire, A. Lartsev, R. Yakimova, A. Asenjo, and O. Kazakova, *IEEE Trans. Magn.*, **49**, 3520 (2013).

161. R. K. Rajkumar, A. Manzin, D. C. Cox, S. R. P. Silve, A. Tzalenchuk, and O. Kazakova, *IEEE Trans. Magn.*, **49**, 3445 (2013).

162. Y.-M. Lin, C. Dimitrakopoulos, K. A. Jenkins, D. B. Farmer, H.-Y. Chiu, A. Grill, P. Avouris, and Ph. Avouris, *Science*, **327**, 662 (2010).

163. Y.-M. Lin, A. V. Garcia, S.-J. Han, D. B. Farmer, I. Meric, Y. Sun, Y. Wu, C. Dimitrakopoulos, A. Grill, P. Avouris, and K. A. Jenkins, *Science*, **332**, 1294 (2011).

164. H. Yang, J. Heo, S. Park, H. J. Song, D. H. Seo, E.-E. Byun, P. Kim, I. Yoo, H.-J. Chung, and K. Kim, *Science*, **336**, 1140 (2012).

165. V. H. Nguyen, J. S. Martin, D. Querlioz, F. Mazzamuto, A. Bournel, Y.-M. Niquet, and P. Dollfus, *J. Comput. Electron.*, **12**, 85 (2013).

166. X. Wang, L. Zhi, and K. Müllen, *Nano Lett.*, **8**, 323 (2008).

167. F. Schedin, A. K. Geim, S. V. Novoselov, E. W. Hill, P. Blake, M. I. Katsnelson, and K. V. Novoselov, *Nat. Mater.*, **6**, 652 (2007).

168. M. Liu, X. Yin, E. Ulin-Avila, B. Geng, T. Zentgraf, L. Ju, F. Wang, and X. Zhang, *Nature*, **472**, 64 (2011).

169. V. V. Cheianov, V. Fal'ko, and B. L. Altshuler, *Science*, **315**, 1252 (2007).

170. A. G. Moghaddan and M. Zareyan, *Phys. Rev. Lett.*, **105**, 146803 (2010).

171. D. Popa, Z. Sun, T. Hasan, F. Torrisi, F. Wang, and A. C. Ferrari, *Appl. Phys. Lett.*, **98**, 073106 (2011).

172. N. Zhan, M. Olmedo, G. Wang, and J. Liu, *Appl. Phys. Lett.*, **99**, 113112 (2011).

# Chapter 5

# Spintronics MR Devices

The field of spintronics has been developed, being triggered by the discovery of giant magnetoresistance (GMR). Interaction between fundamental physics and technological applications has brought about novel concepts and phenomena in the field of spintronics and in solid-state physics. Exchange coupling and the spin valve appeared in the early stage of the development, being followed by tunnel magnetoresistance (TMR), spin current, spin transfer torque, the spin Hall effect, and the inverse spin Hall effect. Recently, various types of torque acting on magnetization have been proposed. In device applications, on the other hand, the GMR effect in spin valve structures has been applied to magnetic sensors, and the TMR of epitaxial ferromagnetic tunnel junctions (FTJs) has been applied to magnetic (magnetoresistive) random access memories. Research of further applications of such phenomena is still in progress in the field of, for example, spin field-effect transistors (spin FETs). In this chapter, we will explain the fundamental phenomena of GMR, TMR, spin transfer torque, and their applications to devices and give a brief introduction of the spin FET.

*Graphene in Spintronics: Fundamentals and Applications*
Jun-ichiro Inoue, Ai Yamakage, and Shuta Honda

ISBN 978-981-4669-56-6 (Hardcover), 978-981-4669-57-3 (eBook)
www.panstanford.com

## 5.1 Introduction

It would not be an exaggeration to say that the field of spintronics started when the phenomenon of giant magnetoresistance (GMR) was discovered in FeCr magnetic multilayers in 1986 [12, 13]. GMR is a phenomenon that the resistivity of the magnetic multilayers reduces dramatically by applying an external magnetic field. The multilayers are thin films composed of an alternating stack of two different types of atomic species. Such structures were first fabricated by using semiconducting materials and were called superlattices. Fabrication of metallic multilayers started about 10 years after the fabrication of superlattices of semiconducting materials. This is because the Fermi wavelength of magnetic metals is much shorter than that of semiconductors, and therefore fabrication of multilayer structures using thin films was quite difficult. The structure of the magnetic multilayers is schematically shown in Fig. 5.1a. Microfabrication of such magnetic structure (Fig. 5.1b) was also performed. Soon after the discovery of GMR, oscillation of exchange coupling between magnetic layers in multilayers was discovered [14, 15]. The oscillation of exchange coupling plays an essential role in the phenomenon of GMR. By using the junction structure of a trilayer made of two magnetic layers separated by one nonmagnetic layer, the effect of GMR has been applied to magnetic sensors for writing/reading heads of hard disc drives (HDDs). The structure is called spin valve [16, 17]. In the next chapter, we will

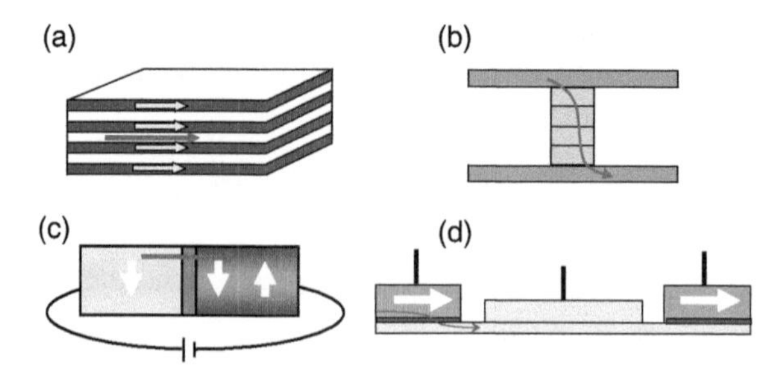


**Figure 5.1** Schematic structures used in the field of spintronics: (a) magnetic multilayers, (b) nanofabricated multilayers, (c) ferromagnetic tunnel junctions, and (d) spin FETs.

explain the phenomenon of GMR, its mechanism, and application to sensors.

Similar magnetoresistance (MR) was discovered for FTJs, which are composed by two ferromagnetic metal layers separated by an insulating barrier layer such as $Al_2O_3$ and MgO. A schematics is shown in Fig. 5.1c. The phenomenon is called tunnel magnetoresistance (TMR). Actually, TMR was discovered prior to GMR for junctions with Ge for the insulating barrier [20] and Ni/NiO/Ni [21]. However, a large TMR effect was first found in junctions Fe/$Al_2O_3$/Fe junctions [18, 19] in which the $Al_2O_3$ layer is amorphous like. Pronounced TMR was reported for junctions Fe/MgO/Fe and FeCo/MgO/FeCo junctions in 2004 [22–24]. Because the junctions is epitaxially fabricated, the MgO layer can be very thin and high MR ratios are realized. TMR has been applied to magnetoresistive random access memory (MRAM) as well as to magnetic sensors. The phenomenon and mechanism of TMR will be explained in Section 5.3. In the application of TMR to MRAM, current control of the magnetization direction of the ferromagnetic junctions is inevitable. A novel method called spin transfer torque or current-induced magnetization switching has been adopted. A brief explanation of this effect will also be explained in the Section 5.3. There are reported other MR effects, such as colossal MR, ballistic MR etc.; however, please refer to suitable references for these MR effects.

It is well known that the silicon technology in the field of nanodevices is now confronted with difficulty to reduce the size of devices, for example, current leakage and energy consumption. One of the purposes of development of MRAM is to solve the latter problem. However, the replacement of silicon RAM with MRAM meets many difficulties. Recently another type of devices is proposed, which combine the semiconductor with ferromagnetic metals. A typical structure of the junction is depicted in Fig. 5.1d. The field-effect transistor (FET) is one of the important devices in the modern silicon technology, in which the current is controlled by a gate contact. The source and drain contacts are made of normal metals. In the proposed devices, the normal metals are replaced with ferromagnetic metals, as shown in Fig. 5.1d. This kind of devices is called a spin FET; the present status of the research of which will be explained briefly in Section 5.4.

There published excellent review articles on these subjects [1–11] and therefore readers are asked to refer to them.

## 5.2 Giant Magnetoresistance

### 5.2.1 *Magnetic Multilayers*

The basic structure of magnetic multilayers are composed of nonmagnetic and ferromagnetic metals. Fe, Co, Ni and their alloys are frequently used for the ferromagnetic layers, while Cr, Ru and noble metals (Cu, Ag and Au) are used for the nonmagnetic layers. The thickness of each layer is 1–10 nm and the number of layers ranges from 3 to about 100. To fabricate high-quality magnetic multilayers, matching of the lattice constants of the constituent metals is important. Magnetic multilayers have the following two important characteristics:

(1) The alignment of the magnetization of the magnetic layers is easily controlled by an external magnetic field, since the coupling between the magnetization of the magnetic layers is weakened by the presence of the nonmagnetic layer between them.
(2) Each layer is thin enough for carrier electrons to feel a change in the magnetization direction of the magnetic layers.

### 5.2.2 *GMR and Exchange Coupling*

The first observation of antiparallel coupling between magnetic layers was reported by Grünberg et al. [12] for Fe/Cr trilayers. They also observed negative MR, that is, a resistivity reduction under an external magnetic field. The magnitude of the MR of Fe/Cr trilayers was observed to be a few percent. Two years later, Fert's group [13] reported MR as large as 40% for Fe/Cr multilayers. This MR was the largest so far observed for magnetic metal films and was called giant MR (GMR). After the discovery of GMR, many experimental works have been performed [25–30].

Figure 5.2a shows magnetization curve of Fe/Cr multilayers [13]. We see that the magnetization increases from zero value

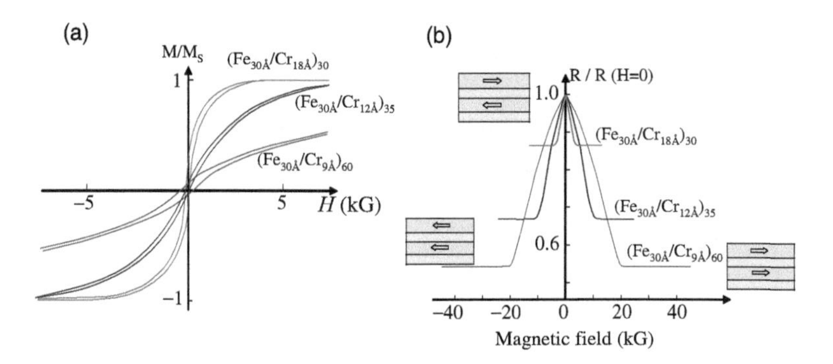


**Figure 5.2** (a) Magnetization curves observed for Fe/Cr multilayers and (b) resistivity change as a function of applied magnetic field $H$. Reprinted (figure) with permission from Ref. [13], Copyright (1988) by the American Physical Society.

to saturated values by an external magnetic field with almost no hysteresis. The result is interpreted that an antiparallel (AP) alignment of magnetization of Fe layers changes to parallel (P) alignment by the external magnetic field. Figure 5.2b shows the corresponding change in resistivity. The resistivity decreases with increasing magnetic field according to the change in the alignment of the magnetization of the Fe layers.

The magnitude of the MR is expressed by the so-called MR ratio, defined as

$$\mathrm{MR} = \frac{\rho_{\mathrm{AP}} - \rho_{\mathrm{P}}}{\rho_{\mathrm{AP}}}, \tag{5.1}$$

or

$$\mathrm{MR} = \frac{\rho_{\mathrm{AP}} - \rho_{\mathrm{P}}}{\rho_{\mathrm{P}}}, \tag{5.2}$$

where $\rho_{\mathrm{AP}}$ and $\rho_{\mathrm{P}}$ are the resistivity in AP and P alignment of the magnetization of the magnetic layers. Since usually $\rho_{\mathrm{P}} < \rho_{\mathrm{AP}}$, the definitions (5.1) and (5.2) are called as pessimistic and optimistic definition, respectively. In experiments, the optimistic definition Eq. (5.2) is usually used, while we use definition Eq. (5.1) for the theoretical results in this chapter. Tables 5.1 and 5.2 show MR ratios observed for several combinations of magnetic and nonmagnetic metals. We see that the MR ratio depends on the combination of metals.

GMR appears when the AP alignment of the magnetization is changed to P alignment by an external magnetic field. Therefore,

**Table 5.1** MR ratios measured for various magnetic multilayers for current parallel to the layer planes

| Multilayer | $\Delta\rho/\rho_P(\%)$ | $\Delta\rho/\rho_{AP}(\%)$ |
|---|---|---|
| Fe/Cr | 108 | 52 |
| Co/Cu | 115 | 53 |
| NiFe/Cu/Co | 50 | 33 |
| FeCo/Cu | 80 | 44 |
| NiFeCo/Cu | 35 | 26 |
| Ni/Ag | 26 | 21 |
| Co/Au | 18 | 15 |
| Fe/Mn | 0.8 | |
| Fe/Mo | 2 | |
| Co/Ru | 7 | |
| Co/Cr | 2.6 | |
| Fe/Cu | 12 | |

**Table 5.2** MR ratios measured for selected magnetic multilayers for current perpendicular to the planes

| Multilayer | $\Delta\rho/\rho_P(\%)$ | $\Delta\rho/\rho_{AP}(\%)$ |
|---|---|---|
| Ag/Fe | 42 | 30 |
| Fe/Cr | 108 | 52 |
| Co/Cu | 170 | 63 |

AP alignment of the magnetization is a prerequisite for GMR. A detailed study of the alignment of magnetization has found that the coupling of magnetization in magnetic layers changes as a function of the nonmagnetic layer thickness [14]. The coupling between magnetic layers is called interlayer exchange coupling [15, 31, 33–35]. Figure 5.3a shows a dependence of the resistivity change of Fe/Cr multilayers as a function of layer thickness [14]. We see that high and nearly zero resistivity change appear periodically. In the multilayers with high resistivity change, the AP alignment of Fe magnetization is realized at zero field, while in those with low resistivity change, the P alignment is realized. This means that the coupling of magnetization in Fe layers oscillates with the thickness of Cr layers. Figure 5.3b shows experimental results of the change

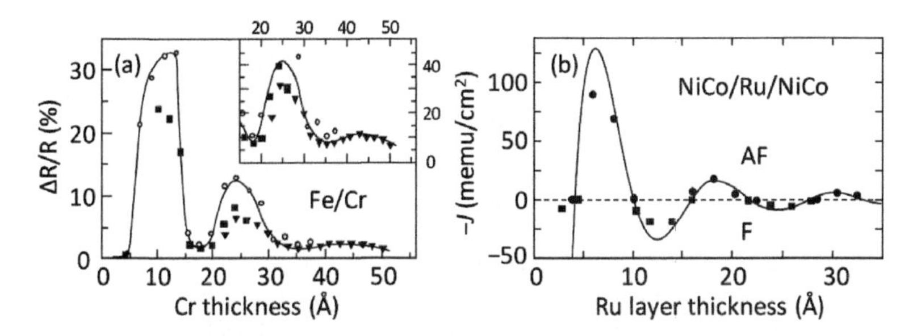


**Figure 5.3** (a) Change in resistivity as a function of Cr thickness in Fe/Cr multilayers (inset: enlargement of the results for thick Cr layers); (b) exchange coupling constant $J$ as a function of Ru layer thickness in NiCo/Ru/NiCo trilayers. Reprinted (figures) with permission from Ref. [14], Copyright (1990), and Ref. [15], Copyright (1991) by the American Physical Society.

in coupling with nonmagnetic layer (Ru) thickness of NiCo/Ru/NiCo multilayers.

The period of the oscillation is rather long and the magnitude decays as the thickness of the nonmagnetic layer increases. The long period of oscillation has been confirmed in various experiments [36–38]. The features are very similar to those of the so-called RKKY interaction between magnetic impurities in metals [39]. The period of the oscillation of the interlayer exchange coupling is determined by the Fermi wave vector $k_F$, as in the RKKY interaction [40–45]. In multilayers, however, the thickness of the nonmagnetic layer changes discretely and therefore $(\pi/a - k_F)$ can also be the period of oscillation, where $a$ is the lattice distance. Since $k_F$ in Cu, for example, is close to $\pi/a$, the period of oscillation of the exchange coupling becomes long. The decay of the magnitude for multilayers is proportional to $L^2$, where $L$ is the nonmagnetic layer thickness, in contrast to $r^3$ for the RKKY interaction.

Magnetic layers in multilayers are usually coupled magnetically (interlayer exchange coupling). The interlayer exchange coupling in Fe/Cr multilayers is rather strong to be controlled by the magnetic field. Multilayers with thicker nonmagnetic layers have nearly zero exchange coupling. Nevertheless, the magnetization direction of the magnetic layers may be controlled by using the difference in the coercive force between magnetic layers of different metals. An

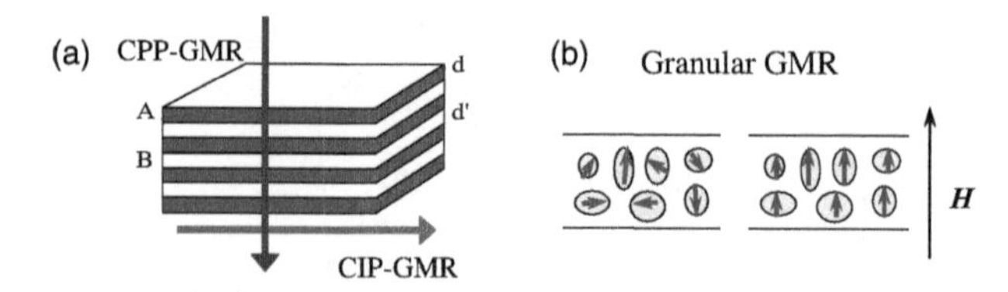


**Figure 5.4** (a) Relation between current direction and layer structures used in CIP-GMR and CPP-GRM. (b) Magnetization alignment of grains with and without external field H in granular magnetic materials.

example is Co/Cu/NiFe multilayers, in which NiFe is a soft magnet with a magnetization easily changed by a weak external magnetic field [46, 47].

The experiments presented so far have used a geometry with the current flowing parallel to the layer planes. GMR with this geometry is called current-in-plane GMR (CIP-GMR). GMR with a geometry with current flowing perpendicular to the planes is called CPP-GMR, as shown in Fig. 5.4a. In CPP-GMR, the resistivity of a sample is too small to be detected, since the layer thickness is usually less than μm and the resistivity of the leads is overwhelming. To make a measurement of sample resistivity possible, several methods have been adopted. One is to utilize superconducting leads [48–50], the second one is to microfabricate the samples [51, 52], and finally to fabricate multilayered nanowire formed by electrodeposition into nanometer-sized pores of a template polymer membrane [53]. In the second and third cases, the resistivity of the sample becomes as large as that of the leads because the resistivity of such systems is governed by the narrow region of the system. Some observed values of CPP-GMR are listed in Table 5.2.

GMR has been observed not only in magnetic multilayers, but also in granular films (Fig. 5.4b) in which magnetized metallic grains of Co or Fe, for example, are distributed in nonmagnetic metals, typically Cu or Ag [54–60]. The size of the grains is of nanoscale and the magnetic moments of the grains are nearly isolated from each other. When an external magnetic field is applied, the random orientation of the magnetic moments of the grains are forced to be parallel, resulting in a decrease of the resistivity, as observed in magnetic multilayers.

## 5.2.3 *Mechanism of GMR*

### 5.2.3.1 Spin-dependent resistivity in ferromagnetic metals: Two-current model

The electronic structure of ferromagnetic metals may be characterized by an exchange splitting of up- and down-spin states, which produces a difference between the up- and down-spin densities of states (DOSs) at the Fermi level. Because the electrical transport is governed by the electronic states near the Fermi level, the resistivity (and conductivity) becomes spin dependent. One might consider that effects of spin mixing due to, for example, spin–orbit interaction (SOI) may eliminate the spin dependence of the electrical resistivity. However, this is not the case because the mean free path of electrons in transition metals is about a few nanometers, which is much shorter than the spin flip length. This means that the electrical resistivity is governed by the spin-conserving scattering. Because of the spin-dependent electronic states near the Fermi level of ferromagnetic metals and spin-dependent scattering potential caused by, for example, impurities, the transport lifetime is spin dependent ($\tau_\uparrow$ and $\tau_\downarrow$) [61]. Then the spin-dependent conductivity $\sigma_s$ ($s = \uparrow, \downarrow$) is given as (Drude theory)

$$\sigma_s = \frac{ne^2\tau_s}{m},$$

where $n$ is the number of electrons contributing to the transport, $e$ the electron charge, and $m$ is the effective mass of transport electrons. It should be noted that both $n$ and $m$ might be spin dependent. Thus the resistivity $\rho = 1/\sigma$ is spin dependent ($\rho_\uparrow$ and $\rho_\downarrow$). Because the up- and down-spin channels are parallel circuit, the total resistivity is given as

$$\frac{1}{\rho} = \frac{1}{\rho_\uparrow} + \frac{1}{\rho_\downarrow}.$$

This is the Mott's two current model for electrical transport in ferromagnetic metals.

Although the measurement of $\rho$ is easy, measurement of $\rho_s$ is difficult. Nevertheless, Fert and Campbell reported details of $\rho_s$ for various ferromagnetic metals with different impurities, the results of which are shown in Fig. 5.5 [62, 63]. The ratio $\rho_\downarrow/\rho_\uparrow$ is called

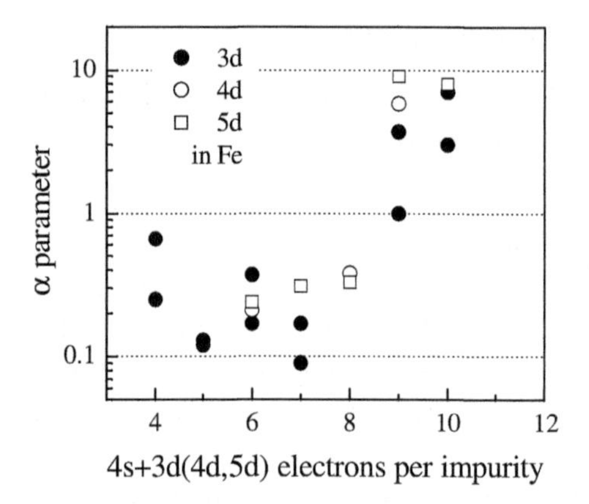


**Figure 5.5** Values of $\alpha$-parameter determined experimentally for 3d, 4d, and 5d transition metal impurities in Fe.

$\alpha$-parameter. The experimental results have been confirmed by a simple analysis of the impurity states and by the first-principles calculations.

#### 5.2.3.2 Spin-dependent resistivity in multilayers

Possible sources of spin-dependent resistivity in multilayers are:

- *Interfacial roughness.* As described, the origin of the spin-dependent resistivity in metals and alloys is the spin dependence of the scattering potentials caused by roughness. The roughness due to random arrangement of atoms also exists in multilayers. In molecular beam epitaxy (MBE) and sputtering fabrication methods, it is impossible to avoid intermixing of atoms at interfaces. The intermixing of magnetic A atoms and nonmagnetic B atoms at an A/B interfaces gives rise to spin dependent random potentials near the interface. The situation is similar to that in ferromagnetic alloys.
- *Roughness within each layer.* As in bulk ferromagnetic metals, there exist scattering sources that give rise spin-dependent resistivity. Schematics of the interfacial and bulk roughness is given in Fig. 5.6a.
- *Band matching/mismatching at interfaces.* The essence of the origin of electrical resistivity is the absence of translational

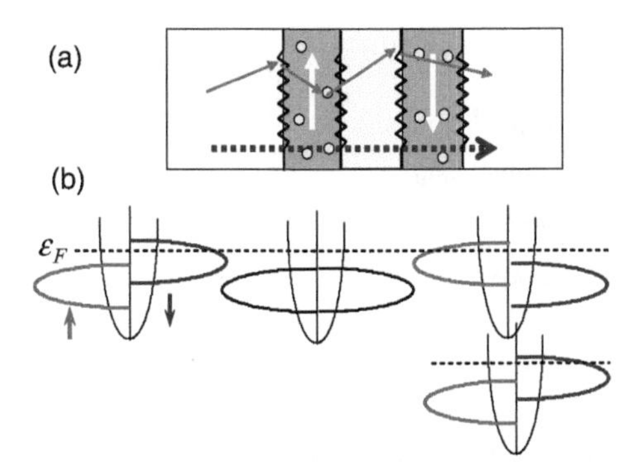


**Figure 5.6** Schematic figure to explain the mechanism of GMR: (a) electron scattering at interfaces and inside of layers and (b) electronic density of states of FM/NM/FM structures.

invariance along the current direction, because the momentum of electrons need not be conserved in this case. When the interfaces are clean, translational invariance parallel to the layer planes is satisfied and there is no electrical resistivity. Even in this case, however, there is electrical resistivity perpendicular to the layer planes, since there is no translational invariance along this direction for thin multilayers. In this case, the difference between the electronic structure of the constituent metals of the multilayers acts as a source of spin-dependent electrical resistivity and gives rise to CPP-GMR. In Co/Cu multilayers, for example, band matching between majority (+) spin states is much better than between minority (−) spin states, as schematically shown in Fig. 5.6b. Therefore, $\rho_+ \ll \rho_-$ is realized. In Fe/Cr multilayers, the opposite relation, $\rho_- \ll \rho_+$, is realized. The spin dependence of the resistivity is the same as that obtained by spin-dependent random potentials.

Thus, both random potentials and matching/mismatching of the electronic structure cause the spin dependence of the resistivity [64, 65, 67–70]. For current flowing perpendicular to the planes, both are crucial for the spin dependence of the resistivity, while the spin-dependent random potential is likely to be more greatly responsible for GMR for current flowing parallel to the planes. As a result, the MR ratio in CPP-GMR is usually larger than that in

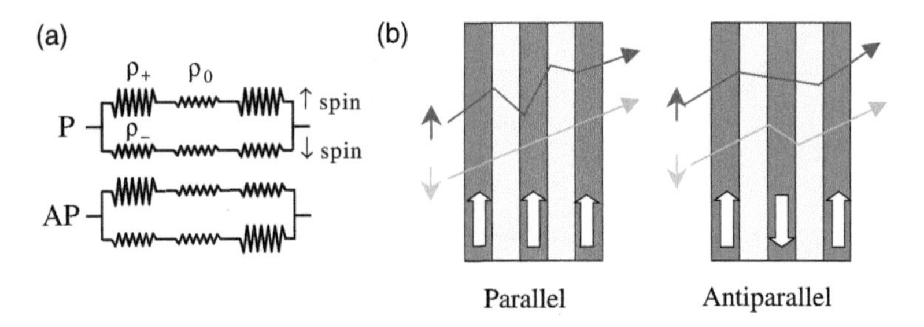


**Figure 5.7** (a) A circuit model to explain the mechanism of GMR. (b) Schematics of electron scattering at interfaces in parallel and antiparallel alignments of magnetic layers.

CIP-GMR as shown in Table 5.2. So far, many theoretical studies on the scattering at interfaces have been performed for CIP-GMR and CPP-GMR [71, 79].

### 5.2.3.3 Phenomenological theory of GMR

The typical length scale of a multilayer is the thickness $L$ of each layer, which is of the order of 1 nm. Since the length scale is shorter than the mean free path and much shorter than the spin diffusion length, Mott's two-current model is applicable to GMR in multilayers as a first approximation. The model is also applicable to granular GMR because the length scale of the sample is the diameter (about 1 nm) of the magnetic grains. In the following we adopt the simplest picture to explain the effect of GMR [64, 65].

In applying the two-current model to magnetic multilayers, the direction of the spin axis, up ($\uparrow$) or down ($\downarrow$), should be defined to deal with the resistivity. Because the magnetization of the magnetic layers is reversed by the magnetic field and the magnetization alignment can be either parallel (P) or antiparallel (AP), $\uparrow$ and $\downarrow$ spin states should be distinguished from the majority ($+$) and minority ($-$) spin states of each magnetic layer. We henceforth use the notation $\uparrow$ and $\downarrow$ spin states as the global spin axes and $+$ and $-$ spin states to express the electronic states of the magnetic metals. For P alignment, $\uparrow$ and $\downarrow$ spin states coincide with $+$ and $-$ spin states, respectively; however, they do not coincide for AP alignment. In the following, we adopt equation (5.1) for the MR ratio.

For simplicity, we consider a case where the current traverses a trilayer composed of a nonmagnetic layer sandwiched by two ferromagnetic layers. Let $\rho_+$, $\rho_-$ and $\rho_0$ be the majority and minority spin resistivities in the ferromagnetic layers and the resistivity in the nonmagnetic layer, respectively. The total resistivity for P and AP alignment may be easily obtained by referring to the equivalent circuits shown in Fig. 5.7a. The resultant MR ratio is given as

$$\mathrm{MR} = \left(\frac{\rho_+ - \rho_-}{\rho_+ + \rho_- + 2\rho_0}\right)^2 . \tag{5.3}$$

The MR ratio increases as the difference between the spin-dependent resistivities $\rho_+ - \rho_-$ increases.

When $\rho_+ + \rho_0$ and $\rho_- + \rho_0$ are rewritten as $\rho_+$ and $\rho_-$, the expression above is written as

$$\mathrm{MR} = \left(\frac{1-\alpha}{1+\alpha}\right)^2 , \tag{5.4}$$

using the $\alpha$-parameter $\alpha = \rho_-/\rho_+$. Thus, a combination of materials, which gives a large value of the $\alpha$-parameter gives rise to a large GMR. The combination of Fe and Cr gives a large GMR In addition, the lattice matching between Fe and Cr layers is good.

We have presented the mechanism of GMR in terms of spin-dependent resistivity and band matching/mismatching between magnetic and nonmagnetic layers. These origins are closely related to each other and present a simple picture for the mechanism of GMR in magnetic multilayers, as shown in Fig. 5.7b. In P alignment of the magnetization, one of the two spin channels has a low resistance (or resistivity) while the other has a high resistance. In AP alignment, both spin channels have high resistance (or resistivity), since both $\uparrow$ and $\downarrow$ spin electrons are scattered at some interfaces. As a result, the resistance in AP alignment is larger than that in P alignment.

### 5.2.3.4 Microscopic theory of GMR

The discussion in the previous section does not take into account the geometry of the multilayers. Therefore, they are not able to distinguish between CIP-GMR and CPP-GMR. To account for the effects of geometry, microscopic theory should be applied to the resistivity, which may be done by adopting the Kubo formula explained in Appendix. Several calculations have been performed so far [72–76], but the results are left for good review articles.

#### 5.2.3.5 Effects of spin flip scattering

So far, we have not taken spin flip or spin diffusion into account. It is easy to include the spin flip process in the phenomenological theory. By adding a electric pass with resistivity $\rho_{\rm sf}$ between up- and down-spin channels at the middle of the circuit shown in Fig. 5.7a, the resistivity in P and AP alignment may be given as

$$\rho_{\rm P} = \frac{2\rho_+\rho_-}{\rho_+ + \rho_-}, \tag{5.5}$$

$$\rho_{\rm AP} = \frac{2\rho_+\rho_- + \rho_{\rm sf}(\rho_+ + \rho_-)}{\rho_+ + \rho_- + 2\rho_{\rm sf}}. \tag{5.6}$$

Here $\rho_{\rm sf}$ indicates the contribution to the resistivity due to spin flip scattering. $\rho_{\rm P}$ does not include $\rho_{\rm sf}$, since the voltage drop at the middle of the circuit is the same for $\uparrow$ spin and $\downarrow$ spin channels. In AP alignment, taking $\rho_{\rm sf} \to 0$ gives $\rho_{\rm AP} = \rho_{\rm P}$ and the MR disappears because the current is shorted. For an infinite $\rho_{\rm sf}$, there is no spin flip effect and the MR ratio decreases with decreasing $\rho_{\rm sf}$. (Note that $2\rho_+ \equiv \rho_\uparrow$ and $2\rho_- \equiv \rho_\downarrow$ according to Fert et al. [63]).

### 5.2.4 *Application of GMR: Spin Valve*

Technological applications of GMR, for example, sensors, require a sharp response of the magnetization direction to the external magnetic field within a few Oe. To achieve such sensitivity, a trilayer structure with an attached antiferromagnet has been designed. The magnetization of the magnetic layer adjacent to the antiferromagnetic layer is pinned by the antiferromagnetism and only the other magnetic layer responds to the external magnetic field. This kind of trilayer is called a spin valve [16, 17, 80–82]. PtMn and FeMn are typical antiferromagnets used in spin valves. An example of GMR in a spin valve type trilayer is shown in Fig. 5.8a.

Top figure of Fig. 5.8c shows the *M–H* curve and resistivity change. The arrows indicate the magnetization alignment. When

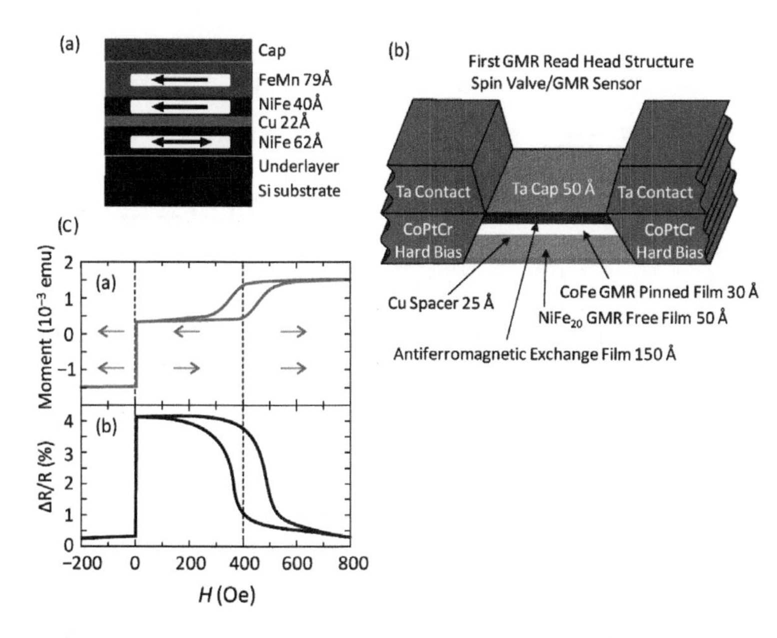


**Figure 5.8** (a) Schematics of spin valve structure, (b) realistic structure of spin valve, and (c) change in magnetization and resistivity with external magnetic field in spin valve. Reprinted (figure) with permission from Ref. [16], Copyright (1991) by the American Physical Society.

the negative magnetic field is applied to the trilayer, the total magnetization is negative, and it becomes zero by applying positive magnetic field. This is due to reversal of magnetization on one of the ferromagnetic layer. Accordingly, the resistivity sharply increases, and there is no hysteresis in this process because of very weak exchange coupling between two ferromagnetic layers. Further increase in the magnetic field, the magnetizations of both ferromagnetic layer become parallel, and the resistivity drops. In this process, there exist a rather large hysteresis because the magnetization of one ferromagnetic layer is strongly pinned by the antiferromagnetic states of pinning layer.

Actual junction structure is rather complicated, as shown in Fig. 5.8b to produce desired lattice structure and performance of the spin valve for magnetic sensors at the reading/writing head of HDD. The spin valve using GMR was developed soon after the discovery of GMR and has been used widely as GMR devices.

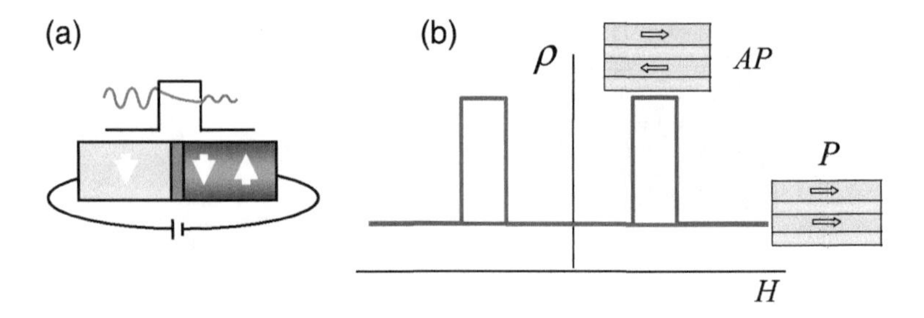


**Figure 5.9** (a) Schematics of ferromagnetic tunnel junction and (b) resistivity change in tunnel junction with applied magnetic field $H$.

## 5.3 Tunnel Magnetoresistance

### 5.3.1 *Ferromagnetic Tunnel Junctions and TMR*

FTJs are made of a thin (about 1 nm thick) nonmagnetic insulator sandwiched between two ferromagnetic electrodes. A schematic figure of such a junction is shown in Fig. 5.9a. The ferromagnetic metals used are predominantly Fe, Co and their alloys, while amorphous $Al_2O_3$ is one of the most stable materials for the insulating barrier. Recently, a single-crystal MgO layer has been used as the barrier in order to generate high TMR ratios. Transmission electron microscopy (TEM) measurements confirm the single crystalline lattice in the MgO barrier.

TMR in FTJs was reported prior to the discovery of GMR [20, 21]. The observed MR ratios, however, were rather small at the time. A large TMR at room temperature was reported for Fe/Al-O/Fe in 1995 [18, 19, 83], and it has attracted considerable attention due to its wide potential application in sensors and memory storage devices in the near future. In the same year, tunnel-type MR was reported for metal-oxide ferromagnetic granular films [84, 85].

The resistivity of the FTJ is reduced when the magnetization alignment of two ferromagnets is changed from AP to P alignment. A schematics of the resistivity change caused by the external magnetic field is shown in Fig. 5.9b. The resistance is high when the magnetization of the two ferromagnetic electrodes is antiparallel, and it is low when the magnetization is parallel. One of the characteristics of TMR is that the external magnetic field required to rotate the magnetization is sufficiently low. This is because there is almost no coupling between the magnetizations of the two electrodes as a result of the insulating barrier inserted between them. The AP and P alignments of the magnetization are realized by using a small difference in the coercive force between the two ferromagnets. The current flows perpendicular to the layer planes, which is similar to CPP-GMR. The resistance in the FTJs is much higher than that in CPP-GMR. This makes it possible to measure the junction resistance without microfabricating the samples. This could be considered to be another characteristic of TMR.

The phenomenon of TMR is similar to GMR, and the MR ratio is defined as

$$\mathrm{MR} = \frac{\rho_{\mathrm{AP}} - \rho_{\mathrm{P}}}{\rho_{\mathrm{AP}}}. \tag{5.7}$$

An alternative definition is often used in experimental studies; in this definition $\rho_{\mathrm{AP}}$ in the denominator is replaced with $\rho_{\mathrm{P}}$.

A large TMR effect has been discovered for FTJ with MgO barrier after theoretical calculations in which nearly 1000 % MR ratio was predicted for the FTJ. The experimental values of the MR ratio are listed in Table 5.3. The largest MR ratio exceeds 1000 % as predicted by theory. However, temperature dependence of the ratio is rather strong, and the MR decreases rapidly with raising temperature

**Table 5.3** Observed MR ratios of ferromagnetic tunnel junctions with MgO barriers. Values at low temperature (LT) and at room temperature (RT) are shown. $Co_2Cr_{0.6}Fe_{0.4}$, $Co_2FeAl_{0.5}Si_{0.5}$, and $Co_2MnSi$ are half-metals (see text). 1) Lee et al., APL 89, 042506 (2007), 2) Ikeda et al., MSJ (2008) 14a2PS-37(B), 3) Marukame et al., APL 90, 012508 (2007), 4) Tezuka et al., JJAP 46, L454 (2007), 5) Ishikawa et al., JAP 103, 07A919 (2008), 6) MSJ(2008) 13pB-11 Andoh G

| Materials | MR ratio (%) | | References |
|---|---|---|---|
| | LT | RT | |
| CoFeB/MgO/CoFeB | 1010 | 500 | [1] |
| CoFeB/MgO(2.1nm)/CoFeB | 1144 | 576 | [2] |
| $Co_2Cr_{0.6}Fe_{0.4}/MgO/Co_{50}Fe_{50}$ | 317 | 109 | [3] |
| $Co_2FeAl_{0.5}Si_{0.5}/MgO/Co_2FeAl_{0.5}Si_{0.5}$ | 390 | 220 | [4] |
| $Co_2MnSi/MgO/Co_2MnSi$ | 683 | 179 | [5] |
| $Co_2MnSi$/MgO(2.5nm)/$Co_{50}Fe_{50}$ | 753 | 217 | [6] |

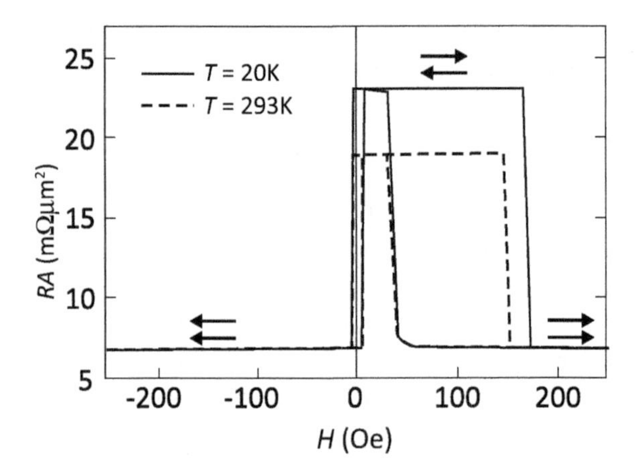


**Figure 5.10** Observed results of area resistivity in Fe/MgO/Fe tunnel junction with spin valve structure as a function of external magnetic field. Reprinted by permission from Macmillan Publishers Ltd: [*Nature Materials*] (Ref. [24]), copyright (2004).

except for junctions made of CoFeB. Recently TMR has been used for spin valve, and the resistivity change is shown in Fig. 5.10 [24].

### 5.3.2 *A Phenomenological Theory of TMR*

The experimental results for TMR can be understood phenomenologically as follows. Let us denote the left and right electrodes as L and R, respectively. When the tunneling process is independent of the wave vectors of tunneling electrons, the tunnel conductance $\Gamma$ is proportional to the product of the densities of states of the L and R electrodes, and is given by $\Gamma \propto \sum_s D_{\mathrm{L}s}(E_\mathrm{F})D_{\mathrm{R}s}(E_\mathrm{F})$, where $s$ denotes the spin [20, 90]. The proportionality constant includes the transmission coefficient of electrons through the insulating barrier. Henceforth, the Fermi energy $E_\mathrm{F}$ is omitted for simplicity.

**Semiclassical theory of tunnel conductance:** Let us consider a junction with two metallic electrodes separated by an insulating barrier. The electronic state of the junction is schematically shown in Fig. 5.11. There is an energy difference $|e|V$ between the left and right electrodes. Tunneling current from left (L) to right (R)

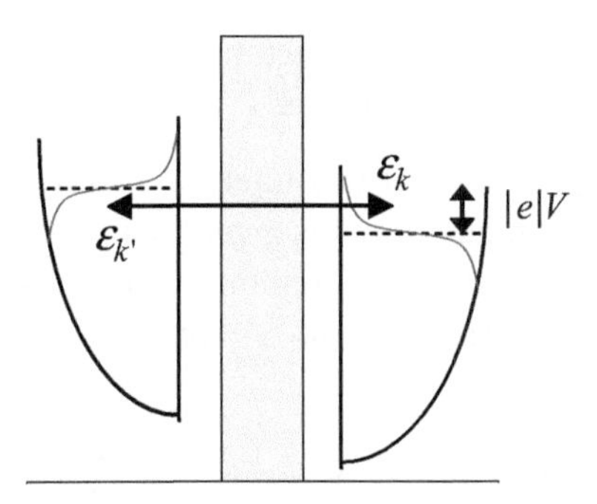


**Figure 5.11** Electron tunneling through insulating barrier sandwiched by two ferromagnetic metals with bias voltage.

electrode and that from R to L electrode are given as

$$J_{L\to R} = 2(-|e|)\sum_{kk'} T_{kk'} f_{k'}(1 - f_k)\delta(\varepsilon_k - \varepsilon_{k'} + |e|V),$$

$$J_{R\to L} = 2(-|e|)\sum_{kk'} T_{kk'}(1 - f_{k'}) f_k\delta(\varepsilon_k - \varepsilon_{k'} + |e|V).$$

Here $T_{kk'}$ is a transmission probability from $k$ to $k'$ state, and a factor 2 indicates spin degeneracy. The total current is then given by

$$\begin{aligned} J &= J_{L\to R} - J_{R\to L} \\ &= -2|e|\sum_{kk'} T_{kk'}\,(f_{k'} - f_k)\,\delta(\varepsilon_k - \varepsilon_{k'} + |e|V). \end{aligned}$$

By making an approximation to replace $T_{kk'}$ with an averaged value $\langle T\rangle$, and using

$$\begin{aligned} \sum_k f_k F(\varepsilon_k) &= \sum_k \int d\varepsilon\delta(\varepsilon - \varepsilon_k) f(\varepsilon)F(\varepsilon) \\ &= \int d\varepsilon D(\varepsilon) f(\varepsilon)F(\varepsilon), \end{aligned}$$

where $D(\varepsilon)$ is the DOS and $F(\varepsilon)$ is an arbitrary function of $\varepsilon$, the current is given as

$$\begin{aligned} J &= -2|e|\langle T\rangle D_L(\varepsilon_F)D_R(\varepsilon_F)\int d\varepsilon d\varepsilon' \left[f(\varepsilon') - f(\varepsilon)\right]\delta(\varepsilon - \varepsilon' + |e|V) \\ &= 2|e|^2\langle T\rangle D_L(\varepsilon_F)D_R(\varepsilon_F)V. \end{aligned}$$

Then we obtain the tunnel conductance as

$$\Gamma = J/V = 2e^2\langle T\rangle D_L(\varepsilon_F)D_R(\varepsilon_F).$$

This mean that the tunnel conductance is proportional to a product of DOS of left and right electrodes.

By using the expression of the tunnel conductance, the conductance for P magnetization alignment is given by $\Gamma_{\mathrm{P}} \propto D_{\mathrm{L}+}D_{\mathrm{R}+} + D_{\mathrm{L}-}D_{\mathrm{R}-}$, and that for AP magnetization alignment is given by $\Gamma_{\mathrm{AP}} \propto D_{\mathrm{L}+}D_{\mathrm{R}-} + D_{\mathrm{L}+}D_{\mathrm{R}-}$. Since the resistivity $\rho_{\mathrm{P(AP)}}$ corresponds to $\Gamma^{-1}_{\mathrm{P(AP)}}$, the MR ratio is given by

$$\text{MR ratio} = \frac{\Gamma_{\mathrm{AP}}^{-1} - \Gamma_{\mathrm{P}}^{-1}}{\Gamma_{\mathrm{AP}}^{-1}} = \frac{2P_{\mathrm{L}}P_{\mathrm{R}}}{1 + P_{\mathrm{L}}P_{\mathrm{R}}}, \tag{5.8}$$

where $P_{\mathrm{L(R)}}$ is the spin polarization of L(R) electrodes and is defined by

$$P_{\mathrm{L(R)}} = \frac{D_{\mathrm{L(R)}+} - D_{\mathrm{L(R)}-}}{D_{\mathrm{L(R)}+} + D_{\mathrm{L(R)}-}}. \tag{5.9}$$

Although the transmission coefficient governs the magnitude of the tunnel conductance, it does not appear in the expression for the MR ratio.

An intuitive picture of the tunneling process explained above is shown in Fig. 5.12. As shown in this figure, in P alignment, majority and minority spin electrons in the L electrode tunnel through the barrier into the majority and minority spin states in the R electrode, respectively. In AP alignment, however, the majority and minority spin electrons in L electrode tunnel into the minority and majority spin states in R electrode, respectively. The difference between the conductances of the P and AP alignments gives rise to TMR.

Equation (5.8) indicates that the spin polarization of the electrodes govern the MR ratio. The spin polarization has

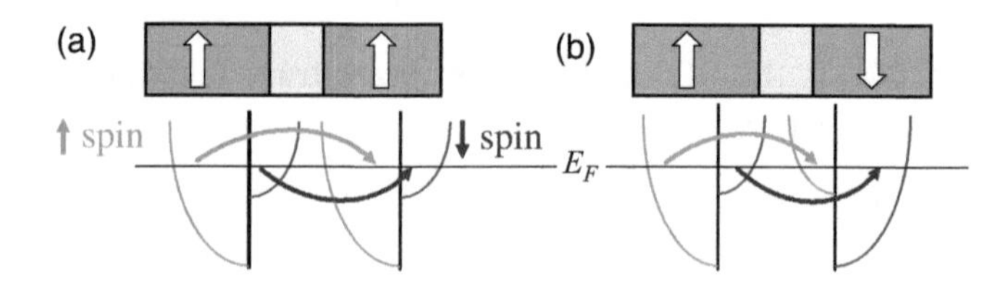


**Figure 5.12** Tunneling process of up- and down-spin electrons for (a) parallel and (b) antiparallel alignments of magnetization in ferromagnetic tunnel junctions.

been experimentally determined by using junctions of ferromagnet/Al/superconductor, or by analyzing the tunneling spectrum obtained using point contacts [91–97]. Measured values of the spin polarization of Fe and Co are 0.3–0.4. Using these values, the experimentally measured MR ratios are explained rather well. For example, the MR ratio measured for Fe/Al-O/Fe junctions is about 0.3 and this value is close to the theoretical value calculated by using the experimental value of $P$ for Fe. However, the MR ratio observed for a single crystal Fe/MgO/Fe junction is 0.7–0.8, which cannot be explained in terms of the spin polarization of Fe.

## 5.3.3 *TMR of Several FTJs*

### 5.3.3.1 Fe/MgO/Fe junctions

Calculations of MR ratios for disorder-free Fe/MgO/Fe FTJs have been performed by using a first-principles method and a realistic tight-binding model [98, 99]. The calculated results show that an extremely high MR ratio may be realized in clean samples. Being inspired by the results, single-crystal Fe/MgO/Fe samples were fabricated [100, 101] and high TMR ratios have been observed [22–24].

The experimentally observed MR ratio is much higher than that predicted by the phenomenological model, and that elucidated using the principles mentioned in the previous section. First, the symmetry of the conduction bands of Fe plays an important role. The conduction band of $+$ spin states of Fe has $\Delta_1$ symmetry, which is composed of $s$, $p_z$, $d_{3z^2-r^2}$ orbitals, and contains the state with $k_\parallel = (0, 0)$, as shown in Fig. 5.13. The $\Delta_1$ band hybridizes with the $s$ and $p_z$ atomic orbitals of MgO. On the other hand, the conduction band of $-$ spin states of Fe has $\Delta_2$ symmetry composed of $d_{x^2-y^2}$ and does not contain the state with $k_\parallel = (0, 0)$, nor hybridize with the $s$ and $p_z$ atomic orbitals of MgO. Therefore, the decay rate of the wave function of the $-$ spin electrons is much larger than that of $+$ spin electrons, and as a result, the transmission coefficient of $+$ spin electrons is much larger than that of $-$ spin electrons. This gives rise to a large MR ratio. The symmetry of the bands and the existence of $k_\parallel = (0, 0)$ state in the conduction bands exert a strong

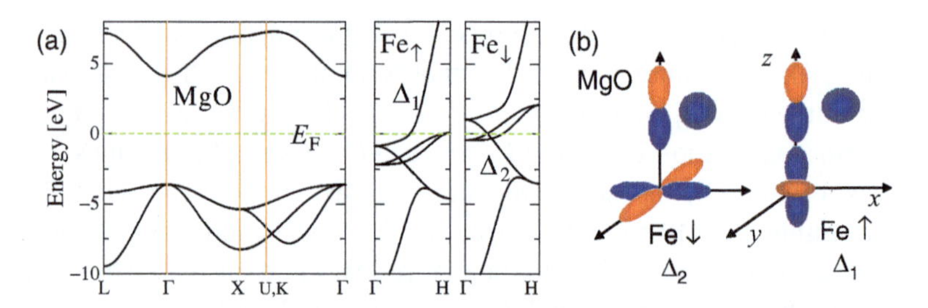


**Figure 5.13** (a) Electronic structure of MgO and up- and down-spin states of Fe calculated in the first-principles method. (b) Arrangement between $p_z$ and $s$ orbitals of MgO layer and $d_{x^2-y^2}$ orbital of Fe layer ($\Delta 2$ symmetry) and that between $p_z$ and $s$ orbitals of MgO and $d_{3z^2-r^2}$ orbital of Fe ($\Delta 1$ symmetry).

influence on the MR ratio. When the MgO layer is thin, the state with $k_{\parallel} \neq (0, 0)$ also contributes to tunneling, and the dependence of the MR ratio on MgO thickness is not simple in general. The calculated results of momentum resolved conductance of up- and down-spin states in P and AP magnetization alignments are shown in Fig. 5.14. The figure shows that up-spin conductance in P alignment shows high conductance near $k_{\parallel} = (0, 0)$ state. The conductance is much higher than those of down-spin conductance in P alignment and of AP alignment, the MR ratio becomes quite large.

When there is roughness in the junction, the momentum $k_{\parallel}$ need no longer be conserved. Then tunneling via the $k_{\parallel} = (0, 0)$ state

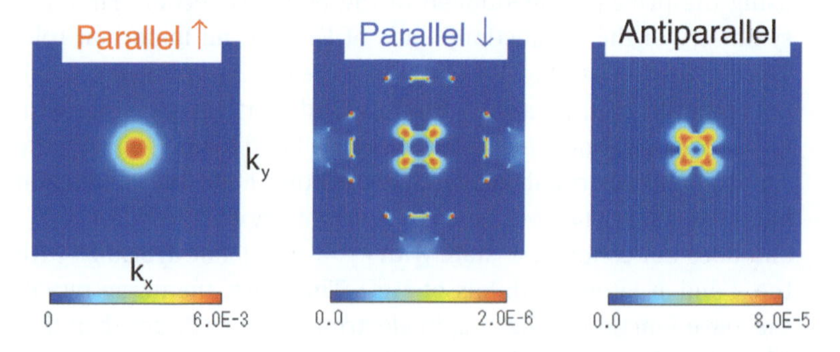


**Figure 5.14** Calculated results of momentum resolved tunneling conductance of Fe/MgO/Fe tunnel junctions in up- and down-spin states of parallel alignment and in antiparallel alignment. Reproduced from Ref. [114] with kind permission from the Magnetics Society of Japan.

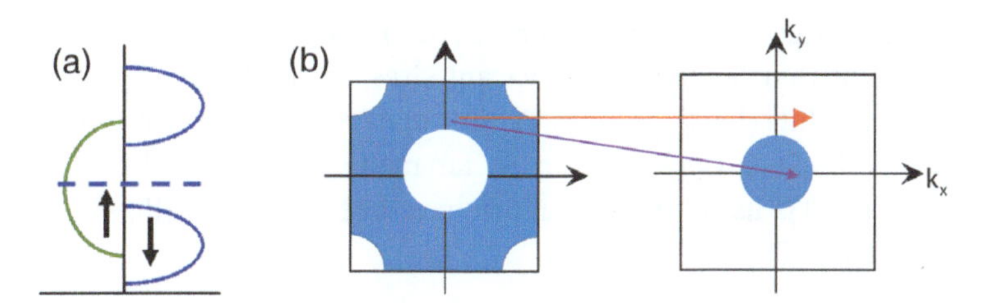


**Figure 5.15** (a) Schematics of density of states of a half-metal. (b) A picture of electron hopping from available states (colored in blue) on left to those on right electrodes.

becomes possible for – spin electrons, and the tunnel conductance of – spin electrons increases resulting in an reduction in the spin asymmetry of the tunnel conductance, which causes the MR ratio to decrease.

Thus we understand how a high MR ratio is realized in clean Fe/MgO/Fe junctions. However, high MR ratios also occur in disordered ferromagnetic electrodes such as FeCoB, which has an amorphous structure [102]. Possible reasons for this may be that B atoms may reside on regular sites in the face-centered cubic (fcc) structure after annealing, and the electronic states of B atoms have energies that are distant from the Fermi energy.

### 5.3.3.2 TMR in FTJ with half-metals

The DOS of metallic ferromagnets are usually spin dependent. When the DOS of either ↑ or ↓ spin state is zero at the Fermi level, and one of two spin states is metallic and the other is insulating, as shown in Fig. 5.15a, the metals are referred to as half-metals. The spin polarization *P* of these half-metals is 100%, and therefore half-metals have potential applicability as magnetoresistive devices. The half-metallic characteristic of the electronic states in metallic ferromagnets was first predicted for NiMnSb in a first-principles band calculation [104]. Later, it was shown in a first-principles band calculation that many Heusler alloys have half-metallic DOS. There are two types of structures in Heusler alloys, namely half-Heusler with an XYZ lattice and full-Heusler with an $X_2YZ$ lattice, where X=Ni, Co, Pt; Y=Cr, Mn, Fe, and Z=Sb, Ge, Al, etc. The basic lattice structure is composed of two nested bcc lattices. The half-Heusler lattice has vacant sites, which are occupied by X atoms in full-Heusler alloys. Typical materials of Heusler alloys are $Co_2FeAl_{0.5}Si_{0.5}$, $Co_2MnGe$ etc. So far many experimental results of TMR of FTJs using Huesler alloys have been reported [86–89, 105, 106, 108, 109].

### 5.3.3.3 Granular TMR

The MR ratio of the granular TMR is given by [110]

$$\mathrm{MR} = \frac{\Gamma^{-1}(0) - \Gamma^{-1}(H)}{\Gamma^{-1}(0)} = \frac{P^2}{1+P^2}, \tag{5.10}$$

which is just half the MR ratio of junction TMR. This is due to the random distribution of the magnetization direction of the grains when no external magnetic field is applied.

The above result is obtained as follows. Consider the two ferromagnetic grains denoted as L and R. The magnetizations of the R grain is canted by an angle $\theta$ with respect to that of the L grain. In this case, the transition probability for an electron on the L grain to hop to the same spin state on the R grain is given by $\cos^2(\theta/2)$, and that to the opposite spin state on the R grain is $\sin^2(\theta/2)$. Thus, the probability of an electron tunneling between L and R grains is given by

$$T_{\uparrow\uparrow} = D_{\mathrm{L}+} D_{\mathrm{R}+} \cos^2(\theta/2) + D_{\mathrm{L}+} D_{\mathrm{R}-} \sin^2(\theta/2), \tag{5.11}$$

$$T_{\downarrow\downarrow} = D_{\mathrm{L}-} D_{\mathrm{R}-} \cos^2(\theta/2) + D_{\mathrm{L}-} D_{\mathrm{R}+} \sin^2(\theta/2), \tag{5.12}$$

where $D_{\mathrm{L}+(-)}$ and $D_{\mathrm{R}+(-)}$ are majority (minority) spin DOS of the L and R grains, respectively. The total tunneling probability is given by

$$T = T_{\uparrow\uparrow} + T_{\downarrow\downarrow} Pp \propto 1 + P^2 \cos\theta, \tag{5.13}$$

where

$$P^2 = \frac{(D_{\mathrm{L}+} - D_{\mathrm{L}-})(D_{\mathrm{R}+} - D_{\mathrm{R}-})}{(D_{\mathrm{L}+} + D_{\mathrm{L}-})(D_{\mathrm{R}+} + D_{\mathrm{R}-})}. \tag{5.14}$$

The spin of the tunneling electron is assumed to be conserved in the tunneling process.

The tunnel conductance in the absence of a magnetic field is given by

$$\Gamma(0) \propto \langle 1 + P^2 \cos\theta \rangle_{av} = 1, \tag{5.15}$$

when an external magnetic field $H$ is applied it is

$$\Gamma(H) = 1 + P^2. \tag{5.16}$$

We then obtain Eq. (5.10). The theoretical result is consistent with the observed one for granular systems except for low temperature [85].

### 5.3.4 *Ingredients of TMR*

In the semiclassical theory described above, the origin of TMR is the spin dependence of DOS. However, as shown below, realistic electronic structures and disorder at interfaces exert a larger effect on determining the magnitude of TMR [98, 99, 111–114]. Here we summarize ingredients for TMR.

#### 5.3.4.1 Role of the transmission coefficient

The magnitude of the tunnel conductance is governed by the transmission coefficient with a prefactor of $\exp(-2\kappa d)$, where $\kappa$ is a decay rate of the wave function and $d$ is the thickness of the insulating barrier. Although $\exp(-2\kappa d)$ is spin independent, it gives rise to a strong dependence on spin in the following way.

In the tunneling process, electrons at the Fermi energy $E_{\mathrm{F}}$ contribute the tunneling process. The energy may be decomposed into two terms as

$$E_{\mathrm{F}} = E(k_{\parallel}) + E(k_z), \tag{5.17}$$

where the first and second terms consist of momentum parallel and perpendicular to the barrier layer, respectively. Because the latter is responsible to the tunneling, the expression of $\kappa$ is given as

$$\kappa = \sqrt{2m(U - E(k_z))}/\hbar \tag{5.18}$$

where $U$ is the height of the barrier potential. We see that when $\kappa d \gg 1$, the electronic state that maximizes $E(k_z)$ (i.e., minimizes

$E(k_\parallel)$) makes the largest contribution to the tunneling conductance. Usually, such a state is $\boldsymbol{k}_\parallel = (0, 0)$, and the tunneling conductance calculated by taking into account only this state agrees with the result obtained using the one-dimensional model. For junctions with thin tunnel barriers, however, the contribution from states with wave numbers besides $\boldsymbol{k}_\parallel = (0, 0)$ becomes large. Thus we find that the electrons injected perpendicular to the barrier planes make the biggest contribution to tunneling. This is a filtering effect for the momentum of a tunneling electron.

Since the electronic states of ferromagnets depend on spin, the tunneling probability becomes spin dependent due to this momentum filter effect, and the decay rate may be written as $\kappa_s = \sqrt{2m(U - E_s(k_z))}/\hbar$. The spin-dependent $\kappa_s$ can produce high TMR ratios.

#### 5.3.4.2 Effects of the Fermi surface

When a junction has no disorder and is translationally invariant along layer planes, the component of the wave vector $k_\parallel$ parallel to layer planes is conserved. This kind of tunneling process is referred to as specular tunneling. On the other hand, when $k_\parallel$ is not conserved, the tunneling process is referred to as diffusive tunneling. In this section, we consider specular tunneling.

Tunnel junctions are usually composed of an insulating barrier sandwiched between two different ferromagnetic metals. Therefore, the Fermi surfaces of the L and R electrodes are generally different. Since $k_\parallel$ is conserved in specular tunneling, only states on the Fermi surface with the same $k_\parallel$ may contribute to tunneling.

In the free-electron model, the state with $\boldsymbol{k}_\parallel = (0, 0)$ is always included in the Fermi surface. However, this state may not be included in the complicated Fermi surfaces of, for example, transition metals. In particular, ferromagnets have spin-dependent Fermi surfaces, and thus the state with $\boldsymbol{k}_\parallel = (0, 0)$ may be included in one spin state, but not in the other spin state. In this case, the tunneling conductance is strongly spin dependent. Furthermore, the tunnel conductance in AP alignment becomes very small because the $\boldsymbol{k}_\parallel = (0, 0)$ state may not contribute to tunneling.

#### 5.3.4.3 Symmetry of the wave function

In specular tunneling, the difference between the symmetries of the wave functions of tunneling electrons in L and R electrodes exerts a strong influence on TMR. The wave function of each state in metals has a specific symmetry, and electrons with a certain symmetry are not able to transfer into a state with a different symmetry. Therefore, electrons on the L electrode can tunnel through the barrier into the R electrode only when the states specified by $k_{\|}$ in the L and R electrodes have the same symmetry; but this becomes impossible when the states are different. When the L and R electrodes are composed of different materials, the symmetry of the wave functions of tunneling electrons play an important role in tunnel conductance. In ferromagnets, the $+$ and $-$ electronic states are different in general. Therefore, when the magnetizations of the two electrodes are AP, the states on the Fermi surfaces of the L and R electrodes can have different symmetries. In this case, the electrons in these states are not able to tunnel through the barrier. The situation makes the tunnel conductance for AP alignment much smaller than that for P alignment, resulting in a large MR ratio.

Figure 5.13a shows the energy–momentum relation along the (001) direction for Fe and MgO [114, 115]. It can be seen from this figure that the symmetry of the $+$ spin band is $\Delta_1$ on $E_F$ for Fe; however, the symmetry does not appear in the $-$ spin band. This situation is realized in Fe/MgO/Fe junctions.

#### 5.3.4.4 Effect of density of states

Surely the spin dependence of the DOS affects the TMR ratio as described in the semiclassical theory. Figure 5.15a schematically shows the half-metallic DOS in which the difference between up- and down-spin DOS becomes the largest. It is however noted that this effect might get blurred due to other effects described in this section.

#### 5.3.4.5 Effect of interfacial states

When a semiconductor (e.g., GaAs) is used as the barrier, interfacial states, called Shockley states, appear within the energy band gap.

Since interfacial states are localized near the interface, they do not contribute to electron transport. For a thin semiconductor barrier, the states at one interface extend up to a few atomic layers inside the barrier and might overlap with those at the other interface. When the Fermi level is located within the energy region where the interfacial states exist, the tunnel conductance may be strongly enhanced by the interfacial states, thus effectively reducing the thickness of the barrier [137].

### 5.3.4.6 Effect of electron scattering

Let us now consider diffusive tunneling, which occurs when some disorder is present in junctions, and the wave vector $k_{\parallel}$ is not conserved. Interfacial roughness and amorphous-like insulators break translational invariance parallel to layer planes. Therefore, the parallel component of the wave vector no longer has to be conserved; that is, the wave vector $k_{\parallel}$ of incident electrons need not coincide with that $k'_{\parallel}$ of the transmitted electrons. In this section, we consider a situation in which the Fermi surfaces of the L electrode differs from that of the R electrode, as shown in Fig. 5.15b. In the specular tunneling case, only states with Fermi wave vectors $k_{\parallel \mathrm{L}} = k_{\parallel \mathrm{R}}$ contribute to tunneling. In the diffusive tunneling case, on the other hand, this restriction is removed, and therefore those states having wave vectors $k_{\parallel} \neq k_{\parallel \mathrm{L}} \neq k_{\parallel \mathrm{R}}$ may contribute as shown in Fig. 5.15b, which increases the tunnel conductance. Even when the Fermi surfaces of the L and R electrodes are the same and do not include the $\boldsymbol{k}_{\parallel} = (0, 0)$ state, tunneling electrons take the $\boldsymbol{k}_{\parallel} = (0, 0)$ state virtually in the tunneling process. The virtual process reduces

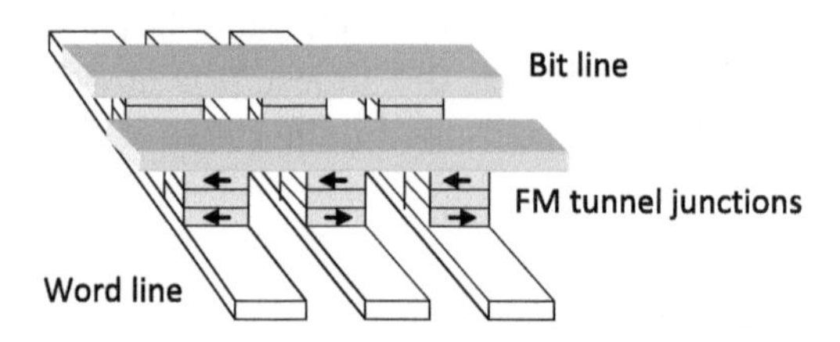


**Figure 5.16** Schematics of device structure of magnetoresistive random access memory (MRAM).

the decay rate $\kappa$ and consequentially increases tunnel conductance [112–114].

### 5.3.5 *Application of TMR: Magnetoresistive Random Access Memory*

The modern memory devices are made of complementary metal-oxide semiconductor (CMOS) random access memory (RAM). The memory is volatile and needs continuous charging up to the memory, which is quite energy consumptive. Therefore, development of nonvolatile memories is indispensable for the information technology. Magnetic random access memory (MRAM) is one of promising candidates for the nonvolatile memory. Immediately after the discovery of TMR effect in FTJs, development of MRAM using FTJs was started in USA, because the P and AP alignments of the magnetization can be the memory unit, the alignment can be controlled by weak magnetic fields, and the alignment can be easily detected by the tunneling current.

The schematic structure of the MRAM using an FTJ is shown in Fig. 5.16. Writing process is done by an effective magnetic field produced by currents on both bit and word lines, and reading process is done by measuring the resistance on the FTJ. Selection of bit and word lines is performed by adopting the CMOS structures used in the semiconductor memory devices. Therefore, the FTJ structure should be mounted on the CMOS structure. Although the memory structure is complicated, small memory devices have already developed and used for certain purposes. Miniaturization of the MRAM of this scheme, however, was hindered by a fact that the current to produce the effective magnetic field to control the magnetization direction increases inversely with decreasing size of the device.

The obstacle has been removed by adopting a novel method to control the magnetization direction, that is, current-induced magnetization switching (CIMS). The current decreases linearly with the size of device in this method. Combined with this method and perpendicular magnetization of ferromagnetic layers, several Mbits of memory has already been fabricated and ready for commercial production. It is noted that the perpendicular

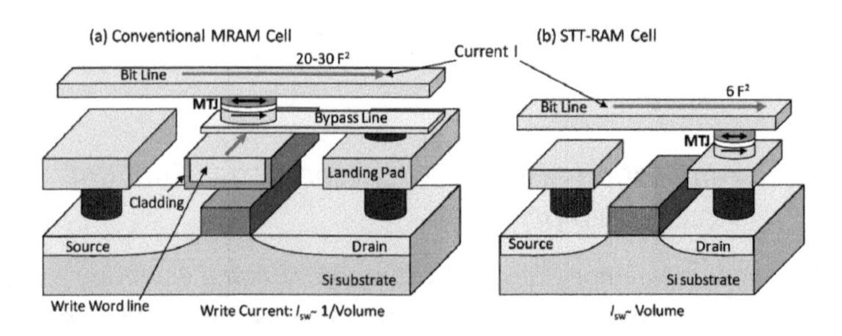


**Figure 5.17** Structures of unit cell of MRAM using domain wall motion [116]. Courtesy of S. Fukami.

magnetization is favorable for both the stability of the magnetization and the switching of magnetization. Because this type of MRAM uses MR, the memory is called magnetoresistive RAM (MRAM). It is also noted that the CIMS is used to control the domain wall. Actually, current control of the domain wall motion is used for MRAM, structures of which are shown in Fig. 5.17 [116–118]. A brief explanation for the CIMS will be given in next section.

In spite of successful fabrication of MRAM, rather high current density is still desirable to control the magnetization direction, and it eventually results in a problem of heat production for high density MRAM. Further research may be desirable to make innovation progress in the field of MRAM.

## 5.4 Current-Induced Magnetization Switching

### 5.4.1 *Spin Transfer Torque*

Electrical current in ferromagnetic metals is usually spin polarized. Although the current itself is conserved, the ratio of the up- and down-spins of the current may not be conserved in general due to presence of spin–orbit interaction and scattering by magnetic impurities. The change in the ratio of up- and down-spin components of the current means a dissipation of spin moments that is a change in the spin angular momentum in time. Because the time dependence of the spin angular momentum produces a torque that may act to change direction of the other magnetization within the ferromagnetic material. In other word, the spin-polarized current

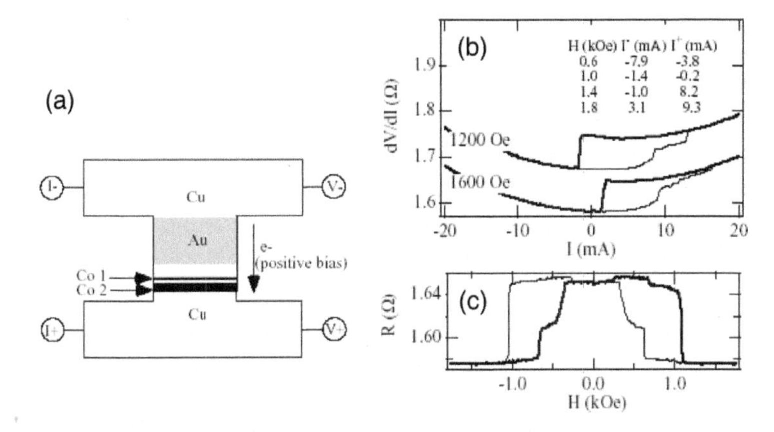


**Figure 5.18** Junction structures used to observed magnetization switching by current in CPP-GMR. (b) Observed results of resistance as a function of current and (c) of applied magnetic field. Reprinted (figure) with permission from Ref. [131], Copyright (2000) by the American Physical Society.

transfers the spin angular momentum to the magnetic moment of the ferromagnet by exerting a torque. The torque is therefore called spin transfer torque (STT) [119, 120].

The CIMS was first observed in CPP-GMR multilayers, as shown in Fig. 5.18a [131]. Figure 5.18c shows the change in resistivity as a function of applied magnetic field. The change shows a hysteresis according to the change in the direction of magnetization on the thinner Co layer. Figure 5.18b shows resistivity change as a function applied current through the junction under certain magnetic field. We find that the magnitude of the resistivity jump observed is the same with that achieved by applying external magnetic field. Therefore, it is concluded that the resistivity change by current is induced by the magnetization switching by the current.

Now, let us consider ferromagnetic junctions made of two FMs, FM1/nonmagnetic metal/FM2, in which FM1 and FM2 have magnetization $\boldsymbol{M}_{\mathrm{p}}$ and $\boldsymbol{M}$, respectively, as shown in Fig. 5.19. Electrons flow from the left to the right electrodes (current flow from the right to the left), and a spin polarization is created by $\boldsymbol{M}_{\mathrm{p}}$ (spin polarizer). The direction of the spin polarization of the current is the same as that of $\boldsymbol{M}_{\mathrm{p}}$. When the two magnetizations are noncolinear, the spins of the current interact with $\boldsymbol{M}$ and the direction of $\boldsymbol{M}$ is

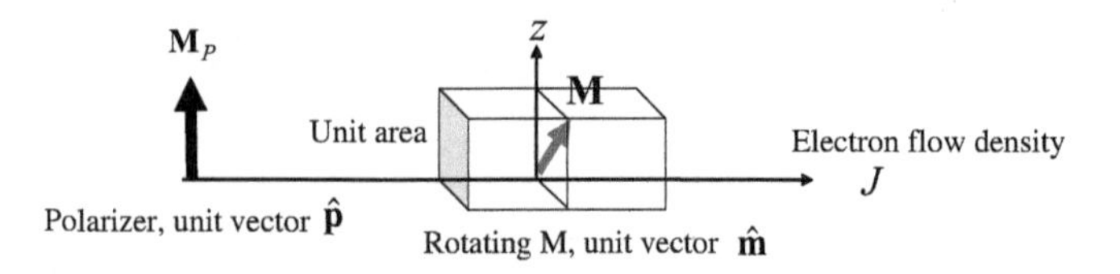


**Figure 5.19** Schematics of magnetization switch of $\boldsymbol{M}$ due to current spin polarized by magnetization $\boldsymbol{M}_{\mathrm{p}}$.

changed. Because the total angular momentum must be conserved, the change in the direction of the spin polarization of the current, that is, the change in the spin angular momentum of the current, must be transferred to the magnetic moment within FM2. The time dependence of the angular momentum is nothing but a torque, and therefore, the change in the spin direction of the current may exert a torque on $\boldsymbol{M}$ resulting in a rotation of $\boldsymbol{M}$.

The magnitude of the spin transfer torque $\boldsymbol{T}$ is estimated as follows way [121]. Let $\boldsymbol{p}$ and $\boldsymbol{m}$ be unit vectors of $\boldsymbol{M}_{\mathrm{p}}$ and $\boldsymbol{M}$, respectively. The current density is given as $J$. The number of electrons that flow through a unit cross section of the junction is then $J/|e|$. These electrons have a spin angular momentum of $(J/|e|)\hbar/2$. Because the direction of the spin polarization of the current changes from $\boldsymbol{p}$ to $\boldsymbol{m}$, the STT acting on $\boldsymbol{M}$ may be given as

$$\boldsymbol{T} = \frac{\hbar}{2|e|} J\,\eta \boldsymbol{m} \times (\boldsymbol{p} \times \boldsymbol{m}), \tag{5.19}$$

where $\eta$ is a parameter representing the spin polarization of the current. The vector $\boldsymbol{T}$ is on a plane spanned by $\boldsymbol{p} \times \boldsymbol{m}$. In general there should exist a component of the torque perpendicular to the plane [122]. These two components may be written as

$$\boldsymbol{T}_{\perp} = g_{\perp} \boldsymbol{m} \times \boldsymbol{p}, \tag{5.20}$$

$$\boldsymbol{T}_{\parallel} = g_{\parallel} \boldsymbol{m} \times (\boldsymbol{p} \times \boldsymbol{m}), \tag{5.21}$$

where $g_{\perp}$ and $g_{\parallel}$ are relevant constants.

The discussion above is based on the fact that the spin-polarized current induces the spin transfer torque. It is quite interesting to note that the temperature gradient produced in the spin-polarized ferromagnet also induces spin transfer torque in FTJs [123].

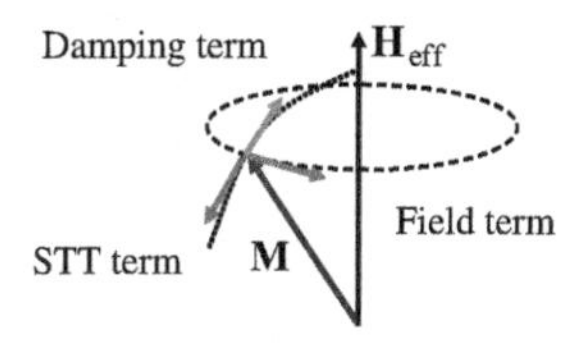


**Figure 5.20** Change in magnetization direction caused by an effective magnetic field and spin transfer torque.

## 5.4.2 *Magnetization Dynamics and Spin Pumping*

The STT may have an influence on dynamics of the magnetization $\boldsymbol{M}$. The magnetization dynamics is analyzed using the Landau–Lifsitz–Gilbert (LLG) equation,

$$\frac{d\boldsymbol{m}}{dt} = -\gamma \boldsymbol{m} \times \boldsymbol{H} + \alpha \boldsymbol{m} \times \frac{d\boldsymbol{m}}{dt}, \tag{5.22}$$

where $\gamma (= g\mu_B)$ is the gyromagnetic ratio and $\alpha$ is the Gilbert damping factor. The first term represents a precession of $\boldsymbol{m}$ around $\boldsymbol{H}$, and the second term gives a damping of direction of $\boldsymbol{m}$ towards $\boldsymbol{H}$. Because the first term is caused by the magnetic field and the second term represents the damping, they are called field term and damping term, respectively.

When a STT is exerted on $\boldsymbol{m}$, $\boldsymbol{T}$ is added on the right-hand side of the LLG equation. The $\boldsymbol{T}_{\parallel}$ term is usually called *torque term* because it exerts a torque on $\boldsymbol{m}$. The $\boldsymbol{T}_{\perp}$ term is called *field term* because $\boldsymbol{p}$ may be considered an effective field caused by magnetization $\boldsymbol{M}_{\mathrm{p}}$ as shown in Fig. 5.20. The most important point is that the sign of $\boldsymbol{T}_{\parallel}$ term can be either positive or negative because it proportional to the current. When the current direction is reversed, the direction of $\boldsymbol{T}_{\parallel}$ is also reversed. This property is the one to use for magnetization reversal by a current. Thus the LLG equation may be generalized as

$$\frac{d\boldsymbol{m}}{dt} = -\gamma \boldsymbol{m} \times \boldsymbol{H} + \alpha \boldsymbol{m} \times \frac{d\boldsymbol{m}}{dt} - \frac{\beta}{M} \boldsymbol{m} \times (\boldsymbol{p} \times \boldsymbol{m})$$

Now, let us consider the physical meaning of the Gilbert damping factor $\alpha$. The factor plays a role in aligning the magnetization in the direction of magnetic field. Because the change in the magnetization direction is nothing but a change in the spin angular momentum, dissipation of the angular momentum occurs in the ferromagnet

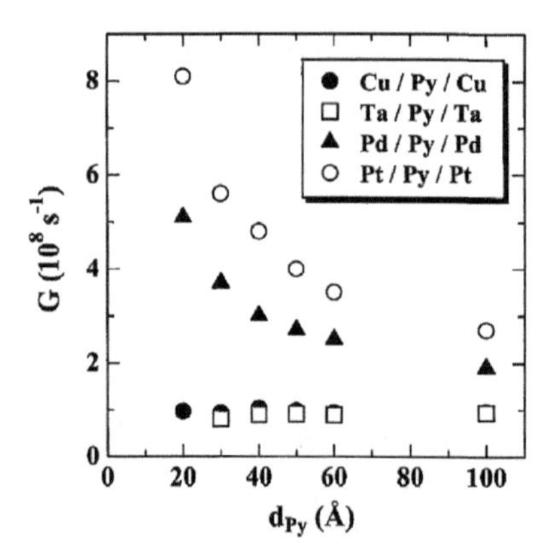


**Figure 5.21** Experimental results of Gilbert damping factor. Reprinted from Ref. [124], Copyright (2001), with permission from Elsevier.

via, for example, the SOI. The value of $\alpha$ thus depends on the ferromagnet species. In magnetic multilayers or ferromagnet/nonmagnet junctions, it also depends on the nonmagnet species. Experimental evidence for the dependence of $\alpha$ on paramagnetic metals is shown in Fig. 5.21 [124]. The value of $\alpha$ depends on species of the paramagnetic metals; it is small for Cu and large for Pt and Pd. The higher value for the latter may be attributed to the large SOI in heavy elements.

The fact that the value of $\alpha$ depends on not only the ferromagnetic metal, but also the nonmagnetic metals indicates that the dissipation of the spin angular momentum is influenced by the nonmagnetic metal. The wave functions of the electrons in the ferromagnetic metal extend into the nonmagnetic metal that is in contact with the ferromagnet. As a result, the effects of the precession of the magnetization of the ferromagnet influence the spin state of the nonmagnetic metal. A change in the spin state of the nonmagnet in turn influences the magnetization dynamics of the ferromagnet, resulting in the dependence of $\alpha$ on the nonmagnetic metal.

One may estimate the spin diffusion length of metals in the measurements of values of $\alpha$. Values of $\alpha$ at room temperature have been estimated 240, 2, 170, and 34 nm for Cu, Ru, Ag, and Au, respectively [125–127].

In STT, the change in the spin current produces magnetization rotation, while in spin pumping, the magnetization rotation creates the spin current. Various aspects of the spin current have been reported in reviews, for example, by Brataas et al. [130].

## 5.5 Spin FET

### 5.5.1 *Proposal of a Spin FET*

As mentioned in Chapter 1, the modern electronics devices are based on CMOS FET in which the current is controlled by gate voltage. Electrode of these FETs are made of nonmagnetic metals. Recently there proposed junction in which the electrodes are made of ferromagnetic metals. A schematic is shown in Fig. 5.22. Using this structure, in addition to the control of current, a role of memory may be added to the junction because both P and AP alignments of magnetization of the electrodes are possible. This type of FET is called spin FET.

The metal-oxide semiconductor (MOS) structure often forms a state of two-dimensional electron gas (2DEG) due to a Schottky barrier at the interface of semiconductor and oxide. Because the electron density varies within the Schottky barrier region, a spin–orbit interaction called Rashba spin–orbit interaction appears due to nonzero gradient of electric field. Spins of electrons injected into 2DEG precess as they transmit within the 2DEG region, and the precession is controlled by gate voltage to match the spin state of electrons injected at the source with the spin state of the drain. This kind of spin FET is called Datta–Das spin FET after their proposal in 1990 [132]. In this spin FET, high coherence of conduction electron

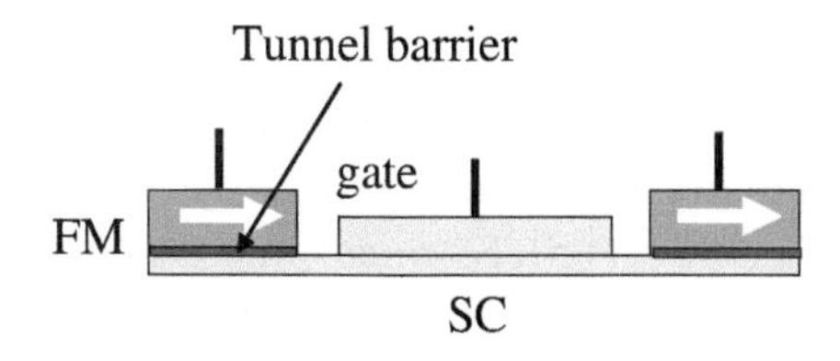


**Figure 5.22** Schematics of device structure of a spin FET.

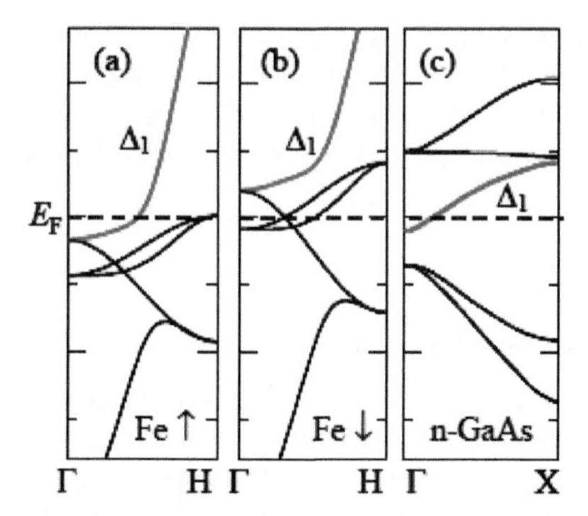


**Figure 5.23** Electronic structure of up- and down-spin states of Fe and that of electron-doped (n-type) GaAs.

spins is desirable. Sugawara and Tanaka proposed spin FET in which the ferromagnetic electrodes are both half-metals [133]. Using half-metals as electrodes, efficient spin injection and detection could be realized. Nearly perfect spin polarization of the half-metal, however, is required. Flatte and Vignale have proposed a unipolar spin diodes and transistors.

In general, there are three basic issues to bc solved to realize the spin FET, spin injection from ferromagnet to semiconductor, spin control in semiconductor and spin detection at the other ferromagnet [134]. Although several concrete structures of such junctions have been proposed and studied experimentally, no actual device has yet been realized. Theoretically, however, the first-principles calculates predicted MR effect for FM/semiconductor/FM junctions in the ballistic transport regime. Figure 5.23 shows electronic structure calculated for Fe and GaAs. The up-spin electronic state at the Fermi energy of Fe is composed of $\Delta_1$ symmetry, while the down-spin state is not. Because the conduction state of GaAs is also of $\Delta_1$ symmetry, the matching of the symmetry is good for up-spin electrons as in Fe/MgO/Fe tunnel junctions. Therefore, high MR ratio should be realized in Fe/GaAs/Fe junctions. Although SOI may reduce the MR ratio, the first-principles calculations give sufficiently high MR ratios [135–137].

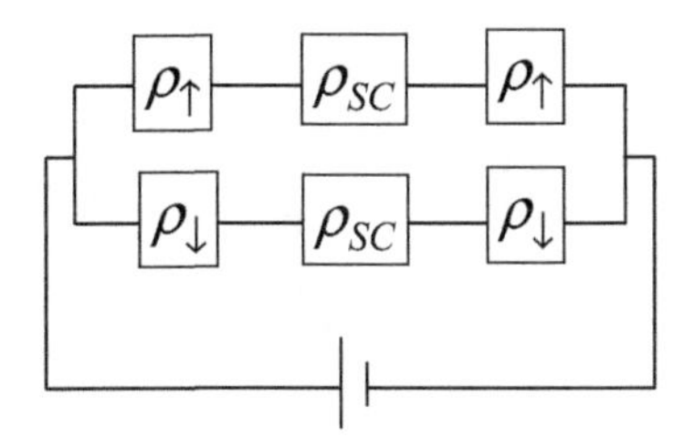


**Figure 5.24** Circuit model to explain the conductivity mismatch in FM/semiconductor/FM junctions.

Nevertheless, successful MR effect has not yet observed, a reason of which has been presented in a concept of "conductivity mismatch" [138], as explained below.

## 5.5.2 *Conductivity Mismatch*

Let us consider an FM/semiconductor/FM junction without SOI. Because of no SOI, the circuit model of the electrical current is given by Fig. 5.24 for P alignment of magnetizations of FM electrodes. The spin polarization defined as

$$\alpha = \frac{I_\uparrow - I_\downarrow}{I_\uparrow + I_\downarrow}$$

may be easily calculated to obtain

$$\alpha = \beta \frac{\rho_{\rm FM}}{\rho_{\rm SC}} \frac{2}{2\frac{\rho_{\rm FM}}{\rho_{\rm SC}} + 1 - \beta^2},$$

where $\beta$ indicates spin polarization of ferromagnetic metal, which is about 0.5, and $\rho_{\rm FM}$ and $\rho_{\rm SC}$ are resistivity of FM and semiconductor, respectively. In the derivation of the formula for $\alpha$, following relations have been used:

$$V = R_\uparrow I_\uparrow = R_\downarrow I_\downarrow,$$
$$R_{\uparrow(\downarrow)} = 2\rho_{\uparrow(\downarrow)} + \rho_{\rm SC},$$
$$\rho_{\uparrow(\downarrow)} = 2\rho_{\rm FM}/\left(1 + (-)\beta\right).$$

Because $\rho_{\rm FM}/\rho_{\rm SC}$ is about $10^{-4}$ in FM/SC/FM junctions, the value of $\alpha$ is nearly zero except for $\beta \approx 1$. Because in the AP alignment, $I_\uparrow = I_\downarrow$, and $\alpha = 0$. Therefore no difference appears in the value

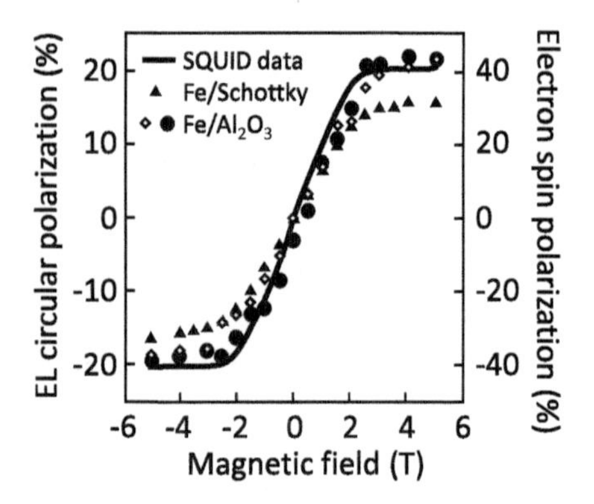


**Figure 5.25** Observed results of electroluminescence circular polarization as a function of magnetic field for Fe/semiconductor junctions. Reproduced with permission from Ref. [143]. Copyright 2004, AIP Publishing LLC.

of $\alpha$ by changing the alignment of magnetizations. Shortly speaking, the conductivity (or resistivity) mismatch between FM and SC is the source of no MR effect in FM/SC/FM junctions.

Immediately after this indication of conductivity mismatch, it has been proposed that this basic obstacle could be removed by inserting an insulating layer between FM and SC layers [139, 140]. Therefore, junctions used later experiments include insulating barrier such as $Al_2O_3$ and MgO between FM and SC layers. When the resistivity of FM matches well with that of SC, there appears no problem, for example, junctions with electrodes made of ferromagnetic semiconductors (GaMn)As show high MR ratios. However, the effect is restricted at low temperature because of low Curie temperature of (GaMn)As.

Fert–Jaffres [140] suggested that suitable interface resistance is necessary for effective spin injection into SC from FM. The spins of injected electrons loose quickly its coherence due to spin–orbit interaction within SC layer. Therefore, when the interface resistance is too high, no sufficient number of electrons is detected at the drain FM, on the other hand, the interface resistance is low, the problem of the conductivity mismatch occurs. Thus they suggested the MR effects on FM/SC/FM appears only for intermediate Interface resistance. In the following we show some experimental results of spin injection and spin detection.

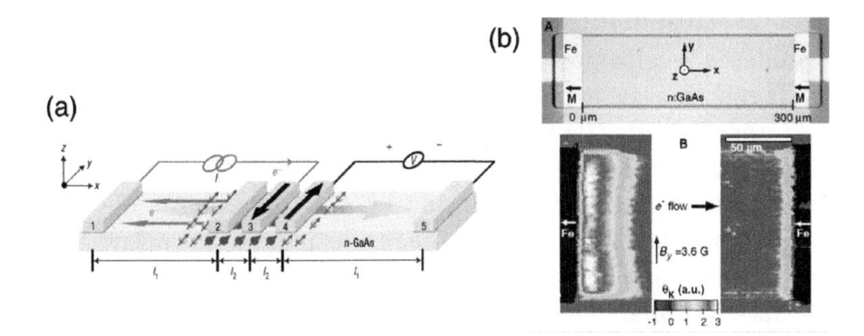


**Figure 5.26** (a) Device structure of nonlocal measurement to detect spin injection into semiconductor. Reprinted by permission from Macmillan Publishers Ltd: [*Nature Physics*] (Ref. [144]), copyright (2007). (b) Imaging of injected spins in Fe/GaAs/Fe junctions obtained by the Kerr effect. From Ref. [145]. Reprinted with permission from AAAS.

## 5.5.3 *Some Experiments of Spin Injection and Detection*

Early experiments for the spin injection into SC from FM have been done by using GaAs for SC to confirm the spin injection by light emission form GaAs or AlGaAs [141, 142]. Because of the spin–orbit interaction of these SCs with direct energy gap, selection

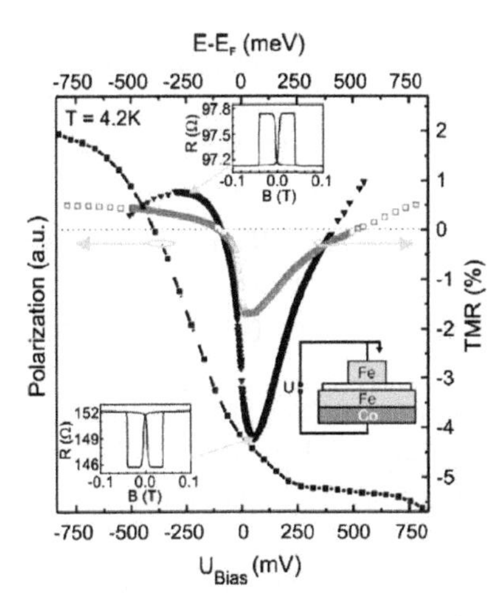


**Figure 5.27** Spin polarization observed in Fe/GaAs/Fe junctions with applied bias voltage. Reproduced with permission from Ref. [146]. Copyright 2006, AIP Publishing LLC.

rule of electron transition between occupied and unoccupied states produces polarization of emitted light. The observed polarization was at most a few percent.

van't Erve et al. [143] observed circular polarization of emitted light as a function of external magnetic field for both Fe/AlGaAs/*p*-GaAs with a Schottky barrier and Fe/Al-O/p-GaAs. The observed results are shown in Fig. 5.25. The results indicate the polarization of light is about 20%, indicating sufficient spin injection from Fe to semiconductors.

Figure 5.26a shows a setup of nonlocal measurements of spin injection [144], and Fig. 5.26b shows an imaging of injected spins using Kerr effect for Fe/GaAs/Fe [145]. These results show a negative spin polarization of injected spins. Figure 5.27 shows results of MR measurement for Fe/GaAs/Fe junction [146]. The results indicate negative MR, that is, the spin polarization of injected spins is negative. The negative spin polarization has been interpreted in terms of effects of interface resonant states appearing in the Schottky barrier of GaAs.

## References

1. T. Shinjo and T. Takada, *Metallic Superlattices*, Elsevier, Amsterdam (1987).
2. P. M. Levy, *Solid State Phys.*, **47**, 367 (1994).
3. M. A. M. Gijs and G. E. W. Bauer, *Adv. Phys.*, **46**, 285 (1997).
4. U. Hartmann (Ed.), *Magnetic Multilayers and Giant Magnetoresistance*, Springer-Verlag, Berlin (1999).
5. M. Ziese and M. J. Thornton (Eds.), *Spin Electronics*, Spinger-Verlag, Berlin, (2001).
6. S. Maekawa and T. Shinjo (Eds.), *Spin Dependent Transport in Magnetic Nanostructures*, Taylor and Francis, New York, (2002).
7. D. L. Mills and J. A. C. Bland (Eds.), *Nanomagnetism*, Elsevier, Amsterdam (2006).
8. Y. B. Xu and S. M. Thompson (Eds.), *Spintronic Materials and Technology*, Taylor and Francis, New York (2006).
9. J. Bass and W. P. Pratt Jr., *J. Phys.: Condens. Matter*, **19**, 183201 (2007).
10. E. Y. Tsymbal and D. G. Pettifor, *Solid State Phys.*, **56**, 113 (2001).

11. T. Shinjo (Ed.), *Nanomagnetism and Spintronics*, Elsevier, Amsterdam (2009).
12. P. Grüberg, R. Schreiber, Y. Pang, M. B. Brodsky, and H. Sowers, *Phys. Rev. Lett.*, **57**, 2442 (1986).
13. M. N. Baibich, J. M. Broto, A. Fert, F. Nguyen Van Dau, F. Petroff, P. Etienna, G. Creuzet, A. Friederich, and J. Chazelas, *Phys. Rev. Lett.*, **61**, 2472 (1988).
14. S. S. P. Parkin, N. More, and K. P. Roche, *Phys. Rev. Lett.*, **64**, 2304 (1990).
15. S. S. P. Parkin and D. Mauri, *Phys. Rev. B*, **44**, 7131 (1991-I).
16. B. Dieny, V. S. Sperious, S. S. P. Parkin, B. A. Gurney, D. R. Wilhoit, and D. Mauri, *Phys. Rev. B*, **43**, 1297 (1991).
17. B. Dieny, V. S. Speriosu, S. Metin, S. S. P. Parkin, B. A. Gurney, P. Baumgart, and D. R. Wilhoit, *J. Appl. Phys.*, **69**, 4774 (1991).
18. T. Miyazaki, and N. Tezuka, *J. Magn. Magn. Mater.*, **151**, 403-410 (1995).
19. J. S. Moodera, L. R. Kinder, T. M. Wong, and R. Meservey, *Phys. Rev. Lett.*, **74**, 3273 (1995).
20. M. Julliere, *Phys. Lett.*, **54A**, 225 (1975).
21. S. Maekawa and U. Gäfvert, *IEEE Trans. Magn.*, **18**, 707 (1982).
22. S. Yuasa, A. Fukushima, T. Nagahama, K. Ando, and Y. Suzuki, *Jpn. J. Appl. Phys.*, **43**, (2004).
23. S. S. P. Parkin, C. Kaiser, A. Panchula, P. M. Rice, B. Hughes, M. Samant, and S.-H. Yang, *Nat. Mater.*, **3**, 862 (2004).
24. S. Yuasa, T. Nagahama, A. Fukushima, Y. Suzuki, and K. Ando, *Nat. Mater.*, **3**, 868-871 (2004).
25. G. Binasch, P. Grüberg, F. Saurenbach, and W. Zinn, *Phys. Rev. B*, **39**, 4828 (1989).
26. J. Barnas, A. Fuss, R. E. Camley, P. Grüberg, and W. Zinn, *Phys. Rev. B*, **42**, 8110 (1990).
27. A. Barthélémy, A. Fert, M. N. Baibich, S. Hadjoudj, and F. Petroff, *J. Appl. Phys.*, **67**, 5908 (1990).
28. D. H. Mosca, F. Petroff, A. Fert, P. A. Schroeder, W. P. Pratt, and R. Laloee, *J. Magn. Magn. Mater.*, **94**, L1–L5 (1991).
29. F. Petroff, A. Barthélémy, A. Hamzić, A. Fert, P. Etienne, S. Lequien, and G. Creuzet, *J. Magn. Magn. Mater.*, **93**, 95–100 (1991).
30. R. Schad, C. D. Potter, P. Beliën, G. Verbanck, V. V. Moshchalkov, and Y. Bruynseraede, *Appl. Phys. Lett.*, **64**, 3500 (1994).
31. S. S. P. Parkin and D. Mauri, *Phys. Rev. B*, **44**, 7131 (1991).
32. S. S. P. Parkin, *Phys. Rev. Lett.*, **71**, 1641 (1993).

33. F. Petroff, A. Barthélémy, D. H. Mosca, D. K. Lottis, A. Fert, P. A. Schroeder, W. P. Pratt, Jr., R. Loloee, and S. Lequien, *Phys. Rev. B*, **44**, 5355 (1991).
34. S. T. Purcell, W. Folkerts, M. T. Jonson, N. W. E. Mcgee, K. Jager, J. aan de Stegge, W. B. Zeper, W. Hoving, and P. Grüberg, *Phys. Rev. Lett.*, **67**, 903 (1991).
35. Z. Q. Qiu, J. Pearson, A. Berger, and S. D. Bader, *Phys. Rev. Lett.*, **68**, 1398 (1992).
36. J. Unguris, R. J. Celotta, and D. T. Pierce, *Phys. Rev. Lett.*, **67**, 140 (1991).
37. J. Unguris, R. J. Celotta, and D. T. Pierce, *Phys. Rev. Lett.*, **69**, 1125 (1992).
38. E. E. Fullerton, S. D. Bader, and J. L. Robertson, *Phys. Rev. Lett.*, **77**, 1382 (1996).
39. Y. Wang, P. M. Levy, and J. L. Fry, *Phys. Rev. Lett.*, **65**, 2732 (1990).
40. P. Bruno, and C. Chappert, *Phys. Rev. Lett.*, **67**, 1602 (1991).
41. D. M. Edwards, J. Mathon, R. B. Muniz, and M. S. Phan, *Phys. Rev. Lett.*, **67**, 493 (1991).
42. M. van Schilfgaarde, and W. A. Harrison, *Phys. Rev. Lett.*, **71**, 3870 (1993).
43. J. Mathon, M. Villeret, R. B. Muniz, J. d'Albuquerque e Castro, and D. M. Edwards, *Phys. Rev. Lett.*, **74**, 3696 (1995).
44. P. Bruno, *Phys. Rev. B*, **52**, 411 (1995).
45. P. Bruno, J. Kudrnovský, V. Drchal, and I. Turek, *Phys. Rev. Lett.*, **76**, 4254 (1996).
46. T. Shinjo, and H. Yamamoto, *J. Phys. Soc. Jpn.*, **59**, 3061 (1990).
47. H. Yamamoto, T. Okuyama, H. Dohnomae, and T. Shinjo, *J. Magn. Magn. Mater.*, **99**, 243–252 (1991).
48. W. P. Pratt, Jr., S.-F. Lee, J. M. Slaughter, R. Loloee, P. A. Schroeder, and J. Bass, *Phys. Rev. Lett.*, **66** (1991).
49. Q. Yang, P. Holody, S.-F. Lee, L. L. Henry, R. Loloee, P. A. Schroeder, W. P. Pratt, Jr., and J. Bass, *Phys. Rev. Lett.*, **72**, 3274 (1994).
50. J. Bass and W. P. Pratt, Jr., *J. Magn. Magn. Mater.*, **200**, 274 (1999).
51. M. A. M. Gijs, S. K. J. Lenczowski, and J. B. Giesbers, *Phys. Rev. Lett.*, **70**, 3343 (1993).
52. M. A. M. Gijs, S. K. J. Lenczowski, R. J. M. van de Veerdonk, J. B. Giesbers, M. T. Johnson, and J. B. Faan de Stegge, *Phys. Rev. B*, **50**, 16733 (1994).
53. L. Piraux, J. M. George, J. F. Despres, C. Leroy, E. Ferain, R. Legras, K. Ounadjela, and A. Fert, *Appl. Phys. Lett.*, **65**, 2484 (1994).
54. A. E. Berkowitz, J. R. Mitchell, M. J. Carey, A. P. Yong, S. Zhang, F. E. Spada, F. T. Parker, A. Hutten, and G. Thomas, *Phys. Rev. Lett.*, **68**, 3745 (1992).

55. J. Q. Xiao, J. S. Jiang, and C. L. Chien, *Phys. Rev. Lett.*, **68**, 3749 (1992).
56. J. Q. Xiao, J. S. Jiang, and C. L. Chien, *Phys. Rev. B*, **46**, 9266 (1992).
57. J. A. Barnard, A. Waknis, M. Tan, E. Haftek, M. R. Parker, and M. L. Watson, *J. Magn. Magn. Mater.*, **114**, L230–L234 (1992).
58. M. Rubinstein, *Phys. Rev. B*, **50**, 3830 (1994).
59. J.-Q. Wang, and G. Xiao, *Phys. Rev. B*, **50**, 3423 (1994).
60. B. J. Hickey, M. A. Howson, S. O. Musa, and N. Wiser, *Phys. Rev. B*, **51**, 667 (1995).
61. N. F. Mott, Proc. R. Soc. London Ser. A, **153**, 699 (1936); **156**, 368 (1936).
62. A. Fert, and I. A. Cambell, *J. Phys. F: Metal Phys.*, **6**, 849 (1976).
63. I. A. Campbell and A. Fert, *Ferromagnetic Materials*, Vol. 3, p. 747, Ed., E. P. Wohlfarth, North-Holland (1982).
64. J. Inoue, A. Oguri, and S. Maekawa, *J. Phys. Soc. Jpn.*, **60**, 376 (1991).
65. J. Inoue, H. Itoh, and S. Maekawa, *J. Phys. Soc. Jpn.*, **61**, 1149 (1992).
66. J. Inoue and S. Maekawa, *J. Magn. Magn. Mater.*, **127**, L249 (1993).
67. H. Itoh, J. Inoue and S. Maekawa, *Phys. Rev. B*, **47**, 5809 (1993).
68. G. E. W. Bauer, *Phys. Rev. Lett.*, **69**, 1676 (1992).
69. G. E. W. Bauer, A. Brataas, K. M. Schep, and P. J. Kelly, *J. Appl. Phys.*, **75**, 6704 (1994).
70. K. M. Schep, P. J. Kelly, and G. E. W. Bauer, *Phys. Rev. Lett.*, **74**, 586 (1995).
71. R. E. Camley, and J. Barnaś, *Phys. Rev. Lett.*, **63**, 664 (1989).
72. Y. Asano, A. Oguri, and S. Maekawa, *Phys. Rev. B*, **48**, 6192 (1993).
73. Y. Asano, A. Oguri, J. Inoue, and S. Maekawa, *Phys. Rev. B*, **49**, 12831 (1994).
74. H. Itoh, J. Inoue, and S. Maekawa, *Phys. Rev. B*, **51**, 342 (1995).
75. K. M. Schep, P. J. Kelly, G. E. W. Bauer, *J. Magn. Magn. Mater.*, **156** 385 (1996).
76. K. M. Schep, J. B. A. N. van Hoof, P. J. Kelly, G. E. W. Bauer, and J. E. Inglesfield, *Phys. Rev. B*, **56** 10805 (1997).
77. K. M. Schep, PhD thesis.
78. T. Valet, and A. Fert, *Phys. Rev. B*, **48**, 7099 (1993).
79. P. Zahn, I. Mertig, M. Richter, and H. Eschring, *Phys. Rev. Lett.*, **75**, 2996 (1995).
80. B. Dieny, *Europhys. Lett.*, **17**, 261 (1992).
81. B. A. Gurney, V. S. Speriosu, J. P. Nozieres, H. Lefakis, D. R. Wilhoit, and O. U. Need, *Phys. Rev. Lett.*, **71**, 4023 (1993).

82. V. S. Speriosu, J. P. Nozieres, B. A. Gurney, B. Dieny, T. C. Huang, and H. Lefakis, *Phys. Rev. B*, **47**, 11579 (1993).
83. T. Miyazaki, and N. Tezuka, *J. Magn. Magn. Mater.*, **151**, 403–410 (1995).
84. H. Fujimori, S. Mitani, and S. Ohnuma, *Mater. Sci. Eng. B*, **31**, 219 (1995).
85. S. Mitani et al., *Phys. Rev. Lett.*, **81**, 2799 (1998).
86. Y. M. Lee, J. Hayakawa, S. Ikeda, F. Matsukura, and H. Ohno, *Appl. Phys. Lett.*, **90**, 212507 (2007).
87. T. Marukame, T. Ishikawa, S. Hatamata, K. Matsuda, T. Uemura, and M. Yamamoto, *Appl. Phys. Lett.*, **90**, 012508 (2007).
88. N. Tezuka, N. Ikeda, S. Sugimoto, and K. Inomata, *Jpn. J. Appl. Phys.*, **46**, L454 (2007).
89. T. Ishikawa, S. Hakamata, K. Matsuda, T. Uemura, and M. Yamamoto, *J. Appl. Phys.*, **103**, 07A919 (2008).
90. J. Bardeen, *Phys. Rev. Lett.*, **6**, 57 (1961).
91. P. M. Tedrow, and R. Meservey, *Phys. Rev. Lett.*, **26**, 192 (1971).
92. R. J. Soulen Jr., J. M. Byers, M. S. Osofsky, B. Nadgorny, T. Ambrose, S. F. Cheng, P. R. Broussard, C. T. Tanaka, J. Nowak, J. S. Moodera, A. Barry, and J. M. D. Coey, *Science*, **282**, 85 (1998).
93. S. K. Upadhyay, A. Palanisami, R. N. Louie, and R. A. Buhrman, *Phys. Rev. Lett.*, **81**, 3247 (1998).
94. S. K. Upadhyay, R. N. Louie, and R. A. Buhrman, *Appl. Phys. Lett.*, **74**, 3881 (1999).
95. D. C. Worledge, and T. H. Geballe, *Phys. Rev. Lett.*, **85**, 5182 (2000).
96. C. Kaiser, S. van Dijken, S.-H. Yang, H. Yang, and S. S. P. Parkin, *Phys. Rev. Lett.*, **94**, 247203 (2005).
97. C. Kaiser, A. F. Panchula, and S. S. P. Parkin, *Phys. Rev. Lett.*, **95**, 047202 (2005).
98. J. Mathon and A. Umerski, *Phys. Rev. B*, **63**, 220403 (2001).
99. W. H. Butler, X.-G. Zhang, T. C. Schulthess, and J. M. MacLaren., *Phys. Rev. B*, **63**, 54416 (2001).
100. S. Yuasa, T. Sato, E. Tamura, Y. Suzuki, H. Yamamori, K. Ando, and T. Katayama, *Europhys. Lett.*, **52**, 344–350 (2000).
101. T. Nagahama, S. Yuasa, Y. Suzuki, and E. Tamura, *Appl. Phys. Lett.*, **79** 4381 (2001).
102. D. D. Djayaprawira, K. Tsunekawa, M. Nagai, H. Maehara, S. Yamagata, N. Watanabe, S. Yuasa, Y. Suzuki, and K. Ando, *Appl. Phys. Lett.*, **86**, 092502 (2005).

103. R. A. de Groot, F. M. Mueller, P. G. van Engen, and K. H. J. Buschow, *Phys. Rev. Lett.*, **50**, 2024 (1983).
104. R. A. de Groot and K. H. J. Buschow, *J. Magn. Magn. Mater.*, **54–57**, 1377 (1986).
105. Y. Sakuraba, J. Nakata, M. Oogane, H. Kubota, Y. Ando, A. Sakuma, and T. Miyazaki, *Jpn. J. Appl. Phys.*, **44**, L1100 (2005).
106. Y. Sakuraba, M. Hattori, M. Oogane, H. Kubota, Y. Ando, A. Sakuma, and T. Miyazaki, *J. Phys. D: Appl. Phys.*, **40**, 1221 (2006).
107. Y. Sakuraba, M. Hattori, M. Oogane, Y. Ando, H. Kato, A. Sakuma, and T. Miyazaki, and H. Kubota, *Appl. Phys. Lett.*, **88**, 192508 (2006).
108. J. Hayakawa, S. Ikeda, Y. M. Lee, F. Matsukura, and H. Ohno, *Appl. Phys. Lett.*, **89**, 232510 (2006).
109. Y. M. Lee, J. Hayakawa, S. Ikeda, F. Matsukura, and H. Ohno, *Appl. Phys. Lett.*, **89**, 042506 (2006).
110. J. Inoue and S. Maekawa, *Phys. Rev. B*, **53**, R11927 (1996).
111. J. M. MacLaren, X.-G. Zhang, W. H. Butler, and X. Wang, *Phys. Rev. B*, **59**, 5470 (1999).
112. H. Itoh, T. Kumazaki, J. Inoue and S. Maekawa, *Jpn. J. Appl. Phys.*, **37**, 5554 (1998).
113. H. Itoh, A. Shibata, T. Kumazaki, J. Inoue, and S. Maekawa, *J. Phys. Soc. Jpn.*, **69**, 1632 (1999).
114. H. Itoh and J. Inoue, *Trans. Mag. Soc. Jpn.*, **30**, 1 (2006).
115. H. Itoh, *J. Phys. D: Appl. Phys.*, **40**, 1228 (2006).
116. S. Fukami, PhD thesis, Nagoya University (2011).
117. S. Fukami, M. Ymanouchi, S. Ikeda, and H. Ohno, *IEEE Trans. Magn.*, **50**, 3401006 (2014).
118. S. Fukami, J. Ieda, and H. Ohno, *Phys. Rev. B*, **91**, 235401 (2015).
119. L. Berger, *J. Appl. Phys.*, **71**, 2721 (1992).
120. J. C. Slonczewski, *J. Magn. Magn. Mater.*, **159**, L1 (1996).
121. J. Z. Sun, *J. Magn. Magn. Mater.*, **202**, 157 (1999).
122. D. M. Edwards, F. Federici, J. Mathon, and A. Umerski, *Phys. Rev. B*, **71**, 054407 (2005).
123. M. Hatami, G. E. W. Bauer, Q. Zhang, and P. Kelly, *Phys. Rev. Lett.*, **99**, 066603 (2007).
124. S. Mizukami, Y. Ando, and T. Miyazaki, *J. Magn. Magn. Mater.*, **226–230**, 1640 (2001).
125. S. Yakata, Y. Ando, T. Mizukami, and T. Miyazaki, *Jpn. J. Appl. Phys.*, **45**, 3892 (2006).

126. Th. Gerrits, M. L. Schneider, and T. J. Silva, *J. Appl. Phys.*, **99**, 023901 (2006).
127. B. Kardasz, O. Mosendz, B. Heinrich, Z. Liu, and M. Freeman, *J. Appl. Phys.*, **103**, 07C509 (2008).
128. Y. Tserkovnyak, A. Brataas, G. E. W. Bauer, and B. I. Halperin, *Rev. Mod. Phys.*, **77**, 1375 (2005).
129. A. Brataas, G. E. W. Bauer, and P. J. Kelly, *Phys. Rep.*, **427**, 157 (2006).
130. A. Brataas, A. D. Kent, and H. Ohno, *Nat. Mater.*, **11**, 372 (2012).
131. J. A. Katine, F. J. Albert, R. A. Buhman, E. B. Myers, and D. C. Ralph, *Phys. Rev. Lett.* **84**, 3149 (2000).
132. S. Datta and B. Das, *Appl. Phys. Lett.*, **56**, 665 (1990).
133. S. Sugahara and M. Tanaka, *Appl. Phys. Lett.*, **84**, 2307 (2004).
134. I. Žutić, J. Fabian, S. Das Sarma, *Rev. Mod. Phys.*, **76**, 323 (2004).
135. O. Wunnicke, Ph Mavropoulos, R. Zeller, and P. H. Dederichs, *Phys. Rev. B*, **65**, 241306 (2002),
136. O. Wunnicke, Ph Mavropoulos, R. Zeller, and P. H. Dederichs, *J. Phys. Condens. Matter*, **16**, 4643 (2004).
137. S. Honda, H. Itoh, J. Inoue, H. Kurebayashi, T. Trypiniotis, C. H. W. Barnes, A. Hirohata, and J. A. C. Bland, *Phys. Rev. B*, **78**, 245316 (2008).
138. G. Schmidt, D. Ferrand, L. W. Molenkamp, A. T. Filip, and B. J. van Wees, *Phys. Rev. B*, **62**, R4790 (2000).
139. E. I. Rashba, *Phys. Rev. B*, **62**, R16267 (2000).
140. A. Fert, H. Jaffres, *Phys. Rev. B*, **64**, 184420 (2001).
141. Y. Ohno, D. K. Young, B. Beschoten, F. Matsukura, H. Ohno, and D. D. Awschalom, *Nature*, **402**, 790 (1999).
142. R. Fiederling, M. Keim. G. Reuscher, W. Ossau, G. Schmidt, A. Waag, and L. W. Molenkamp, *Nature*, **402**, 787 (1999).
143. O. M. van't Erve, G. Kioseoglou, A. T. Hanbicki, C. H. Li, B. T. Jonker, R. Mallory, M. Yasar, and A. Petrou, *Appl. Phys. Lett.*, **84**, 4334 (2004).
144. X. Lou, C. Adelmann, S. A. Crooker, E. S Garlid, J. Zhang, K. S. M. Reddy, S. D. Flexner, C. J. Palmstrom, and P. A. Crowell, *Nature Phys.*, **3**, 197 (2007).
145. S. A. Crooker, M. Furis, X. Lou, C. Adelmann, D. L. Smith, C. J. Palmstrøm, and P. A. Crowell, *Science*, **309**, 2191 (2005).
146. J. Moser, M. Zenger, C. Gerl, D. Schuh, R. Meier, P. Chen, G. Bayreuther, W. Wegscheider, D. Weiss, C.-H. Lai, R.-T. Huang, M. Kosuth, and H. Ebert, *Appl. Phys. Lett.*, **89**, 162106 (2006).

# Chapter 6

# Magnetoresistive Graphene Junctions: Realistic Models

So far we have reviewed various aspects of graphene junctions with nonmagnetic electrodes and magnetoresistive junctions with magnetic electrodes. In electronics and/or spintronics applications of graphene, detailed study on the transport phenomena of graphene with electrodes is inevitable. In this chapter we study the magnetoresistive properties of graphene junctions with magnetic electrodes, typically with iron and nickel electrodes, by using the same numerical method applied to investigate the GMR and TMR. We will show that the junctions produce characteristic feature of magnetoresistance (MR) distinct from GMR and TMR. Because of the translational symmetry along the current direction is broken, we start with model calculations for junctions using the single-orbital tight-binding (TB) model, the results of which may give an intuitive understanding of the characteristic properties of the MR, and the model is easy to include the effects of disorder. We then proceed into the realistic junctions using the full-orbital TB model. Before going to the calculated result we present a short introduction to the subject of MR of graphene junctions.

*Graphene in Spintronics: Fundamentals and Applications*
Jun-ichiro Inoue, Ai Yamakage, and Shuta Honda

ISBN 978-981-4669-56-6 (Hardcover), 978-981-4669-57-3 (eBook)
www.panstanford.com

## 6.1 Introduction

The magnetoresistance (MR) in ferromagnetic junctions is one of the most interesting phenomena for spintronics devices. The junctions basically consist of ferromagnetic electrodes, for which transition metal ferromagnets are used, and nonmagnetic spacer inserted between two electrodes. Various materials are used for the, nonmagnetic metals such as Cu, insulators MgO and $Al_2O_3$, etc., semiconductors such as Si and GaAs, and organic materials, including polymers. Because of the variety of the electronic structures of these materials, the feature of MR depends rather crucially on the spacer materials. The typical difference appears in junctions with nonmagnetic spacers and with insulating spacers: the MR of the former is called giant magnetoresistance (GMR) and that of the latter is called tunnel magnetoresistance (TMR). On the other hand, no successful result of MR has been achieved for ferromagnetic junctions with semiconductor spacers, as mentioned in the previous chapter. The graphene may be one of the organic materials and could be used for spacer materials for ferromagnetic junctions. The characteristic feature (Dirac cone or zero-gap semiconductor) of the electronic states of graphene near the Fermi level is much more different from the electronic states of metals and insulators. Therefore, it is interesting to study how the conductance and MR behave in metal/graphene/metal junctions and what is the role of the Dirac points (DPs). Effects of the interface and those of disorder in the junction may affect the MR strongly as studied for GMR and TMR. These issues are also important subjects to be studied.

The characteristics of the lattice structure, constituent elements, and electronic structure of graphene may give rise to various advantages in junction efficiency: gate controllability in junctions, long spin diffusion length caused by the weak spin–orbit interaction, and the high mobility due to the linear dispersion relations. Because of these features of electrical transport properties distinguished from those in conventional metals and semiconductors, the graphene is expected to be an alternative to silicon in the contemporary electronics, that is, the metal/graphene/metal junctions could be field-effect transistors (FETs). When two ferromagnetic metal (FM) electrodes are attached to the graphene spacer, the junction may also work as a magnetic memory. Thus the FM/graphene/FM may work as both logic and memory device, and has a huge potential for application to next-generation devices.

Despite several proposals of novel MR devices using graphene and graphene NR, the experimental and theoretical results reported previously for the MR effect in two-terminal graphene junctions are not so exciting. The MR ratios observed in FM/graphene/FM junctions are at most 10% [1–3] and theoretical results by Brey and Fertig [14] indicate that the MR ratio in FM/graphene/FM junctions cannot be large. Because the reported MR ratios are not large enough for device applications and no clear mechanism for the MR has been presented, theoretical study of MR in two-terminal graphene (G) junctions with FM electrodes (FM/G/FM junctions) is desirable to clarify the mechanism and to provide a guiding principle for junction design. So far, we have performed numerical calculations of the MR effects of FM/graphene/FM junctions using realistic models for the electronic structures of the junctions, and reported that the MR ratio can be large [4–7]. The large MR was attributed partly to the mode-matching/mode-mismatching and partly to the change in the electronic states near the DP at the interface. The mechanism of large MR is distinct from the mechanisms of the giant MR in magnetic multilayers and of tunnel MR in magnetic tunnel junctions [8]. Recently, Dulbak et al. [9] reported a well-behaved MR effect Co/graphene/Co junctions with $Al_2O_3$ at the interfaces.

In this chapter, we first present numerical results of the conductance and MR for model junctions using the single-orbital TB model for the electrodes. Two types of lattice structures of the electrode have been adopted: one is a square lattice (SL) and the other is a triangular lattice (TL). Effects of disorder on the conductance an MR of graphene junctions are also studied for these junctions. We then report the results calculated for graphene junctions with realistic electrodes for which the full-orbital TB model was adopted.

We demonstrate that a sufficiently high MR ratio could be achieved in these junctions when certain conditions are satisfied, and clarify possible mechanisms for the high MR ratio. A simple mechanism of MR is the matching/mismatching model for conduction channels, and the other one is a spin-dependent shift of effective DPs caused by band mixing between the graphene and the electrodes, that is, the MR originates from a conflict between the DP shift in the up- and down-spin states in the antiparallel alignment of the FM magnetization. The MR ratio is independent of the graphene length because of a characteristic dependence of the tunneling conductance via states near the DP. The mechanism of the MR effect is different from those previously described [10, 11] in that it is caused by contact with the electrodes. We further show that a huge MR effect may appear in junctions with electrodes made of FeCo, FeCr, and FeV alloys. The result is attributed to a suppression of conducting states in the majority spin state near the junction contacts, which indicates matching between DPs and conduction channel of electrodes becomes worse in the spin state.

## 6.2 Magnetoresistance of Graphene Junctions: Model Calculations

In this section, we presented numerical study of the spin-dependent conductance and MR effect for graphene junctions the electrodes of which have a simple lattice. First, we give results for artificial graphene junctions without electrodes, and move to junctions with electrodes with an SL and a TL. The electrodes of the junction are in contact with the zigzag edge of the graphene. The electrode with the TL is useful because we can take the area of the contact region arbitrarily. Conductance of the junctions are calculated by using Kubo formula and recursive Green's function method in a simple TB model.

### 6.2.1 *Artificial Graphene Junctions*

We consider artificial graphene junctions in which two doped graphene sheets with half-infinite length are used as electrodes instead of metallic electrodes. The doped graphene is of n-type, which is realized by shifting the electronic states by 1 eV. Both armchair and zigzag contacts are adopted as shown in Figs. 6.1a and 6.2a. The electronic states of the graphene are characterized by $p$-orbital hopping as usual. The hopping integrals at the interface are

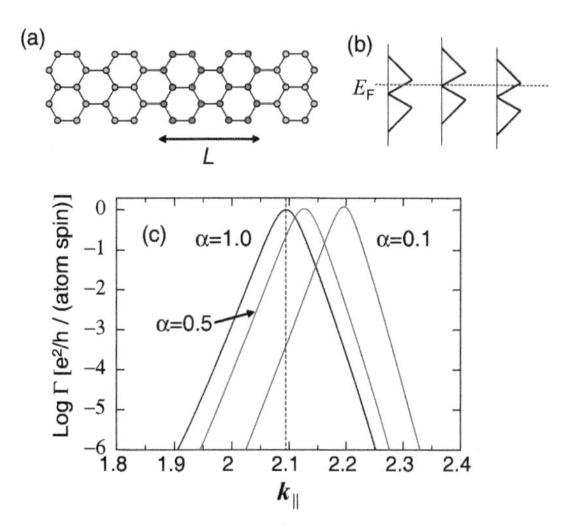


**Figure 6.1** (a) A model of graphene junction attached with a doped graphene electrode at zigzag edges and (b) its density of states. (c) Calculated results of momentum-resolved conductance of graphene junctions with three values of the hopping integral $\alpha$ at the interface of graphene and doped graphene.

multiplied by a factor $\alpha$. This kind of model junctions seems to be similar to *p*-*n* junctions; however, here we would like to study the momentum-resolved conductance and its dependence on the factor $\alpha$. The DP appears at $k = 0$ and $k \approx 2.1$ $(1/a)$ for the armchair and zigzag-edge contacts, respectively.

The calculated results show that the conductance of junctions with zigzag-edge contacts shifts with $\alpha$ keeping the maximum value of $\Gamma$ unchanged, as shown in Fig. 6.1c. The maximal value of $\Gamma$ is the same with the quantized conductance $e^2/\hbar$ per atom per spin. The shift of the maximal value of $\Gamma$ may be attributed the change in the electronic structure in the junction as presented in Chapter 3, and suggest that the DP shifts effectively due to the junction structure. On the other hand, as shown in Fig. 6.2c, the conductance of junctions with armchair-edge contacts decreases with shifting of $k_{\parallel}$ keeping the conductance $\Gamma$ at $k_{\parallel} = 0$ unchanged even the value of $\alpha$ is varied.

It is noted that the values of $\Gamma$ of junctions with armchair-edge contacts show an oscillation with a three-monochair (MC) period,

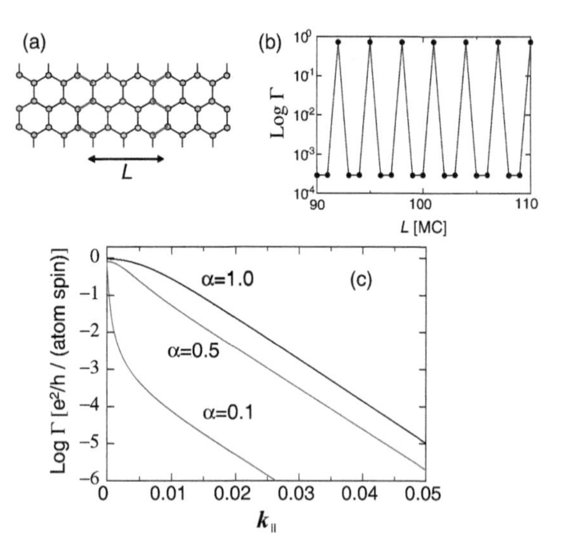


**Figure 6.2** (a) A model of graphene junction attached with a doped graphene electrode at armchair edges. (b) Calculated results of conductance as a function of the graphene length $L$. (c) Calculated results of momentum-resolved conductance of graphene junctions with three values of the hopping integral $\alpha$ at the interface of graphene and doped graphene.

as shown in Fig. 6.2b. The oscillation is essentially same with that appearing in graphene nanoribbon (GNR) with armchair edges.

## 6.2.2 *Square-Lattice/Graphene/Square-Lattice Junctions*

### 6.2.2.1 Junction structure and model

Here we calculate the conductance and MR of junctions in which SL electrodes are contacted with graphene zigzag edges. Two types of junction structures are considered, one is lateral junctions and the other is overlayer junctions, as shown in Figs. 6.3a and 6.3b, respectively. The former lateral junctions have been used in previous analytical studies; however, this type of junction is unrealistic because the hopping integrals at the contact become zero due to $s$ and $p$ orbitals on the electrodes and graphene sheet. So we consider here the overlayer structure. The width and length are denoted as $W$ and $L$, respectively. We assume the periodicity along $y$ direction,

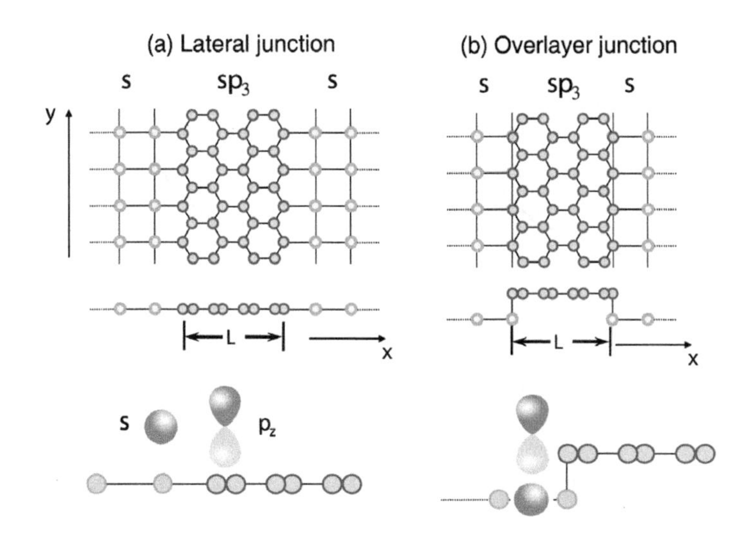


**Figure 6.3** Lattice structures of square-lattice/graphene/square-lattice junctions for (a) lateral junction and (b) overlayer junctions with arrangements of the *s* orbital on the square lattice and the $p_z$ orbital on graphene.

therefore $W = \infty$. The value of $L$ are taken from a few MCs to 1000 MCs. The left and right electrodes are semi-infinite.

The hopping integral on graphene sheet is given by $pp\pi$ and that on the electrodes is given by $ss\sigma$. At the contact, it is given by $\alpha \times sp\sigma$ where $\alpha$ is a proportionality constant. These parameter values are determined according to textbooks [12, 13]. Atomic potentials on $s$ orbital on the electrodes and $p$ orbital on the graphene are taken as $\epsilon_s$ and $\epsilon_p$. These values are listed on Table 6.1.

**Table 6.1** Values of band parameters used in the tight-binding model for graphene junctions

| Energy levels (eV) | Hopping integrals (eV) |
|---|---|
| $\epsilon_s = -8.550$ | $ss\sigma = -4.926$ |
| $\epsilon_p = 0$ | $sp\sigma = 6.474$ |
| | $pp\sigma = 5.700$ |
| | $pp\pi = -2.850$ |

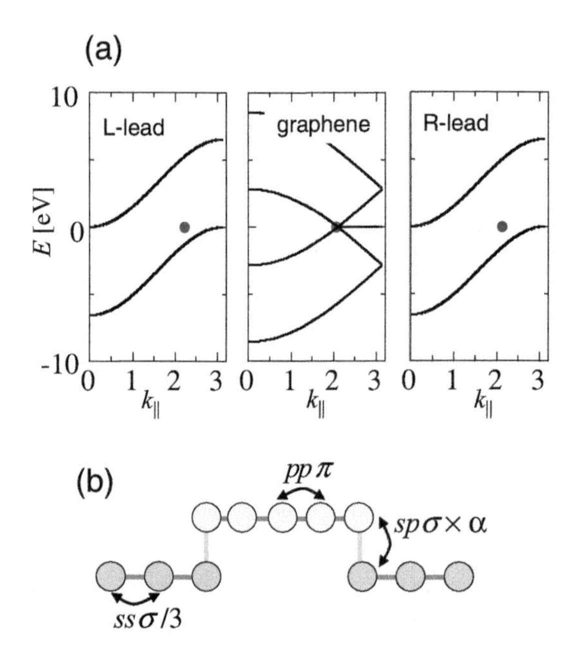


**Figure 6.4** (a) Schematic figures of electronic structures as a function momentum parallel to the interface direction and (b) definition of hopping parameters.

#### 6.2.2.2 Conductance and momentum-resolved conductance

Schematics of the energy dispersion relations are shown in Fig. 6.4 as a function of momentum $k_{\parallel}$ along the contact direction ($y$ direction). Because the periodicity of the junction is same with that the SL along only $y$ direction, $k_{\parallel}$ is defined between $-\pi/a$ and $\pi/a$. Dots in the figure indicate the position of the DP $K'$, which is denoted $\bar{K}'$ because it is a projected momentum along to $k_{\parallel}$.

Figure 6.5 shows the calculated results of $\Gamma$ as a function of graphene length $L$ for undoped and doped graphene (n-type) shown by black and red curves, respectively. The n-doped graphene is realized by shifting the energy states by $-0.5$ eV. The conductance of the undoped graphene decreases with $1/L$ in accordance with the analytical results. The length dependence may be understood in terms of contribution from tunneling conductance via infinitesimally small energy gap near the DP, as explained in

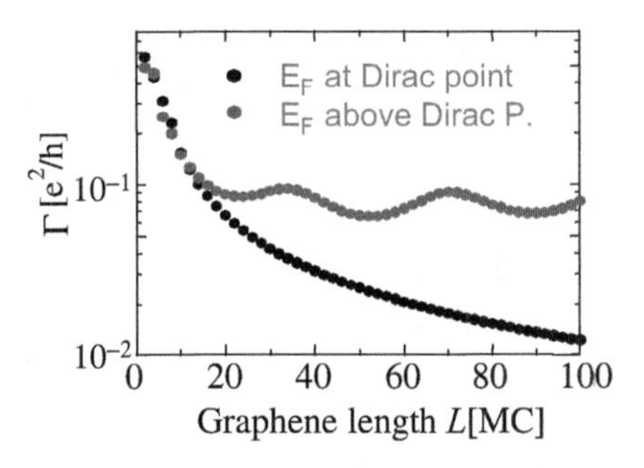


**Figure 6.5** Calculated results of conductance Γ of overlayer junctions with undoped and doped graphene as a function of graphene length.

Chapter 5. The oscillation period of the conductance is given by the Fermi wavelength of the doped graphene, as explained below.

Figure 6.6a shows schematics of the electronic structure of doped graphene. The position of the Fermi energy is denoted by a dotted line, which crosses the energy bands of the graphene. Because the energy bands of the graphene sheet of the junction consist of bundle of many energy branches, as shown in Fig. 3.7 in Chapter 3, and

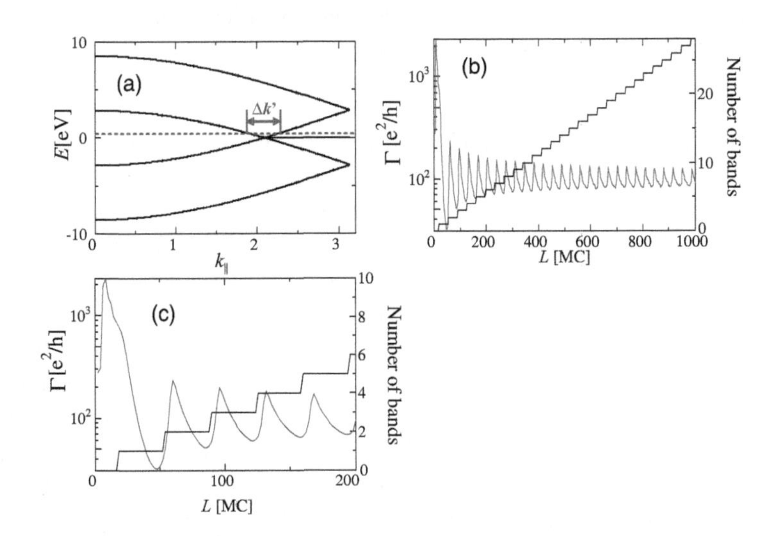


**Figure 6.6** (a) Electronic states along $k_{||}$ in graphene. (b) Calculated results of conductance as a function of graphene length $L$ and (c) enlargement of the result shown in (b).

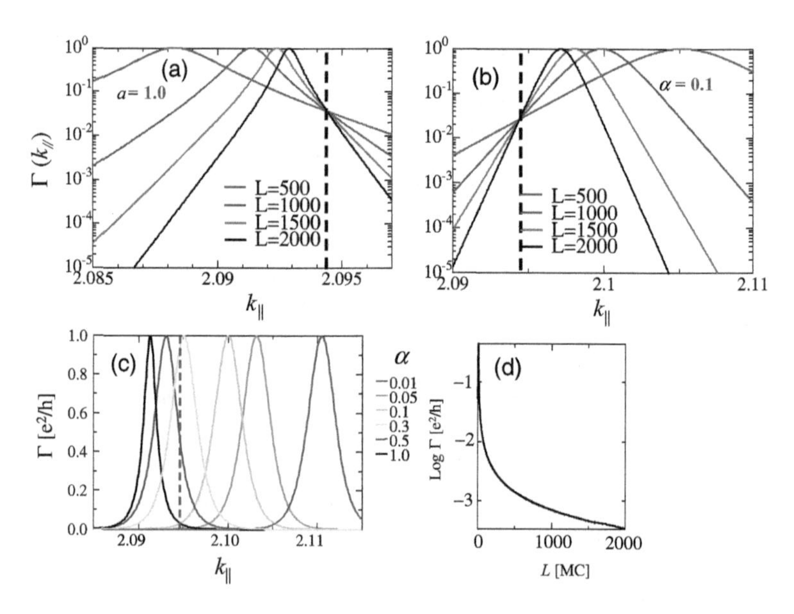


**Figure 6.7** Calculated results of momentum-resolved conductance for square-lattice/graphene/square-lattice junctions with hopping integral (a) $\alpha = 1.0$ and (b) $\alpha = 0.1$ with several values of graphene length. (c) Calculated results of momentum-resolved conductance for several values of $\alpha$ and (d) conductance as a function of graphene length $L$.

the number of the branches increases with increasing graphene length $L$. Figure 6.6b shows the calculated values of $\Gamma$ and number of crossing branches of the energy bands, where the value of $\alpha$ is 0.1. The enlarge scale of the results are shown in Fig. 6.6c. We find the number of branches increases with $L$ and correspondingly the values of $\Gamma$ oscillate as a function of $L$. The oscillation period is obtained as $\lambda = 35.6$ MCs for the band shift −0.5 eV. This length corresponds to the length of $\lambda \times 3a/4\sqrt{3} = 15.4a$ where $a$ is the lattice constant of the graphene sheet. The Fermi wavelength, on the other hand, may be defined by a half of $k_{\rm F} = \Delta k'/2$, where $\Delta k'$ is shown in Fig. 6.6a. The value of $\Delta k'$ is obtained to be $0.4084/a$. Because $\lambda_{\rm F} = \pi/k_{\rm F}$, $\lambda_{\rm F} = 15.38a$, which agree with the value $15.4a$ derived from the numerical results. Thus we conclude the oscillation of the conductance is governed by the Fermi wavelength of the doped graphene.

Results of the momentum-resolved conductance $\Gamma(k_{\parallel})$ (per spin) calculated for junctions with constant values of $\alpha = 1.0$ and 0.1 are shown in Figs. 6.7a and 6.7b, respectively, for several graphene length, $L$. We see two characteristic features in the figures. The conductance $\Gamma(k_{\parallel})$ is $e^2/h$ at a certain value of $k_{\parallel}^0$ near $\bar{K}'$, and decays exponentially away from $k_{\parallel}^0$. The momentum $k_{\parallel}^0$ may be called the effective DP. The effective DP shifts with $L$, and direction of the shift is opposite for $\alpha = 1.0$ and 0.1. The shift of the effective DP may be related to the electronic structure of the junction shown in Fig. 3.14 in Chapter 3. The effective DP may be a momentum close to $k_{\parallel}$ at which the lower conduction band and upper valence band meet with together. The shift of the effective DP may also be seen in the calculated results of the energy dispersions of finite-size graphene junctions.

It is also interesting to note that the value of $\Gamma(k_{\parallel})$ at $\bar{K}'$ is smaller than $e^2/h$, but is independent of $L$. Physical reason for the independent value of $\Gamma(\bar{K}')$ on $L$ is yet unknown; however, it may be related to the features of the shift of the effective DP with $L$ observed in the numerical results: one is that $|\bar{K}' - k_{\parallel}^0| \propto 1/L$, and the other is that the half-width of the function $\Gamma(k_{\parallel})$ changes also with $1/L$. The latter result may correspond with the result that the integrated conductance $\Gamma \propto 1/L$, as shown in Fig. 6.7d. The effective DP also shifts when the magnitude of hopping integral at the junction contact is varied, as shown in Fig. 6.7c.

The most important point to be noticed in these results is that there appears nonzero conductance at $k_{\parallel}$ away from the effective DP. The nonzero values of the conductance is attributed tunneling of carrier electrons through the energy gap near the effective DP. Because of tunneling, the conductance at $k_{\parallel}$ away from the effective DP decreases quickly as increasing graphene length $L$. In spite of several characteristic features of $\Gamma(k_{\parallel})$, the conductance $\Gamma$ integrated over $k_{\parallel}$ obeys the well-known result presented already, that is $\Gamma \propto 1/L$. Thus, the finite value of conductance of undoped graphene junctions with finite length is attributed to the tunneling through the energy gap near the effective DP.

Finally, let us mention on the numerical accuracy. As shown in Fig. 6.7, the momentum-resolved conductance is finite only narrow range of the momentum. Therefore, a sufficiently fine mesh of the

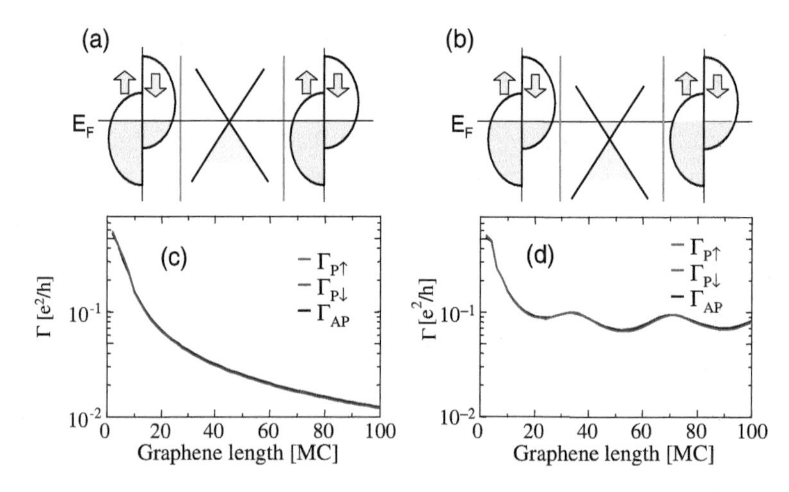


**Figure 6.8** Schematic figures of density of states for (a) undoped and (b) doped graphene and calculated results of conductance of up- and down-spin states in parallel alignment and those in antiparallel alignment as a function of graphene length for (c) undoped and (d) doped graphene.

momentum is desired to obtain converged results. Sampling of a small number of $k$-points results in inaccurate results.

### 6.2.2.3 Magnetoresistance

The MR effect may be studied by introducing exchange splitting of the $s$-bands of the electrodes. Because of the exchange splitting, the momentum-resolved conductance is spin dependent; however, the integrated conductance shows almost no spin dependence. Figures 6.8a and 6.8b present calculated results of the conductance of parallel (P) and antiparallel (AP) alignments of the magnetization of the spin-polarized electrodes for undoped and doped graphene junctions, respectively. Here $\Gamma_{\rm AP} \equiv \Gamma_{\rm AP\uparrow} = \Gamma_{\rm AP\downarrow}$. For both cases, we find almost no difference exists, resulting vanishingly small MR ratio. The results consistent with those obtained Brey and Fertig [14]. However, we will show later that large MR effects may appear when a specific condition is satisfied.

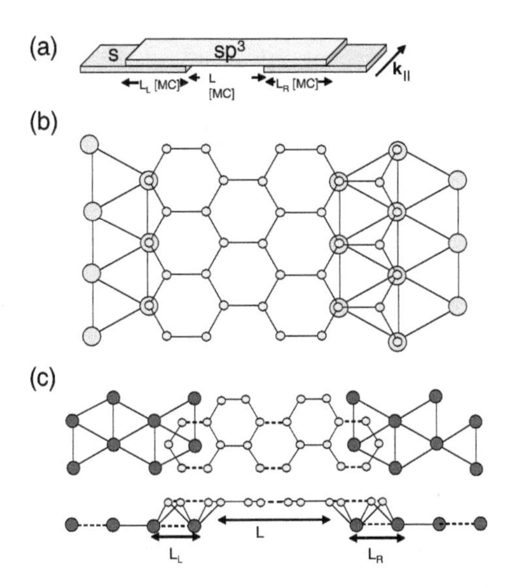


**Figure 6.9** (a) Structure of an overlayer graphene junction with triangular-lattice electrodes. (b) and (c) Two types of lattice structures of triangular/graphene junctions.

## 6.2.3 *Triangular-Lattice/Graphene/Triangular-Lattice Junctions*

TL electrodes have an advantage in that the area of contact region of graphene and electrodes can be taken arbitrarily (Fig. 6.9a). Figures 6.9b and 6.9c show two types of contact of a TL with the zigzag-edge graphene: one is TL sites are just below the carbon atoms of the graphene, and the other is the TL sites are located below the center of the hexagon of the graphene lattice. The length of the overlap region of left and right electrodes are denoted $L_{\rm L}$ and $L_{\rm R}$, respectively, as shown in Fig. 6.9a. Here, we describe calculated results mainly for the first type of the junction structure. The values of the band parameters are the same with those used for junctions with SL electrodes.

We first show results of the momentum-resolved conductance $\Gamma(k_{\|})$ and integrated conductance $\Gamma$ for $L_{\rm L} = L_{\rm R} = 1$ in Fig. 6.10a. Feature of these results is the same with that for junctions with SL electrodes, that is, $\Gamma(k_{\|}) = e^2/h$ at a certain momentum $k_{\|}$ and

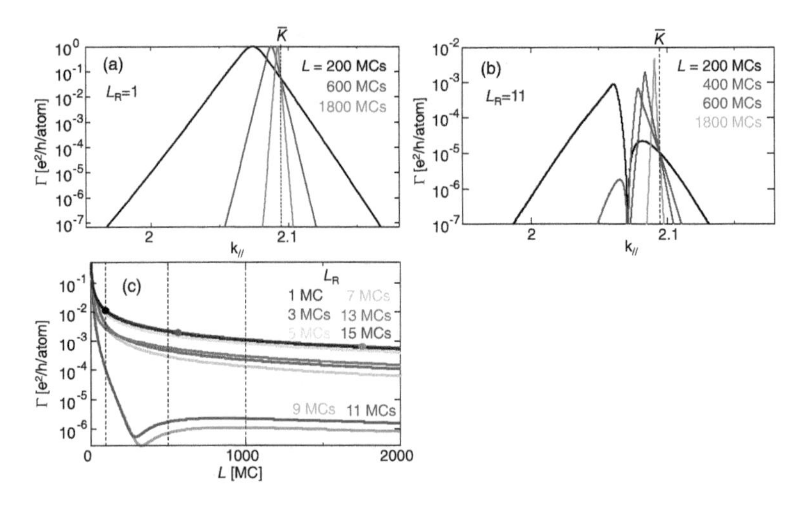


**Figure 6.10** Calculated results of momentum-resolved conductance for graphene/triangular-lattice junctions with (a) one MC overlap region at interface and (b) 11 MC overlap region at the interface. (c) Calculated results of conductance as a function of graphene length for several values of the size of overlap regions.

decays quickly away from the momentum, and $\Gamma \propto 1/L$, as shown in Fig. 6.10c. However, when $L_L = 1$ and $L_R = 11$, both $\Gamma(k_\parallel)$ and $\Gamma$ behaves differently from the results for $L_L = L_R = 1$, as shown in Fig. 6.10b. There appears a dip in $\Gamma(k_\parallel)$, and $L$ dependence of $\Gamma$ is nonmonotonic. Thus, there is a strong dependence of the conductance on the overlapping region between electrodes and graphene. Figure 6.11a shows the calculated results of $\Gamma(k_\parallel)$ for junctions with several values of $L_L = L_R$ for junctions with $L = 1000$ MCs, and Fig. 6.11b are calculated $\Gamma$ plotted as a function of $L_L = L_R$. We find that $\Gamma$ shows oscillation as a function of $L_L (= L_R)$. Values of $\Gamma$ calculated for junctions with shorter graphene $L = 100$ oscillate more rapidly.

Because the band mixing at the contact region should modify the electronic structure of graphene as well as those of electrodes, the strong dependence of $\Gamma$ on $L_L (= L_R)$ must be related to the change in the electronic structure at the contact region. Calculated results of the energy dispersion relation for finite-size graphene junctions with TL electrodes are shown in Fig. 3.18 in Chapter 3.

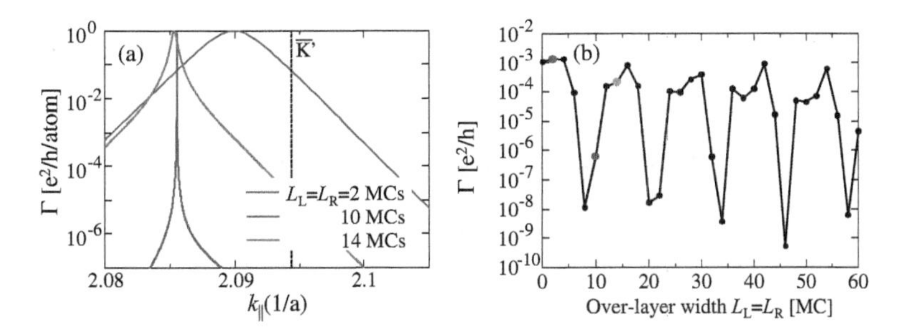


**Figure 6.11** (a) Calculated results of momentum-resolved conductance for graphene/triangular-lattice junctions with several values of the size of overlap regions. (b) Calculated results of the momentum-resolved conductance as a function of size of the overlap region at interfaces.

The length of the graphene layer and that of TL electrodes are 20 MCs and 50 MCs, respectively. The overlap region was taken to be $L_L(= L_R) = 0$ and 4, where zero overlap means that the TL is separated from the graphene sheet by $\sqrt{3}a/2$ where $a$ is the atomic distance of the TL. The results are essentially same with those obtained for graphene junctions with SL electrodes. With increasing the overlapping region, however, there appear several dispersions near $E = 0$. These dispersions have rather large intensity in the wave functions, and become flat near DP. These dispersions may be responsible to the rather anomalous dependence of $\Gamma(k_{\|})$ and $L$ on $L_L(= L_R)$.

To clarify the anomalous behavior in $\Gamma(k_{\|})$, we compare the calculated results of momentum-resolved local density of states (DOS) calculated at certain sites of the junction with $\Gamma(k_{\|})$. Figure 6.12 shows the momentum-resolved local DOS at the edge site of graphene and at a certain site of a triangular electrode ($L_L = L_R = 6$) shown in the inset. The calculated results of $\Gamma(k_{\|})$ is also plotted. We find that dips appear in the momentum-resolved local DOS at the TL, and that the local DOS at the graphene edge shows a peak at the DP. The $\Gamma(k_{\|})$ also shows a peak at the DP. Thus, there exists a good correspondence between the peaks in the local DOS and those in $\Gamma(k_{\|})$. A rather complicated structure of the local DOS near the DP may be responsible to the anomalous feature in $\Gamma(k_{\|})$.

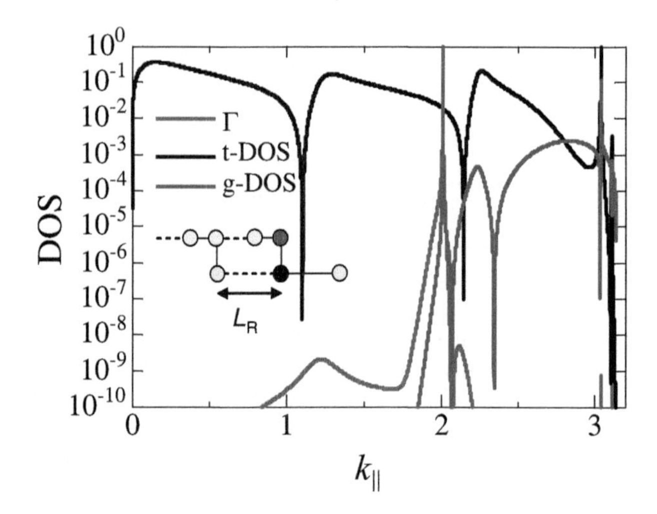


**Figure 6.12** Calculated results of momentum-resolved density of states for graphene/triangular-lattice junctions.

The complicated electronic structure at the contact regions of the graphene junctions with TL electrodes may results in rather unpredictable dependence of the conductance on the junction structures; region of the overlayer, thickness of the TL, hopping integrals at the contact, etc. Figure 6.13a shows calculated results of conductance Γ as a function of $L_R$ with fixed value of $L_L = 1$ for $L = 100$, 500, 1000 and 2000 MCs. Γ shows an oscillatory dependence of $L_R$, but tends to converge for longer $L_R$. However, the convergence is slow for longer $L_R$. Figure 6.13b shows dependence of Γ on the hopping integral between graphene and the TL. We find the results depend on the hopping integral only weakly. Figures 6.13d, 6.13e, and 6.13f show dependence of Γ on the thickness of the TL, the structure of which is shown in Fig. 6.13c. We see that the oscillation of Γ with respect to $L_R$ persists for thicker TLs; however, the amplitude of the oscillation tends to decrease. The oscillatory feature of Γ remains also for various values of the atomic potentials on the TL. So we may conclude that the feature of the conductance may depend on the lattice structures rather than the parameters which characterize the electronic states. Similar results may be obtained for another type of contact of graphene with the TL.

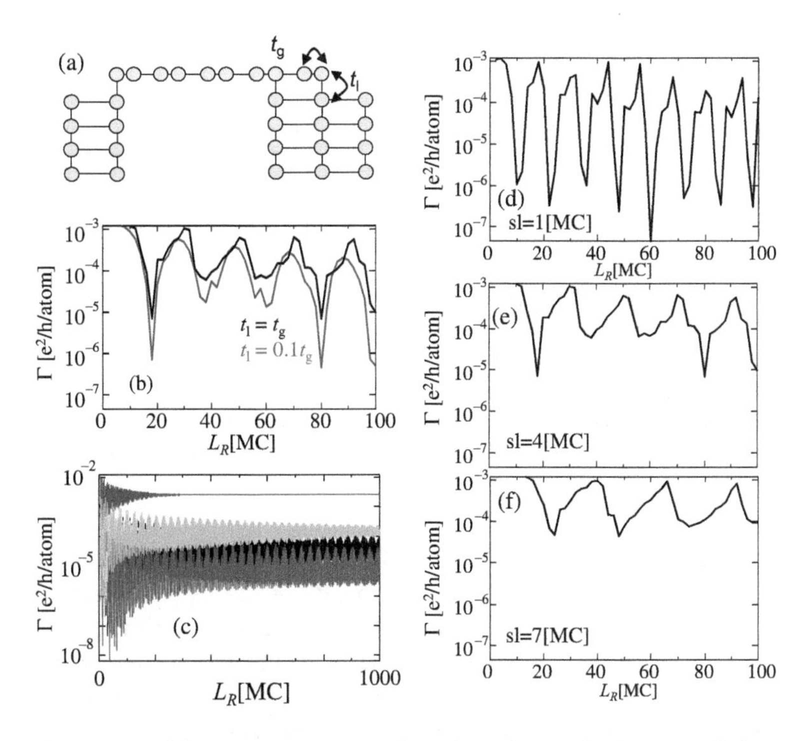


**Figure 6.13** (a) Junction structure of graphene/triangular lattice with four atomic layers. Calculated results of conductance as a function of overlap region for (b) narrow overlap region and (c) wide overlap regions and those with a different number of atomic layers of the triangular lattice and their results with an enlarged scale: (d) one atomic layer, (e) four atomic layers, and (f) seven atomic layers.

## 6.2.4 *Mode-Matching/Mode-Mismatching Model for MR*

As described in the previous section, the MR effect is rather weak for graphene junctions with electrodes the electronic structure is composed by *s*-like orbitals [14]. Nevertheless, they predicted large MR effect for an exceptional case. The large MR effect may be attributed matching/mismatching effect of the DP with the conductive states of the electrodes. Here we give a simple explanation of the effect, with some numerical calculations for graphene junctions with a spin-polarized SL model with an *s* orbital.

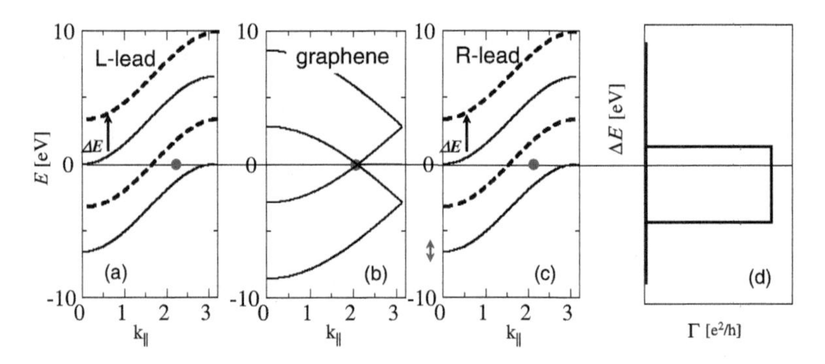


**Figure 6.14** Schematic figure of electronic structures of (a) L electrode, (b) graphene, and (c) R electrode. A square lattice is assumed for the electrodes. (d) Expected conductance change when the energy bands of L and R electrodes are shifted by $\Delta E$.

Figures 6.14a, 6.14b, and 6.14c, respectively, show schematics of relative position of the energy bands of the left (L) electrode, graphene and the right (R) electrode as a function of the momentum $k_{\|}$ parallel to the zigzag edge for ferromagnetic graphene junctions. The electrodes consist of *s* orbitals on the SLs, and have energy bands of cosine type, as shown by solid curves in Fig. 6.14. The DP is formed at $k_{\|} = 2\pi/3$ in the bulk graphene and shown by dots in the figure. The Fermi level is taken to be 0 eV.

In this band structure of the junction, the DP is located within the energy bands of the L and R electrodes, and therefore the junction show finite conductance. The situation is unchanged even if the energy bands of the electrodes are shifted by a few eV. The shift of the energy bands is $\Delta E$ and shown by broken curves in the figure. In the ferromagnetic junctions with parallel (P) magnetization alignment, the situation is similar to that shown in Fig. 6.14.

However, the situation is changed when the magnetization of one of the electrodes is reversed. As a result, a large MR effect may appears. For junctions with antiparallel (AP) magnetization alignment, the relative position of the energy bands are schematically given in Figs. 6.15a, 6.15b, and 6.15c, where blue and red curves indicate the up- and down-spin states, respectively, with an exchange splitting $\pm\Delta_{\rm ex}/2$. In the situation of the energy bands shown in these figures, no current flows because the DP is outside

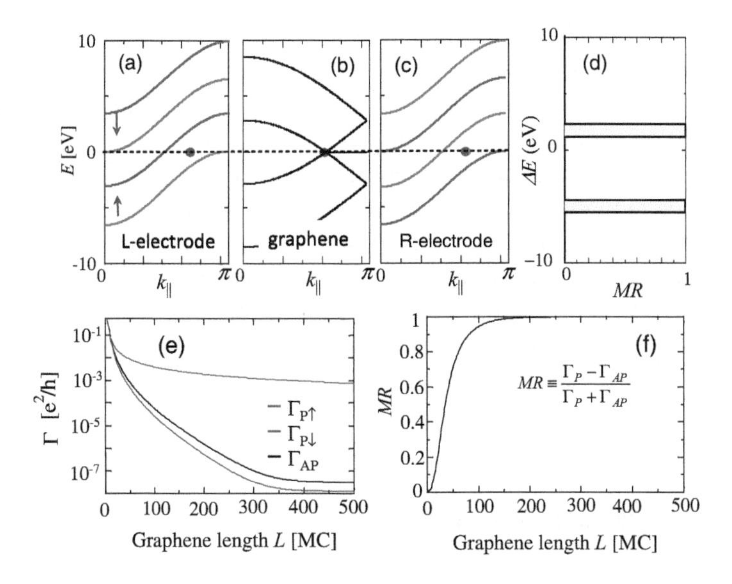


**Figure 6.15** (a–d) Mechanism of appearance of a large MR effect due to The matching/mismatching model. (c) Calculated results of conductance as a function of graphene length for up- and down-spin states in parallel magnetization alignment and those for antiparallel alignment. (d) Calculated results of the MR ratio as a function of graphene length.

the up- or down-spin band of the electrodes. Because the current is finite for the P alignment of the magnetization, we find large MR effect appears.

Now we estimate the MR ratio by shifting the energy bands of both the left and right electrodes by $\Delta E$. When the momentum of the DP is included in both the majority and minority spin bands, $\Gamma_{\mathrm{P}} = \Gamma_{\mathrm{AP}}$ resulting in zero MR ratio. The MR ratio is also zero when the momentum of the DP is not included in both the majority and minority bands, because $\Gamma_{\mathrm{P}} = \Gamma_{\mathrm{AP}} = 0$ in this case. The MR ratio is nonzero only when the momentum of the DP is included in either the majority or minority spin band. In this case, $\Gamma_{\mathrm{P}} \neq 0$ but $\Gamma_{\mathrm{AP}} = 0$, and the MR ratio is one, that is,

$$\mathrm{MR} \equiv \frac{\Gamma_{\mathrm{P}} - \Gamma_{\mathrm{AP}}}{\Gamma_{\mathrm{P}} + \Gamma_{\mathrm{AP}}} = 1.$$

Such a state may be called the *half-metallic like* state. Thus, the MR ratio becomes one for two narrow regions of $\Delta E$, as shown in

Fig. 6.15d. This is the simple mechanism of the MR in the mode-matching/mode-mismatching model.

Examples of calculated results of $\Gamma_{P\sigma}(\sigma = \uparrow, \downarrow)$, $\Gamma_{AP}$, and MR ratio as a function of graphene layer length $L$ are shown in Fig. 6.15. The conductance decreases with $1/L$ and tends to be constant which is due to finite value of imaginary energy $i\delta$ introduced to make the numerical calculation feasible. For $L < 100$, tunneling conductance is dominated, and therefore the effect of matching/mismatching is relatively weak and the MR ratio is smaller than 1, but it tends to be 1.0 with increasing $L$. Similar results of the MR effects are obtained for the TL/graphene junctions .

## 6.3 Effects of Disorder

Effects of disorder in graphene sheet on the transport properties are themselves interesting to be studied because of the characteristic feature of the electronic state near the Fermi level. As described already, the disorder introduced into undoped graphene increase the conductivity as a result of quantum effect. On the other hand, semiclassical description is efficient for the transport in highly doped graphene. Several theoretical approaches have been reported: most studies have been done using direct calculation for finite-size graphene ribbons, semiclassical approach using Monte Carlo technique has also been adopted [15] and an effective medium method has been examined for conductivity for infinite graphene sheet. As for the last method, one should take an average of the conductivity given by a product of two Green's function. Because the average of the product of two Green's function is not in general equal to a product of two averaged Green's function, so-called vertex corrections should properly be evaluated. The vertex corrections would not vanish for materials with linear dispersion relation, which is the case for graphene, and therefore careful analysis is necessary. Such analysis for graphene has been done by, for example, Wang et al. [16]. In this article, we report results obtained by direct calculations of conductance for finite-size graphene ribbons including imperfections.

When the junction includes disorder, the momentum perpendicular to the current direction, need not be conserved, and the condition of the mode-matching/mode-mismatching will be relaxed. As a result, the MR ratios in disordered junctions may be different from those in junctions without disorder. To quantify the effect of the disorder on the MR, it is necessary to perform real space simulations for finite-size graphene junctions. Here we will perform such simulations for junctions made of the graphene nanoribbon (GNR) and ferromagnetic electrodes by adopting a single-orbital TB model. We will show that the large MR effect caused by the mode-matching/mode-mismatching at the momentum of the DP is much reduced by the disorder. On the other hand, when the electronic state of the electrodes is half-metallic, the MR effect develops by the disorder, even though the DP does not match the conduction state of the electrodes.

### 6.3.1 *Models and Method of Calculation*

We deal with graphene junctions with SL electrodes attached at the zigzag edges of the graphene nanoribbon (GNR). The length and width of the GNR are $L \leq 1000$ and $W = 100$ MCs, respectively. We adopt a single-orbital TB model, $p_z$ and $s$ orbital for the graphene and the electrode, respectively. The nearest neighbor hopping is assumed for both the GNR and electrodes as before. The hopping integral of the GNR is $pp\pi = -2.850$ eV, and that of the SLs is assumed to be $t_s = ss\sigma/3$, where $ss\sigma = -4.926$ eV is the hopping integral for $s$ orbital in the graphene [12]. The value of the hopping integral between the GNR and the electrodes is assumed to be $sp\sigma = 6.474$ eV which is the hopping integral between $s$ and $p$ orbitals in the graphene. The exchange splitting $\Delta_{ex}$ of FM electrodes is assumed to be 1.0 eV, and the majority (minority) spin states is shifted by $+(-)0.5$ eV.

As for disorder in the GNR, we introduce short range $\delta-$function-type random potentials with magnitude $\pm v$ on carbon atoms. The value of $v$ is assumed to be 2.33 eV corresponding to the potential of boron impurities in the graphene. To fix the Fermi level at the energy of DP, we introduce same amount of impurities with $+v$ and $-v$. An average of the conductance $\Gamma$ is taken over 10 samples with different impurity distributions. Here, the magnitude of $\upsilon$ is not essential in the simulation for the results, and the effect of disorder may be scaled by $(\upsilon/pp\pi)^2$.

The conductance is calculated by using the linear response theory (Kubo formalism) and the recursive Green's function method [17, 18]. In order to make the numerical calculation feasible, we introduce a small imaginary part $\eta = 10^{-8}$ eV to the energy in the Green's function. The MR ratio is defined as $MR = (\Gamma_{\mathrm{P}} - \Gamma_{\mathrm{AP}})/\Gamma_{\mathrm{P}}$, where $\Gamma_{\mathrm{P}}$ and $\Gamma_{\mathrm{AP}}$ are the conductances in parallel (P) and antiparallel (AP) alignment of magnetizations of the electrodes. In this definition, the maximum value of the MR ratio is 1.

### 6.3.2 *Conductance in Nonmagnetic Disordered GNR Junctions*

To study the effects of disorder on the MR and conductance, we should perform simulations for junctions with finite-size GNRs. In such junctions, an energy gap may open in the graphene sheet, and the conductance decays exponentially with increasing graphene length. This makes it difficult to study the effect of disorder on the MR caused by the mode matching/mismatching mechanism for junctions with long GNRs. However, it has been reported that there exists a kind of threefold periodicity with respect to the width of the infinite armchair GNR and that the conductance remains finite even for long GNRs in one case out of three [19]. We have also confirmed that the periodicity exists for junctions with finite width $W$. When $W = 101$ and 102, the conductance of the junction decays exponentially for $L \gtrsim 500$, while when $W = 100$, the conductance remains finite even for $L > 500$, as shown in Fig. 6.16b.

Figure 6.16c,d shows calculated results of conductance $\Gamma$ for disordered junctions as a function of graphene length $L$. The hopping integral at the contact is taken to be $\alpha \times sp\sigma$, and $\alpha = 0.1$ and 1.0 are taken for the results in Figs. 6.16c and 6.16d, respectively. Results are average over 10 samples with 5% doping of impurities. We find that the conductance increases for long graphene junctions as compared with the undoped junctions. Nevertheless, the threefold periodicity remains in the disordered junctions. The increase in the conductance due to disorder is a feature of electron (hole)-undoped

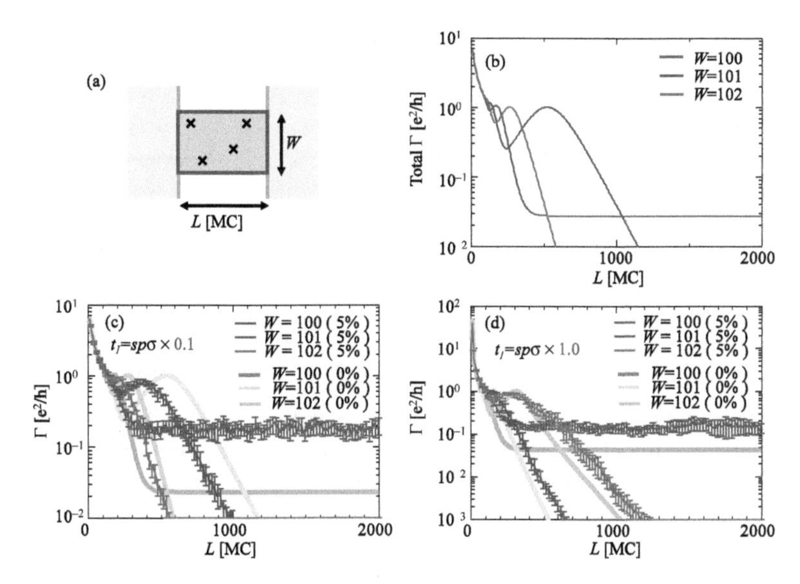


**Figure 6.16** (a) Schematic figure for junctions with impurities denoted by crosses and calculated results of conductance Γ as a function of graphene length $L$ (b) without impurity, (c) with impurities for $\alpha = 0.1$, and (d) with impurities for $\alpha = 1.0$. In (c) and (d), results of Γ without impurity are presented for comparison.

graphene junctions. In Fig. 6.17, we show results of conductance as a function of impurity concentration for undoped and n-doped graphene junctions. We find that the conductance of the n-doped graphene junctions deceases with increasing impurities, which is the same feature with that seen in usual metals. The increase in the

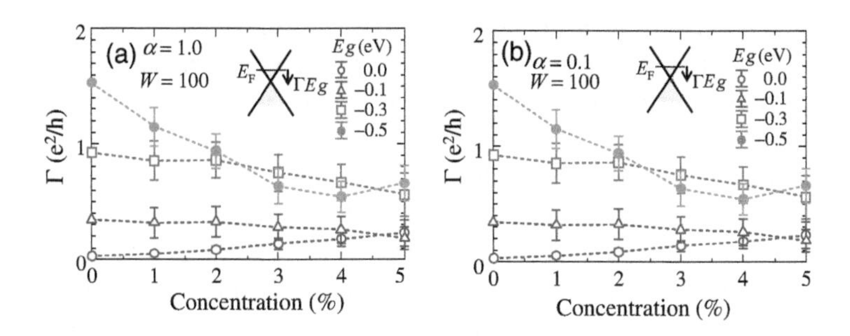


**Figure 6.17** Calculated results of conductance as a function of impurity concentration. (a) $\alpha = 1.0$ and (b) $\alpha = 0.1$.

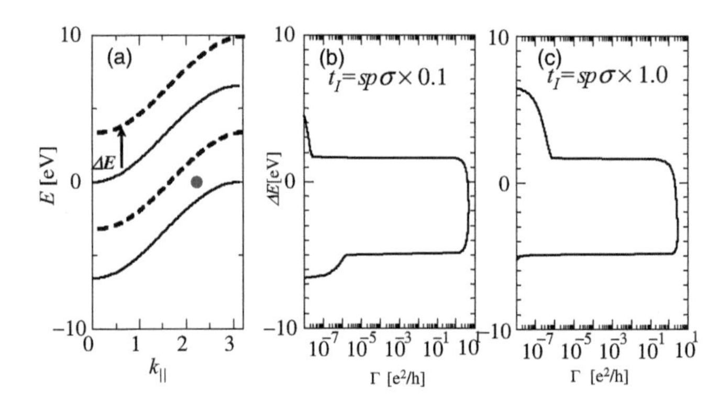


**Figure 6.18** (a) A schematic figure of shift of the energy band of electrodes by $\Delta E$. (b) and (c) Calculated results of conductance $G$ with energy band shift $\Delta E$ for graphene junctions with impurities by adopting a different choice of the hopping integral at the interface.

conductance with impurity concentration in the undoped graphene junctions may be attributed to increase in the conduction channel due to disorder scattering. In general, the tunnel conductance increases with increasing disorder because of the relaxation of the momentum conservation law [17, 18]. In graphene junctions, a similar increase in the conductance appears when the Fermi energy is located at the DP, but the conductance is decreased by the disorder when the Fermi level is located within the conduction or valence band of the graphene [20]. The localization in the latter case [21, 22] has not been taken into account in the present calculations because the GNR length is shorter than the localization length.

To clarify the role of disorder on the mode-matching/mode-mismatching model for MR, we calculate the conductance by shifting the energy bands of the electrodes by $\Delta E$. Figure 6.18a shows the energy band of the nonmagnetic electrodes. The position of the DP of the graphene sheet is shown by black dot. When the DP is located within the energy band of the electrodes, the value of $\Gamma$ is nonzero, but it becomes vanishingly small with increasing (or decreasing) $\Delta E$, and becomes zero with further increase (or decrease) of $\Delta E$, as shown in Fig. 6.18b,c. Here the width and length of the graphene NR are $W = 100$ and $L = 500$, respectively. The small but nonzero

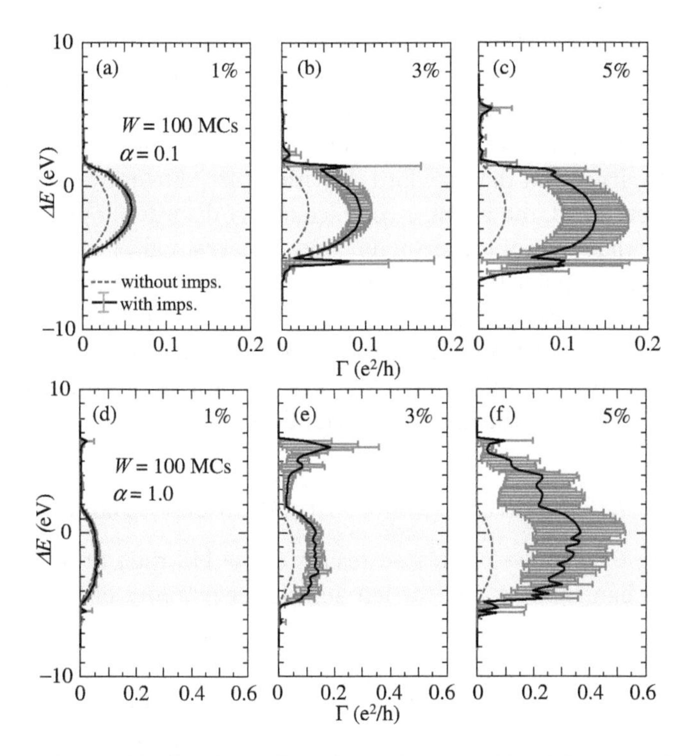


**Figure 6.19** Calculated results of conductance $\Gamma$ of impurity-doped graphene junctions with (a–c) $\alpha = 0.1$ and (d–f) $\alpha = 1.0$ as a function of the energy band shift of $\Delta E$. Impurity concentrations are 1, 3, and 5 atom percents. Averaged values of $\Gamma$ are shown by solid curves, which are compared with $\Gamma$ of the junction without impurity. Horizontal lines indicate the mean square deviation.

values of $\Gamma$ are attributed to mixing of graphene bands with the energy bands of the electrodes.

Figure 6.19 shows calculated results of conductance for disordered junctions with a GNR with $W = 100$ and $L = 500$ as a function of the band shift $\Delta E$ of both the left and right electrodes. Results for the impurity concentrations 0, 1.0, 3.0 and 5.0 atomic percent are presented in Figs. 6.19a, 6.19b, and 6.19c, respectively, by taking the hopping integral between the graphene and electrodes to be $0.1\times sp\sigma$, whereas in Figs. 6.19d, 6.19e, and 6.19f the hopping integral is taken to be $1.0\times sp\sigma$. The conductance of junction without disorder is shown by red broken curve in these figures for comparison. The shaded region indicates the values of $\Delta E$ for which $\Gamma \neq 0$ in the clean junction. The averaged values of $\Gamma$ calculated for ten samples with different impurity distributions are shown by solid curves and the standard deviations are shown by error bars. As mentioned, the average conductance of disordered junctions is larger than that of the clean junctions. Interestingly, $\Gamma$ can be large in regions outside the region where the matching condition between the DP and the energy bands of the electrodes is satisfied. The results are attributed to the relative position of the Fermi energy and energy bands of the electrodes and to the disappearance of the translational invariance along the junction (current) direction of the system.

### 6.3.3 *MR in Ferromagnetic Disordered GNR Junctions*

Figure 6.20 shows calculated results of the MR ratio as a function of the band shift $\Delta E$ of the left and right electrodes of a graphene

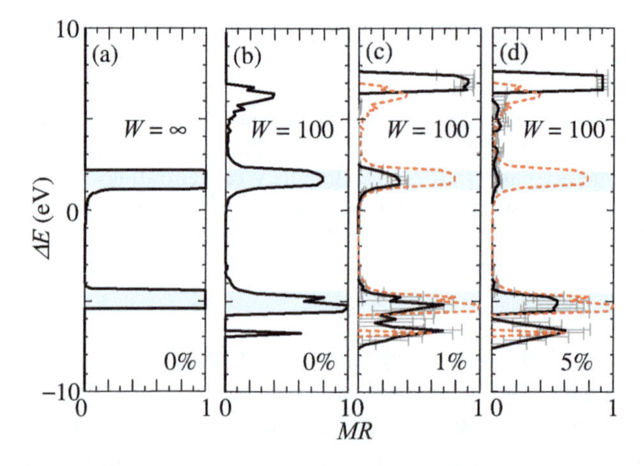


**Figure 6.20** (a) Calculated results of the MR ratio for a clean FM/graphene/FM junction with infinite width. Calculated results of the MR ratio of FM/GNR/ FM junctions for impurity concentrations (b) 0, (c) 1.0, (d) 5.0 atomic percents, respectively, as a function of the energy band shift $\Delta E$ of the electrodes. The width and length of GNR are $W = 100$ and $L = 500$ chains, respectively. The conductance shown in (b) is repeated for comparison by a red broken curve in (c) and (d).

junction with $L = 500$. Figure 6.20b–d shows the results for junctions with GNR ($W = 100$) including 0, 1.0, and 5.0 atomic percent of impurities, respectively. For comparison, results of the MR ratio for a clean junction with $W = \infty$ is presented in Fig. 6.20a. The blue shaded regions show the areas in which the mode-matching/mode-mismatching mechanism gives rise to the perfect MR effect, that is, the MR ratio is 1.

We find several characteristics in the results shown in Fig. 6.20b–d. First, the MR ratio is almost zero in the region between two shaded regions, that is, in the region of $-4.5 \lesssim \Delta E \lesssim 1.0$ eV. Nevertheless the conductance increases in this region. Secondly, the perfect MR ratio in the shaded regions decreases with increasing impurity concentration. Finally, the MR ratio increases substantially near $\Delta E \approx \pm 7$ eV.

The first point is easily understood by noting that the mode-matching condition holds irrespective to the impurity concentration between the DP and the conduction state of the electrode. In this $\Delta E$ region, the conductance in the parallel and antiparallel alignment, $\Gamma_{\mathrm{P}}$ and $\Gamma_{\mathrm{AP}}$, respectively, increases by nearly the same amount with increasing impurity concentration, and MR $\approx 0$.

The second and third characteristics may be understood in terms of the position of the energy band in the electrodes. To quantify the second and last characteristics, we plotted the MR ratios at $\Delta E =$ 1.6 and -6.8 as a function of the impurity concentration in Fig. 6.21. Solid circles show results for $\Delta E = 1.6$. The results indicate that the MR ratio decreases rapidly with increasing impurity concentration. The energy band of the electrodes in P alignment of this case and the position ($E$ and $k_{\|}$) of the DP are shown by a dot in the left inset. In this case, the MR ratio should be one for clean junctions with $W = \infty$ due to mode-matching/mode-mismatching mechanism. In other words, the minority spin ($\downarrow$) band at $k_{\|} = 2\pi/3$ is located above the Fermi energy, while the majority spin ($\uparrow$) band is located below the Fermi energy at the momentum, and therefore $\Gamma_{P\uparrow} \neq 0$ but $\Gamma_{P\downarrow} = \Gamma_{\mathrm{AP}} = 0$. The energy state at $k_{\perp} = 2\pi/3$ may be identified to be half-metallic in the momentum space (half-metallic-like state). It is noted that, when $W = 100$, the MR ratio is about 0.7 even for the clean GNR junction. This is attributed to the disappearance of the translational invariance along the zigzag-edge contact. By

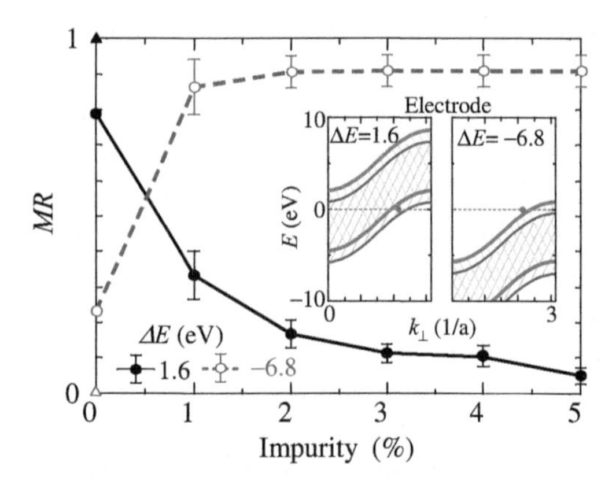


**Figure 6.21** Calculated results of the MR ratio of FM/GNR/FM junctions as a function of impurity concentration for $\Delta E = 1.6$ eV (solid circles) and $-6.8$ eV (open circles). The width and length of GNR are $W = 100$ and $L = 500$ chains, respectively. Error bars denote the standard deviation of the MR ratios. Insets show relative position of the energy bands of the ferromagnet electrode and Dirac points of the graphene.

introducing impurities in GNR, the disorder scattering makes $\Gamma_{P\downarrow}$ and $\Gamma_{\text{AP}}$ nonzero, and makes the MR ratio smaller. The decrease in MR ratio for $\Delta E \approx -0.5$ eV is understood similarly.

Open circles in Fig. 6.21 indicates that the MR ratio increases with increasing impurity concentration. The relative position of the energy bands of the electrodes and position of the DP is shown in right inset. As shown in the inset, the electronic state of the electrode is half-metallic with negative spin polarization, that is, there are minority spin states at the Fermi energy but no state exists in the majority spin state. When we look at $k_{\parallel} = 2\pi/3$, however, the Fermi level is located above both the majority and minority spin bands. This means that there is no conducting states in clean junctions, and no MR appears. However, disorder included in the GNR makes it possible that the scattering of electron from $k_{\parallel} = 2\pi/3$ to momentum near the zone boundary, and makes $\Gamma_{P\downarrow}$ in the P alignment finite keeping $\Gamma_{P\uparrow}$ and $\Gamma_{\text{AP}}$ zero. Therefore, the MR ratio increases rapidly with increasing impurity concentration. The large momentum change in the scattering process is caused by the short-range potentials of impurities. The momentum change may be small for long-range impurity potentials, and the increase in the MR ratio is smaller than that shown in Fig. 6.21. It is noted that nonzero MR ratio for clean junctions is also attributed to the disappearance of the translational invariance along the contact. A similar situation occurs for $\Delta E \approx 7$ eV. Thus, the half-metallicity of the electrode is crucial for the increase in the MR ratio.

We have studied the effects of electron scattering by disorder on the MR in FM/GNR/FM junctions. The large MR ratio obtained in the mode-matching/mode-mismatching mechanism for clean junctions is attributed to the half-metallic feature in the momentum space at $k_{\perp} = 2\pi/3$ in the electrode (half-metallic-like state). In this case, because the disorder scattering relaxes the conservation of the momentum of electrons and makes both $\Gamma_{\mathrm{P}}$ and $\Gamma_{\mathrm{AP}}$ increase, the MR ratio decreases with increasing impurity concentration. On the other hand, when the energy state of the electrodes is half-metallic and the momentum state at $k_{\perp} = 2\pi/3$ is outside the energy band, the conductance is zero in clean junctions, but it becomes finite due to the disorder scattering, and the MR ratio increases with increasing disorder. The results obtained above may be useful to interpret the effects of roughness in realistic ferromagnets/graphene junctions.

## 6.4 Realistic Graphene Junctions

### 6.4.1 *Introduction*

Because the electronic structure of the graphene, a gapless semiconductor, is quite different from that of conventional nonmagnetic metals, semiconductors, and insulators, FM/G/FM junctions may exhibit to a novel MR mechanism distinct from those of giant MR and tunnel MR in the field of spintronics [8]. The MR has been observed in conventional two-terminal graphene junctions [1, 2, 39, 40] as well as in nonlocal measurement [3]. Nevertheless, MR in two-terminal junctions remains to be studied, because the reported MR ratios are not large enough for device applications, and no clear mechanism for the MR has been presented. Therefore, theoretical study of MR in two-terminal graphene (G) junctions with FM electrodes (FM/G/FM junctions) is desirable to clarify the mechanism and to provide a guiding principle for junction design. Furthermore, several difficulties should be solved to realize graphene FET: a band gap should be introduced to realize the gate controllability [23–25] roughness caused by ad atoms might be avoided [26] graphene–metal contact should be well controlled, and so on. Recent first-principles calculations have indicated the importance of the electronic state at graphene/metal (G/M) contacts [27–30].

In this section, we review our theoretical results for graphene junctions with realistic electrodes, one is made of ferromagnetic body-centered cubic (bcc) iron and the other is ferromagnetic face-centered cubic (fcc) nickel. We adopt zigzag-edge contact as done in the previous paper, because the $K'$ point, one of two DPs is included in the momentum states parallel to the junction contact. This enable us to analyze the property of the electrical conductance of junctions in detail. The junction structures with electrodes of bcc iron and fcc nickel may be considered to be a generalization of the simple junction structures tested in the previous section, junctions with SL electrodes and TL electrodes, respectively. Finite overlapping region is realized by adopting electrodes with an fcc Ni (111) lattice and a TL, as studied in the previous section.

The electronic structures of the junction will be calculated by using a realistic TB model, and the conductance and MR of the junction will be calculated by using the recursive Green's function method. We will make the junction more realistic by introducing the potential profile proposed by the first-principles calculation near the contact region of the graphene and ferromagnetic electrodes. The realization may make it possible to predict the MR properties of the junction in detail. Effects of roughness is considered in view of the results calculated for FM/GNR/FM junctions with TL as well as SL electrodes.

**Bcc Fe/graphene/bcc Fe junctions:** Calculated results for these junctions may suggest three mechanisms for MR in FM/G/FM junctions: (1) spin-dependent shift in the effective DP due to spin-dependent band mixing between the graphene and electrode, (2) a simple model of mode-matching/mode-mismatching of the DP with conduction states in the electrodes, and (3) spin-dependent change in the electronic structure near DPs. It was shown that the mechanism (1) is important in bcc Fe/G/bcc Fe(001) junctions, while mechanism (3) contributes to the MR most strongly in junctions with electrodes made of bcc Fe alloys. The MR ratio is independent of the graphene length because of a characteristic dependence of the tunneling conductance via states near the DP. The mechanisms of the MR effect is different from those previously described [10, 11]. We further show that a huge MR effect may appear in junctions with electrodes made of FeCo, FeCr, and FeV alloys. The result is attributed to a suppression of conducting states in the majority spin state near the junction contacts, which indicates matching between DPs and conduction channel of electrodes becomes worse in the spin state. It should be noted, however, that the calculated results have been obtained for junctions with a rather unrealistic contact structure between the electrodes and graphene: the electrode is in contact with the graphene at only one zigzag chain, and that roughness and defects in junctions have been ignored.

**Fcc Ni alloy/graphene/fcc Ni alloy junctions:** The origin of the larger MR effect has been attributed to a spin-dependent change in the electronic states near the DP caused by band mixing between the graphene and electrodes, that is, $p_z - s$ mixing for the up-spin state and $p_z - d$ mixing for the down-spin state. Effects of overlapping region at the contact between graphene and electrode, roughness and magnitude of the band mixing at the contact are studied in detail. It is shown that the MR ratio increases with increasing overlapping region due to an enhancement of the spin asymmetry in the change of the electronic states near the DP. However, the MR ratio decreases with increasing roughness and decreasing band mixing between the graphene and electrodes. We further show that FM/G/FM junctions with Ni-rich alloy electrodes show high MR ratios.

**Ni/graphene/Ni junctions with more realistic contact:** In this study, we make a full survey of the effects of the contact in fcc Ni/G/fcc Ni (111) junctions on the conductance and MR in terms of the size of the contact region, distance between graphene and metal electrodes, and potential change at/near the contact caused by electron or hole dopants. The material dependence, that is the effects of alloying on the MR, is also studied. The effects of roughness in the junction are discussed briefly. However, the effects of band-gap formation in, for example, double-layer graphene, and gate controllability of the MR will be studied in future. The calculated results suggest the necessary conditions for realizing high MR ratios in FM/G/FM junctions : band mixing between graphene and Ni should be rather weak, the effect of dopants on the MR is small for Ni electrodes, and a contact region a few tens of nanometers in size is sufficient to obtain high MR ratios. We further show that Ni-Co and Ni-Cu alloys are useful for electrode materials. The increase in MR has been attributed to matching/mismatching of the DP with the band structure of fcc Ni at the contact. The importance of the *d* orbitals of fcc Ni is also emphasized.

This section is organized as follows. First, we explain the model and the method of the calculation. We present the calculated results of the conductance and MR ratios next for bcc Fe/Graphene/bcc Fe and fcc Ni/Graphene/fcc Ni(111) lateral junctions. Further consideration on the effects of contact size, distance between the graphene and the electrodes and the potential profile near the contact of fcc Ni/graphene/fcc Ni(111) lateral junctions will be presented later.

### 6.4.2 *Model and Method of Calculation*

Schematics of lateral bcc Fe/graphene/bcc Fe (001) junctions with overlayer structure are shown in Fig. 6.22a. We adopt a zigzag-edge contact in the junction, as shown in Fig. 6.22b, because the contact structure is easily modeled, as shown in the figure. It is also expected that the zigzag contact may strongly affect the electronic states of the graphene. Figure 6.22c shows a side view of the junction, where an Fe thin film with three atomic layers is used for the electrode, and Fig. 6.22d is the junction structure of graphene junctions with The SL electrode, results for which will be compared with those calculated Fe/graphene/Fe junctions. A periodic boundary condition is used for the junction width. The graphene length $L$ is taken to be less than $\sim$20,000 monochains (MCs) typically 1000 MCs, where the number

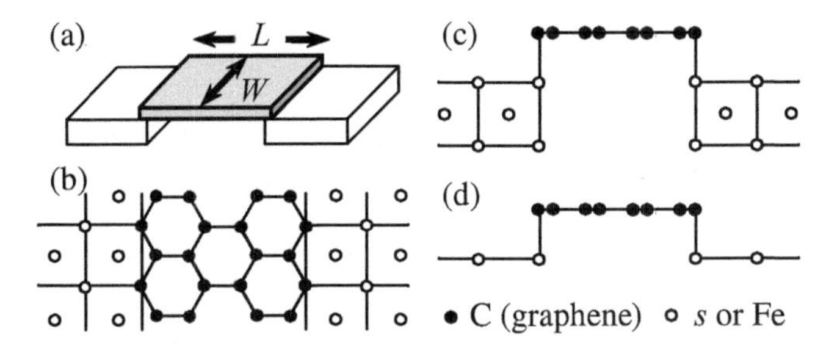


**Figure 6.22** (a) Schematic drawing of a junction, (b) top view of a junction with electrodes made of bcc Fe, (c) side view of a bcc-Fe/graphene/bcc-Fe junction, and (d) side view of a junction with electrodes made of a square lattice.

is taken in such a way that one zigzag chain corresponds to two MCs. No roughness is included in the junction, and therefore the momentum $k_{\parallel}$ parallel to the contact line is conserved.

Schematic figures of fcc Ni/G/fcc Ni(111) lateral junctions with zigzag-edge contacts are shown in Fig. 6.23a. The electrodes are made of four atomic layers of Ni. TL/G/TL junctions have the same structure with that fcc Ni/G/fcc Ni(111) junction except that the electrodes are made of one atomic layer. A periodic boundary condition is used along the edge direction ($y$ direction) for junctions without disorder. The graphene length $L$ (along the $x$ direction) is considered to be less than ~5000 MCs which corresponds about 500 nm. Most of the calculation are performed for $L = 1000$ MCs. The overlapping (contact) region $a$ between the electrode and graphene sheet shown in Fig. 6.23b is less than 100 MCs. The finite-size of the contact region may weaken the difference between the zigzag and armchair-edge contacts [30].

Figure 6.23c shows the Brillouin zone of the graphene, and $K$ and $K'$ are the DPs. As the energy dispersion of the graphene is projected onto momentum $k_{\parallel}$ parallel to junction contacts, a DP exists at $k_{\parallel} = 2\pi/3a_{\mathrm{G}} \simeq 2.094/a_{\mathrm{G}}$, as shown in Fig. 6.23d. Hereafter the momentum $k_{\parallel} = 2\pi/3a_{\mathrm{G}}$ is denoted $\bar{K}'$ point. The edge state of the zigzag edge of the graphene is also shown at $E = 0$.

Full-orbital TB models of the $sp^3$ orbitals and $sp^3d^5$ orbitals are adopted for the graphene and fcc Ni electrode, respectively, although the $p_z$ orbital is essential for the graphene layer. The band

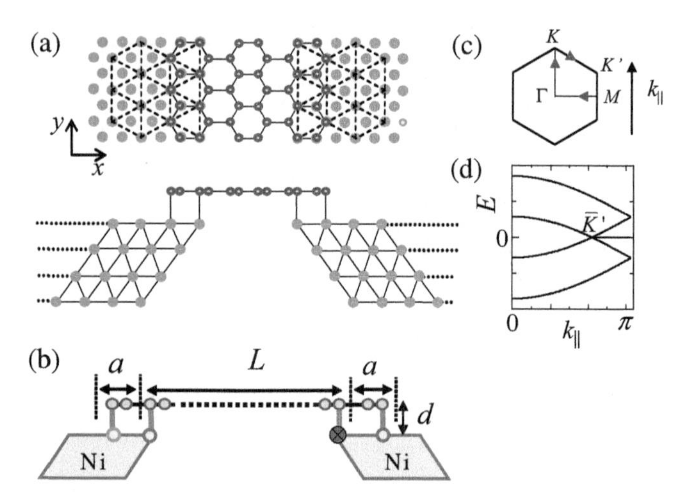


**Figure 6.23** (a) Schematic illustrations of fcc Ni/graphene/fcc Ni (111) junctions with zigzag-edge contacts in which FM electrodes have an fcc lattice. (b) Definitions of the graphene length *L*, size *a* of the contact region in units of MCs (see text), and distance *d* between graphene and Ni layers. Here $a = 2$ MCs and $L = 8$ MCs. The crossed circle indicates the site where the local density of states is calculated. (c) Brillouin zone of graphene and (d) schematics of the electronic structure projected on the momentum parallel to the junction contact.

parameters are taken from textbooks [12, 13]. The overlapping integrals of the graphene layer are $ss\sigma = -5.291$ eV, $sp\sigma = 6.953$ eV, $pp\pi = -3.061$ eV, and $pp\sigma = -6.122$ eV. The overlapping integrals between the graphene and the Ni electrodes are determined by a geometrical average of $ss\sigma$, $sp\sigma$, $pp\sigma$, and $pp\pi$ for graphene and fcc Ni metal. Because no *d* orbitals exist in graphene, the overlapping integrals $sd\sigma$, $pd\sigma$ and $pd\pi$ are determined in such a way that $sd\sigma$ is given by a geometrical average of $ss\sigma$ of the graphene and $dd\sigma$ of the Ni metal. The atomic potential of the *s* and *p* orbitals are taken as −8.55 and 0.0 eV, respectively, for the carbon atoms in graphene, and those of the *s*, *p* and *d* orbitals are taken from the textbook [12, 13]. The geometrical approximation for the overlapping integrals between Ni and graphene could be reasonable unless the distance *d* between Ni and graphene layer is much shorter than the atomic distance $d_{\mathrm{Ni}}$ of Ni.

A single-orbital TB model, however, is adopted for SL/G/SL and TL/G/TL junctions to facilitate numerical calculations for junctions with finite width and with impurities as presented in the previous section. For these junctions, the nearest-neighbor hopping in SL and TL is taken to be $t_s = ss\sigma/3$. The hopping parameters between graphene and the electrodes are assumed to be $\alpha\times$ (hopping integrals of graphene), where $\alpha$ is a parameter, for SL/G/SL and TL/G/TL junctions. A similar parameter $\alpha$ defined as $\alpha\times$ (hopping integrals of fcc Ni) may be used for the interface hopping in fcc Ni/G/fcc Ni(111) junctions.

When disorder is included in the junction, we deal with a graphene nanoribbon with a typical width (along the $y$ direction) $W \approx 100$ MLs. Impurities are randomly distributed in the graphene layer, and the impurity potentials are of $\delta$-function type with magnitude $\pm V$. The value of $V$ is 2.33 eV corresponding to the potential difference between carbon and boron or nitride. To avoid a possible shift in the Fermi level due to impurity potentials, we assumed that the same amount of impurities with $+V$ and $-V$ are introduced. An average of the conductance is taken over 10 samples with different impurity distributions.

The conductance of the junction is calculated by using the Kubo formula with recursive Green's function method which are explained in Appendix. Details are described in Itoh and Inoue [17]. Conductance $\Gamma_{P(AP)\sigma}$ of $\sigma$-spin state is calculated for parallel (P) (antiparallel (AP)) alignment of electrode magnetizations, as shown in Fig. 6.24, and the MR ratio is defined as

$$MR = \frac{\Gamma_P - \Gamma_{AP}}{\Gamma_P + \Gamma_{AP}}. \tag{6.1}$$

with $\Gamma_{P(AP)} = \Gamma_{P(AP)\uparrow} + \Gamma_{P(AP)\downarrow}$. Because we adopt symmetric junctions with respect to left and right electrodes, $\Gamma_{AP\uparrow} = \Gamma_{AP\downarrow}$.

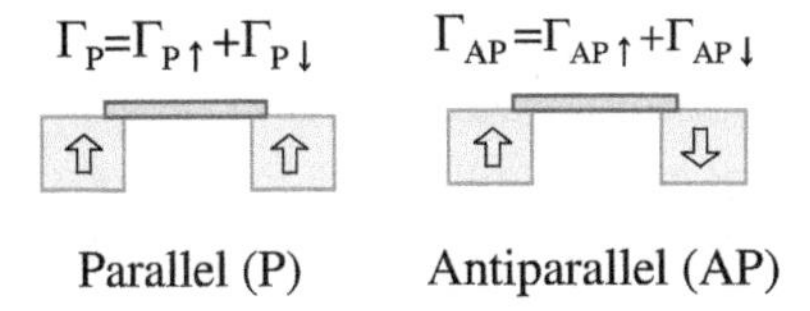


**Figure 6.24** Parallel and antiparallel alignments of magnetization in graphene junctions with ferromagnetic electrodes.

### 6.4.3 *BCC Fe/Graphene/BCC Fe Junctions*

#### 6.4.3.1 Calculated results

We have shown using simple graphene junctions that a large MR occurs due to mode-matching/mode-mismatching mechanism. In this mechanism, $\Gamma$ is finite only in a certain energy range of the band. When either $\Gamma_\uparrow$ or $\Gamma_\downarrow$ is finite, large MR effect appears, and the MR ratio is either 1 or 0. For graphene junctions with ferromagnetic electrodes, such situation may occur because the energy band shift is spin dependent due to the exchange splitting of the ferromagnetic states. Thus the MR effect appears only when the $K'$ point is included in either up- or down-spin conduction state of the electrode. In other words, a matching/mismatching of the conduction pass between the graphene and the leads is essential for finite MR in this case.

The simple mechanism explained above may not work for electrodes with realistic FMs because the spin splitting of bands in realistic FMs is not so large as demonstrated above. So we study the relationship between the conductance and electronic states near the DP putting an emphasis on the role of the band mixing between the graphene and the electrodes.

Results for the $k_\parallel$-resolved conductance $\Gamma_{k_\parallel}$ calculated for graphene junctions with SL electrodes show that the shift and shape of $\Gamma_{k_\parallel}$ depend on various values of parameters. We attribute the shift of the effective DP to band mixing between the $s$ band of the electrodes and the $\pi$ band of the graphene, after calculating the details of the electronic states in finite-size junctions. The shape of $\Gamma_{k_\parallel}$ has a finite width, and the width is determined by the *tunneling of electrons* through the graphene and is related to the tunneling probability, which is $T(k_\parallel, L) \propto \exp\left[-cL(k_\parallel - k_{\text{DP}})\right]$ near the effective DP. By integrating $T(k_\parallel, L)$ over $k_\parallel$, we get the well-known result $\Gamma \propto 1/L$ [31, 32].

Now we propose a possible mechanism for MR in graphene junctions using spin-polarized $s$, $p$ and $d$-bands in Fe electrodes. Since band mixing at the contacts is generally spin dependent, we expect the shift of the DP to become spin dependent and MR to appear. We show below that MR does appear in Fe/G/Fe junctions.

The structure of an Fe/G/Fe junction is shown in Fig. 6.22b,c. The bcc Fe layer is a thin film with three atomic layers. Figure 6.25a

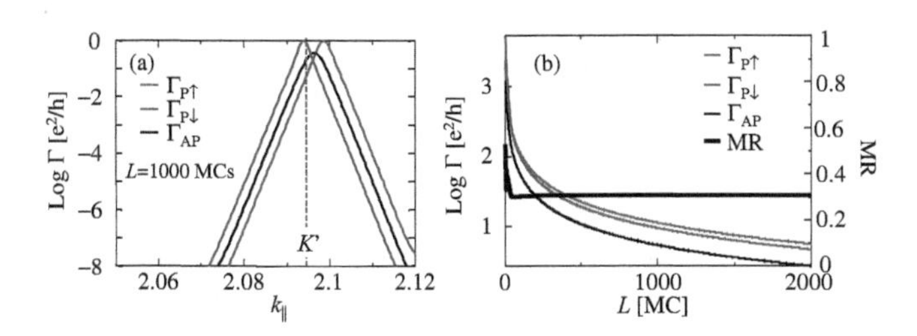


**Figure 6.25** (a) Momentum-resolved conductance $\Gamma_{k_\parallel}$ for parallel and antiparallel alignment of Fe magnetization and (b) Γ and MR ratio as functions of graphene length *L*.

shows the calculated results of the $\Gamma_{k_\parallel}$ with P and AP alignments of Fe magnetization in a junction with $L = 1000$ MCs. We find that the up (↑) spin conductance in the P alignment $\Gamma_{k_\parallel P\uparrow}$ shifts from the $K'$ point more than the down (↓) spin conductance $\Gamma_{k_\parallel P\downarrow}$ does. The shift in $\Gamma_{k_\parallel}$ is attributed to spin-dependent band mixing at the contact, as explained above. In the AP alignment, on the other hand, the maximum value of $\Gamma_{\mathrm{AP}}$ is smaller than $e^2/h$. (Note that Γ is plotted in logarithmic scale.) This is because the spin dependence of the hopping integrals of the left and right contacts is reversed in the AP alignment, and the ↑ and ↓ spin shifts of the DP point conflict with each other in the graphene junction.

As a result, MR occurs in FM/G/FM junctions, as shown in Fig. 6.25b, in which the *L* dependence of $\Gamma_{P\uparrow(\downarrow)}$, $\Gamma_{\mathrm{AP}}$, and the MR ratio are presented. The conductance decays in proportion to $1/L$, as expected, and the spin dependence is maintained independently of *L*; therefore, the MR ratio is constant except for very short *L*. The latter result is attributed to the *L* dependence of $\Gamma_{k_\parallel}$: both the shift of the DP and the half-width of $\Gamma_{k_\parallel}$ are proportional to $1/L$. This is a characteristic of the linear dispersion near the DP.

The MR obtained for Fe/G/Fe junctions is rather moderate because it originates from the spin dependence of the hopping integrals at the contacts. It would be interesting, however, to examine whether a large MR ratio can be realized in any ferromagnetic alloys due to spin-dependent matching/mismatching of the conductive states between graphene and electrodes, as explained for the simple MR mechanism. In Fe-based ferromagnetic alloys, a rigid band model may be applicable: the electronic structure of an Fe alloy is given

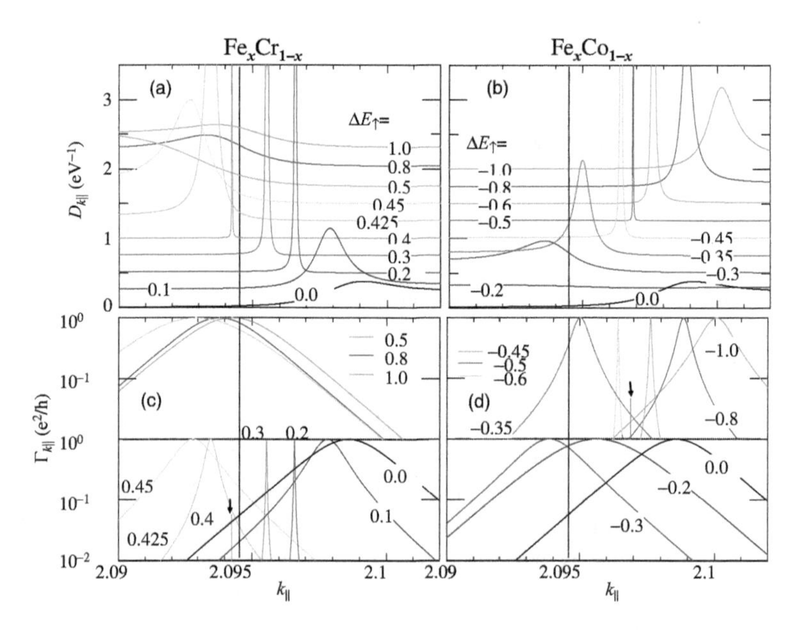


**Figure 6.26** Calculated momentum-resolved local density of states of Fe at an interface obtained by shifting the up-spin band of Fe for (a) $\Delta E_\uparrow > 0$ and (b) $\Delta E_\uparrow < 0$. (c) and (d) show the corresponding conductance values.

by shifting the up-spin Fe bands by $\Delta E_\uparrow$ (in eV) while keeping the down-spin bands unchanged. For example, FeCo alloy corresponds to $\Delta E_\uparrow < 0$, and FeCr or FeV corresponds to $\Delta E_\uparrow > 0$.

Calculated results of $k_\parallel$-resolved local DOS, $D_{k_\parallel}(E = E_F)$, at the contact with the electrode are shown in Figs. 6.26a and 6.26b for $\Delta E_\uparrow > 0$ and $<0$, respectively. We find that the shape of $D_{k_\parallel}$ changes systematically with $\Delta E_\uparrow$ and becomes extremely sharp around $\Delta E_\uparrow = -0.5$ and $0.2 - 0.4$. The corresponding results for $\Gamma_{k_\parallel\uparrow}$ are shown in Fig. 6.26c,d in a logarithmic scale. As expected, $\Gamma_{k_\parallel\uparrow}$ shifts with $\Delta E_\uparrow$ since the hopping parameter at the contact depends on the band structure at $E_F$, and $\Gamma_{k_\parallel\uparrow} \propto e^{-cL(k_\parallel - k_{DP})}$. However, $\Gamma_{k_\parallel\uparrow}$ becomes extremely small around $\Delta E_\uparrow = -0.5$ and 0.2–0.4, as pointed by arrows in the figure. Because of a strong correlation between $D_{k_\parallel}$ and $\Gamma_{k_\parallel\uparrow}$, the change in $\Gamma_\uparrow(k_\parallel)$ shown in Fig. 6.26c,d may be attributed to a change in the local DOS at the contact. $D_{k_\parallel}$ of the leads nearly vanishes for $\Delta E_\uparrow \approx -0.5$ and 0.2–0.4 in the region of $k_\parallel$ near $K'$.

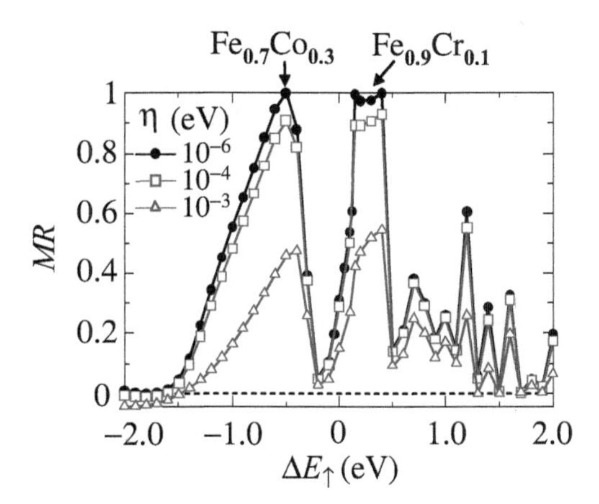


**Figure 6.27** Calculated results of MR as a function of the shift in the up-spin Fe bands. $\eta$ is an imaginary part of the energy (in eV) and shows lifetime broadening.

The sharp peaks in $D_{k_\parallel}$ may be caused by mixing with the graphene states.

The negligibly small local DOS at certain momentum states may result in huge MR because the matching of the conductive states in the graphene and the electrodes becomes worse. Figure 6.27 shows the calculated MR ratio as a function of $\Delta E_\uparrow$. We see that a large MR occurs for $0.15 < \Delta E_\uparrow < 0.4$ and at around $\Delta E_\uparrow \approx -0.5$. The former corresponds to $Fe_{0.9}Cr_{0.1}$ alloy and the latter to $Fe_{0.7}Co_{0.3}$. Since these alloys are ferromagnetic with well-defined magnetic moments, and the electronic states of down-spin electrons are not very different from those of Fe, large MR is expected to occur in graphene junctions.

Figure 6.28 shows calculated results of Γ and MR of doped (n-type; see Fig. 6.28a) graphene junctions with bcc Fe electrodes. The conductance oscillates with graphene length as reported in the previous section, and the oscillation phase depends on magnetization alignment as shown in Fig. 6.28b. As a result, the MR ratio also oscillates with graphene length, as shown in Fig. 6.28d. The oscillation period is proportionally to $1/\Delta k_\parallel$, as shown in Fig. 6.28c.

Several issues remain to be studied; effects of doping and roughness [19, 20, 33] on the conductance and MR, transport properties

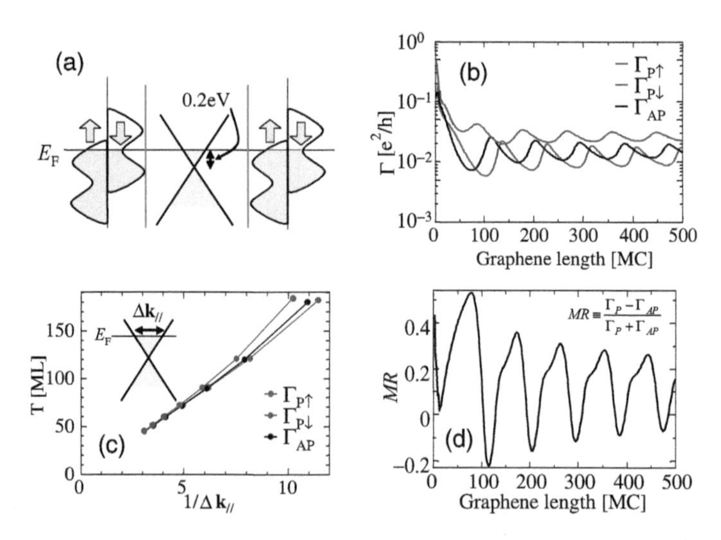


**Figure 6.28** (a) Schematics of density of states of junctions. (b) Calculated results of conductance in parallel and antiparallel alignments for Fe/graphene/Fe junctions with doped graphene as a function of graphene length. (c) Oscillation period shown in (b) and calculated results of the MR ratio as a function of graphene length.

of graphene junctions with an energy gap [23, 24, 34], such as a double-layer graphene [35], realistic graphene/metal contacts [27], and effects of hydrogen absorption. Roughness may give rise to nonconservation of $k_{\parallel}$, breaking the matching/mismatching condition of the conductive states. In fact, a lifetime broadening effect decreases the MR, as shown in Fig. 6.27. Some of these issues will be discussed in next sections.

#### 6.4.3.2 Summary

Closing this section, we may conclude that MR appears in bcc Fe/G/bcc Fe junctions with zigzag-edged contacts due to a spin-dependent shift of the DPs caused by spin-dependent band mixing between the graphene and the electrodes. The MR is independent of graphene length because of a characteristic dependence of the tunneling conductance near a DP. We further suggest that the MR can be quite large for FM electrodes made of certain ferromagnetic transition metal alloys because their electronic structure is strongly modified at the interface. The electronic states of both electrodes and graphene near the contacts are crucial for the MR effect.

## 6.4.4 *FCC Ni/Graphene/FCC Ni Junctions*

### 6.4.4.1 Effects of interlayer distance on MR

Recent first-principles calculations show that G/M contacts can be divided into two groups; one consists of those with short interlayer distance $d$ (strong coupling or chemical contact) between the graphene and the metal, and the other consists of those with long $d$ (weak coupling or physical contact) [27–29]. Contacts with ferromagnetic Ni and Co belong the first type, and the distance $d$ is about 2.1 Å [29]. Because the values of $d$ obtained in the first-principles calculation have not yet strongly confirmed by experiments [36] we treat $d$ as a parameter in our simulations and study its effect on the MR of the graphene junctions.

Figure 6.29a shows the dependence of the overlapping integrals on the distance $d$ after taking the geometrical average; $\gamma_1, \gamma_2$ and $\gamma_3$ are the ratios $pp\sigma$(G-Ni)/$pp\sigma$(Ni-Ni), $sp\sigma$(G-Ni)/ $sp\sigma$(Ni-Ni) and $pd\sigma$(G-Ni)/$pd\sigma$(Ni-Ni), respectively, for $\downarrow$ spin state. We see that they have different dependence on $d$ and that $pd\sigma$ depends most strongly on $d$.

The closed circles in Fig. 6.29b show the calculated results of the MR ratio as a function of the distance $d$ between the graphene and fcc Ni layers for junctions with contact size $a = 50$ MCs and $L = 1000$ MCs. The calculated MR ratio become small when $d$ becomes shorter than the lattice constant of fcc Ni $d_{\mathrm{Ni}} = 2.485$ Å. The MR ratio is about 0.3 for the equilibrium distance proposed by the first-principles calculation, $d = d_{\mathrm{eq}} \approx 2.1$ Å. When the distance becomes larger than $\approx 5.0 d_{\mathrm{Ni}}$, the MR ratio also decreases. Large MR ratios appear only for intermediate distances $d_{\mathrm{Ni}} \lesssim d \lesssim 5.0 d_{\mathrm{Ni}}$, in which region the hopping integrals are $10^0$–$10^{-2}$ times those in fcc Ni metal. The open squares in Fig. 6.29b represent the MR ratios calculated by multiplying the overlapping integrals of fcc Ni for given distance $d$ by an average factor $\gamma \equiv (\gamma + \gamma_2 + \gamma_3)/3$. Both curves show the same dependence on $d$. The dependence of MR on the interlayer distance for $d_{\mathrm{Ni}} \lesssim d$ is rather independent of the average of the

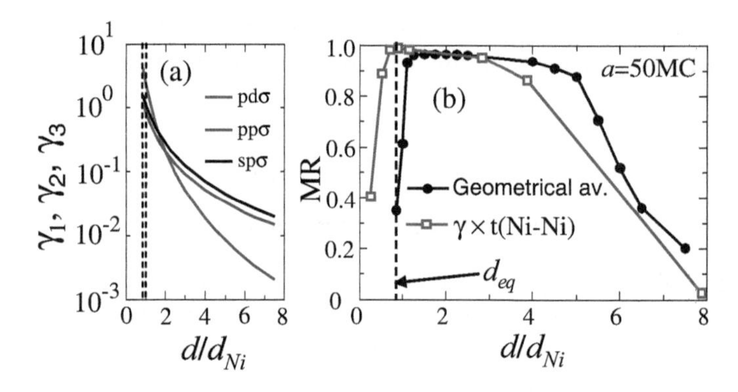


**Figure 6.29** (a) Dependence of band parameters on distance $d$ between graphene and Ni layers. $\gamma_1, \gamma_2$, and $\gamma_3$ are ratios of band parameters at the graphene/fcc Ni contact with those in fcc Ni (see text). (b) MR ratios calculated as a function of distance $d$ (solid circles) and calculated by multiplying the band parameters of Ni by a constant (open squares).

hopping parameters. We also examined MR dependence on $\alpha$ which characterize the hopping parameter at the interface being defined by $\alpha\times$ (hopping integrals of fcc Ni). The MR ratio begins to decrease when $\alpha$ is smaller than $10^{-1}$–$10^{-2}$ in consistent with results shown in Fig. 6.29b.

The results shown in Fig. 6.29b may be interpreted as follows. When $d$ is large, the DP, which is only the path of electrical conduction in the graphene sheet, would not be affected by the spin-dependent electronic structure of Ni, and the MR ratio becomes small. On the other hand, when $d$ is short, band mixing between the DP and the Ni bands becomes large to produce large high spin dependence in the electronic state near the DP. The MR ratios appear only for contacts with slightly insulating characteristics. The results indicate the importance of the contact properties to the MR ratio at least within the linear response regime.

### 6.4.4.2 Effects of contact size

A finite-size of the contact region is inevitable in simulations of the conductance of realistic junctions. Figure 6.30 shows the calculated results of the MR ratios, $\Gamma_{P\uparrow}$, $\Gamma_{P\downarrow}$ and $\Gamma_{AP\uparrow}$ $(= \Gamma_{AP\downarrow})$ as a function of the contact size $a$ for $d = 0.85d_{Ni}$, $d_{Ni}$, and $1.25d_{Ni}$. Here, the

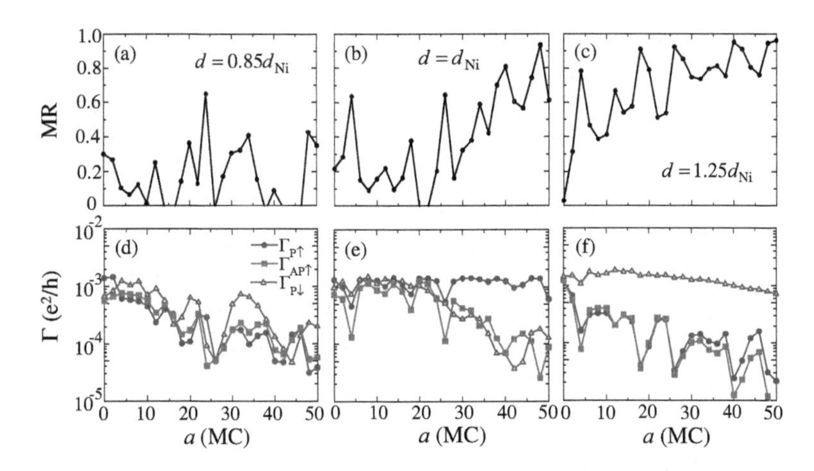


**Figure 6.30** Calculated MR ratios as a function of the contact size $a$ for (a) $d = 0.85d_{\mathrm{Ni}}$, (b) $d = 1.0d_{\mathrm{Ni}}$, and (c) $d = 1.25d_{\mathrm{Ni}}$, where $d_{\mathrm{Ni}}$ is the lattice constant of fcc Ni metal. (d), (e), and (f) show the corresponding spin-dependent conductance, $\Gamma_{\mathrm{P}\uparrow}$ and $\Gamma_{\mathrm{P}\downarrow}$ for parallel alignment and $\Gamma_{\mathrm{AP}\uparrow}$ (= $\Gamma_{\mathrm{AP}\downarrow}$) for antiparallel alignment.

distance $d = 0.85d_{\mathrm{Ni}}$ is predicted to be a stable distance by the first-principles band calculation. In addition, the atomistic structure at the contact is such that the C and Ni atoms are vertically aligned, as shown in Fig. 6.23a.

We find that the MR ratio for the contact with $d = 0.85d_{\mathrm{Ni}}$ cannot be sufficiently large irrespective of the contact size. The results shown in Fig. 6.29b are those for $a = 50$ MCs. The conductances $\Gamma_{\mathrm{P}\uparrow}$, $\Gamma_{\mathrm{P}\downarrow}$, and $\Gamma_{\mathrm{AP}\uparrow}$ show a similar dependence on $a$. With increasing distance $d$, larger MR ratios are realized for wide contacts (large $a$), as shown in Fig. 6.30b. When $d = 1.25d_{\mathrm{Ni}}$, the MR ratio can be large even for narrow contacts as shown in Fig. 6.30c. The large MR ratios are attributed to the difference between the resistance of the P and AP alignments of the magnetization, that is, $\Gamma_{\mathrm{P}\downarrow} \gg \Gamma_{\mathrm{P}\uparrow}$, and $\Gamma_{\mathrm{AP}\uparrow}$, as shown in Fig. 6.30f.

Negative MR can be seen at some values of $a$. When the magnetization alignment is AP, the degree of the mixing of states at left contact is different from that at right contact. Because the (effective) DP forms a coherent state within the graphene layer, a subtle change in the conductance may be caused by an interference of states at left and right contacts via the DP, and $\Gamma_{AP\uparrow(\downarrow)} \approx \Gamma_{P\uparrow}$. Therefore the magnitude of the conductance in AP alignment could be larger than the P alignment resulting in negative MR. The change in the conductance with the size of the contact region is related to the change in the momentum-resolved DOS near the DP at the contact. However, no systematic correlation between the MR ratio and parameter values of $a$ and $d$ could be identified.

The results shown in Fig. 6.30 suggest that large MR ratios may be obtained irrespective of the size of the contact region once a suitable interlayer distance between the graphene and the FM electrodes is adjusted.

#### 6.4.4.3 Effects of doping at contact

Recent theoretical and experimental studies have shown that electron or hole doping occurs in graphene sheets at the contact with a metal [27–30, 36]. The degree of doping depends on the type of metal; Sc, Ti, Al, Au, and Pt produce a large potential shift in graphene, whereas Co and Ni produce rather small potential shifts, -0.002 and 0.009 eV, respectively. Barraza-Lopez et al. [30] proposed a potential profile near the contact and performed TB calculations of the conductance and obtained a good agreement with the results obtained in the first-principles calculation. We adopted the proposed potential profile and introduced a potential change in the graphene sheet as shown in Fig. 6.31a. We calculated the conductance and MR ratio for junctions with various values of the potential height $\Delta E$ as defined in the figure. A tiny magnetic moment might be induced on graphene sheet at the contact with a ferromagnetic metal, but the effect on the conductance could be negligible.

Figure 6.31b,c shows the spin-dependent conductance as a function of contact size $a$ for $d = 1.25d_{\mathrm{Ni}}$. The values of $\Delta E$ are $-0.1$ eV and 0.1 eV in Figs. 6.31a and 6.31b, respectively. No great change appears in the conductance compared with the results for $\Delta E = 0.0$ shown in Fig. 6.30f. These results may indicate that the effects of charge or hole doping on the MR is weak for junctions with Ni and Co, because the predicted magnitudes of $\Delta E$ are $-0.002$ and 0.009 eV for Co and Ni, respectively [29].

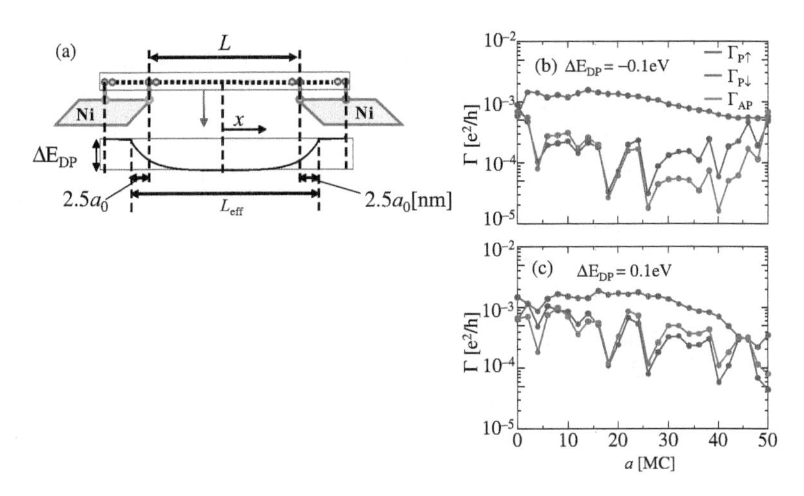


**Figure 6.31** (a) Potential profile of Ni/graphene/Ni junctions and (b) calculated results of conductance as a function of the overlap region with electron-doped graphene and (c) hole-doped graphene.

#### 6.4.4.4 Material dependence

We also consider the dependence of the MR ratio on the electrode materials. Because only states near the DPs may contribute to the transport at FM/G/FM junctions, the conductance and MR ratio could be sensitive to the electronic structures of the electrodes. Here we study the effect of alloying of the electrodes by adopting a simple model of a rigid band shift of the down-spin bands in Ni for Ni-Co and Ni-Cu alloys.

The calculated results of MR ratios for fcc Ni alloy/G/fcc Ni alloy (111) junctions are presented in Fig. 6.32a–c as a function of the band shift $\Delta E_{\downarrow}$, for $d = 0.85d_{\mathrm{Ni}}$, $d_{\mathrm{Ni}}$, and $1.25d_{\mathrm{Ni}}$, respectively, with $a = 50$ MCs and $L = 1000$ MCs. Ni-Co (Ni-Cu) alloys correspond to positive (negative) regions of $\Delta E_{\downarrow}$. Lifetime broadening $\eta$ is also taken into consideration to estimate the effect of roughness.

We first discuss the results for negligible values of $\eta$. When $d = 0.85d_{\mathrm{Ni}}$, high MR ratios appear at $\Delta E_{\downarrow} \gtrsim 0.4$. As the distance $d$ increases, the region of high MR ratio spreads and shifts to negative values of $\Delta E_{\downarrow}$. The results indicate that Ni-Co (Ni-Cu) alloys show high MR ratios when the contact distance is short (long). This behavior is closely related to band mixing between the $p_z$ orbital of

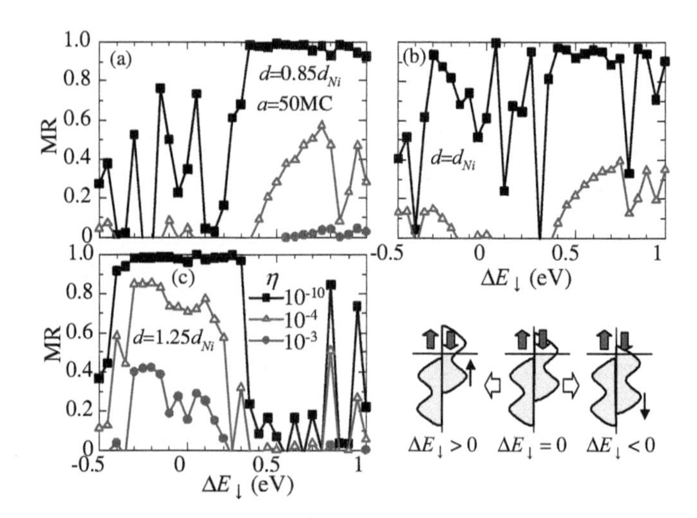


**Figure 6.32** Calculated MR ratio as a function of the shift $\Delta E_\downarrow$ of the down-spin band of Ni for (a) $d = 0.85d_{Ni}$, (b) $d = d_{Ni}$, and (c) $d = 1.25d_{Ni}$. $\eta$ is a broadening factor of the energy levels, which corresponds to the effect of roughness. (d) Schematics of $\Delta E_\downarrow$ of the down-spin band of Ni (rigid band shift).

the graphene sheet and the $s$ and $d$ orbitals of FM electrodes, as will discussed in the next section.

However, the MR ratio decreases with increasing $\eta$. This may be attributed to the sensitivity of the electronic states near the DP to the roughness. In our previous study, $\eta = 10^{-4}$ eV corresponds to impurity concentrations of 5% in the graphene sheet.

### 6.4.4.5 Effects of roughness

In Fig. 6.32 the effect of roughness was taken into account by introducing a factor $\eta$ of lifetime broadening. However, it is not clear how the magnitude relates with the roughness, for example, concentration of impurities. Figure 6.33a shows results of $\Gamma$ calculated for fcc Ni/G/fcc Ni with $\eta = 10^{-3}$ and $10^{-4}$ eV as a function of the size of the contact region. For comparison, Fig. 6.33b shows $\Gamma$ calculated for TL/G/TL junctions with 1 and 5% of impurities. From the comparison, we may expect that 5% of impurities may correspond to $\eta \approx 10^{-4}$ eV.

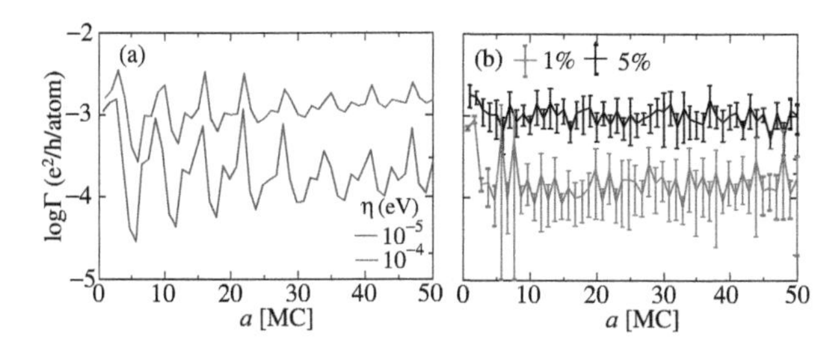


**Figure 6.33** (a) Calculated results of conductance as a function of the overlap region in a Ni/graphene/Ni junction with lifetime broadening. (b) Results calculated for triangular-lattice/graphene/triangular-lattice junctions with 1% and 5% of impurities.

Figure 6.34a shows $\eta$ dependence of $\Gamma$ and MR, which indicates that $\Gamma$ tends to increase with increasing $\eta$ as studied previously, and MR decreases with $\eta$ because a difference between $\Gamma_{P\uparrow}$ and $\Gamma_{P\downarrow}$ becomes small. Because the MR ratios are rather strongly reduced by introducing lifetime broadening in fcc Ni/G/fcc Ni(111) junctions, as shown in Fig. 6.34a, the effect of roughness on the MR ratio could be rather large in realistic junctions. Actually a decrease in the MR effect in tunnel junctions is a rather common feature in tunnel junctions, as studied Fe/MgO/Fe tunnel junctions [18, 37]. Figure 6.34b shows the calculated MR ratio as a function of the size of the contact region. We find that the simple argument on the effect of roughness might not hold for junctions with a narrow contact regions, which may be due to some additional size effects near the contacts. Figure 6.34c shows the dependence of the MR ratio on the hopping integral (a ratio $\alpha$) between the electrodes and graphene with a finite lifetime broadening $\eta = 10^{-4}$ (eV). The MR ratio tends to decrease with decreasing $\alpha$, that is, increasing interlayer distance between the electrodes and graphene, in accordance with previous results shown in Fig. 6.29b.

### 6.4.5 *Mechanism of MR*

To make the mechanism of MR in fcc Ni/G/fcc Ni junctions clearer, we calculated the momentum ($k_{\parallel}$) resolved conductance as was done for bcc Fe/G/bcc Fe junctions [5]. The results are shown

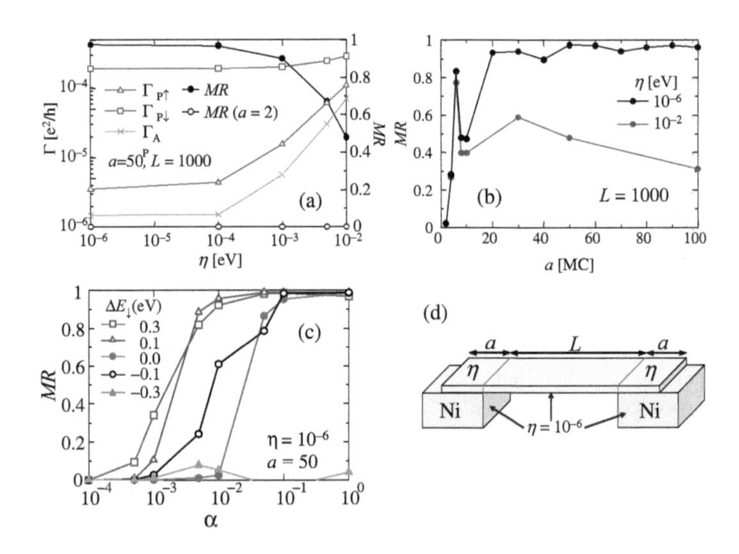


**Figure 6.34** (a) Calculated results of conductance and MR ratio for Ni/graphene/Ni junctions as a function of lifetime broadening, (b) those of MR as a function overlap region, and (c) those of MR as a function hopping parameter at the interface. (d) Schematics how the lifetime broadening $\eta$ is introduced in the junction.

in Fig. 6.35. Figures 6.35a, 6.35b, and 6.35c represent the $k_{\parallel}$ resolved conductances $\Gamma_{P\uparrow}$ and $\Gamma_{P\downarrow}$ for P alignment and $\Gamma_{AP\uparrow} = \Gamma_{AP\downarrow}$ for AP alignment calculated for Ni alloy junctions with $\Delta E_{\downarrow} = -0.40$, $-0.25$, and 0.35 eV, respectively. The corresponding MR ratios are ~0.04, 0.65, and 0.79, respectively, as given in the figures. The contact distance and contact size are $d = 0.85d_{Ni}$ and $a = 10$ MCs, respectively, and $L = 1000$ MCs. We identified these results as typical patterns of the $k_{\parallel}$ resolved conductance after calculating the conductance for various parameter values.

The $k_{\parallel}$ resolved conductances $\Gamma_{P\uparrow}$, $\Gamma_{P\downarrow}$ and $\Gamma_{AP\uparrow}$ shown in Fig. 6.35a exhibit a maximum value $e^2/h$ (quantized conductance) at a momentum slightly different from the $\bar{K}'$ point and decay exponentially. These features are the same as those calculated for bcc Fe/G/bcc Fe junctions [5]. The momentum at which the conductance shows the maximum value may be called an effective DP. The deviation of the effective DP from the $\bar{K}'$ point is related to band mixing between the graphene and the Ni electrodes, as

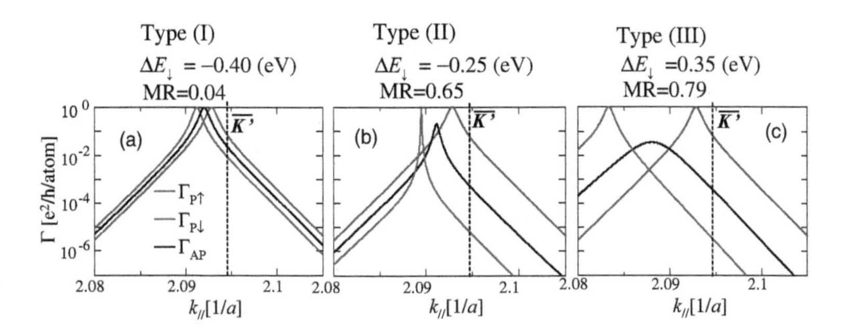


**Figure 6.35** Calculated results of momentum-resolved, spin-dependent conductance in units of $e^2/h$ per atom in parallel ($\Gamma_{P\uparrow}$ and $\Gamma_{P\downarrow}$) and antiparallel ($\Gamma_{AP\uparrow} = \Gamma_{AP\downarrow}$) alignment for junctions with $d = 0.85d_{Ni}$ and a shift in the down-spin band of Ni of (a) $\Delta E_\downarrow = -0.40$, (b) $\Delta E_\downarrow = -0.25$, and (c) $\Delta E_\downarrow = 0.35$ eV. Corresponding MR ratios are also shown. Broken lines show the position of the $\bar{K}'$ point.

discussed below. The exponential decay of the conductance is because of electron tunneling via the infinitesimally small tunnel barrier near the DP [5]. Because the shapes of $\Gamma_{P\uparrow}$, $\Gamma_{P\downarrow}$ and $\Gamma_{AP\uparrow}$ shown in Fig. 6.35a are almost the same, the conductance integrated over $k_{\|}$ shows almost no spin dependence, therefore the MR vanishes.

The calculated results shown in Fig. 6.35b differ from those in Fig. 6.35a in which the peak of $\Gamma_{P\downarrow}$ is extremely sharp, and the maximum value of $\Gamma_{AP\uparrow}$ is less than $e^2/h$. As a result, the integrated conductance becomes spin dependent, and the MR ratio is rather large. In Fig. 6.35c, the shift of the effective DP is much larger for $\Gamma_{P\downarrow}$ than for $\Gamma_{P\uparrow}$, and $\Gamma_{AP\uparrow}$ shows a broad peak with a much smaller value than $e^2/h$. In this case, the MR ratio becomes large because $\Gamma_P \gg \Gamma_{AP}$. Notable features of the $k_{\|}$ resolved conductance include the shift of the effective DP and the sharpness (or broadness) of the conductance peak. The qualitative features are irrelevant to the junction structures studied so far (SL/G/SL, TL/G/TL, bcc Fe/G/bcc Fe, and fcc Ni/G/fcc Ni junctions); however, quantitatively they depend strongly on the contact parameters and the electronic structure of the electrodes. Therefore, we infer the band mixing at the contact plays a central role in the $k_{\|}$ resolved conductance.

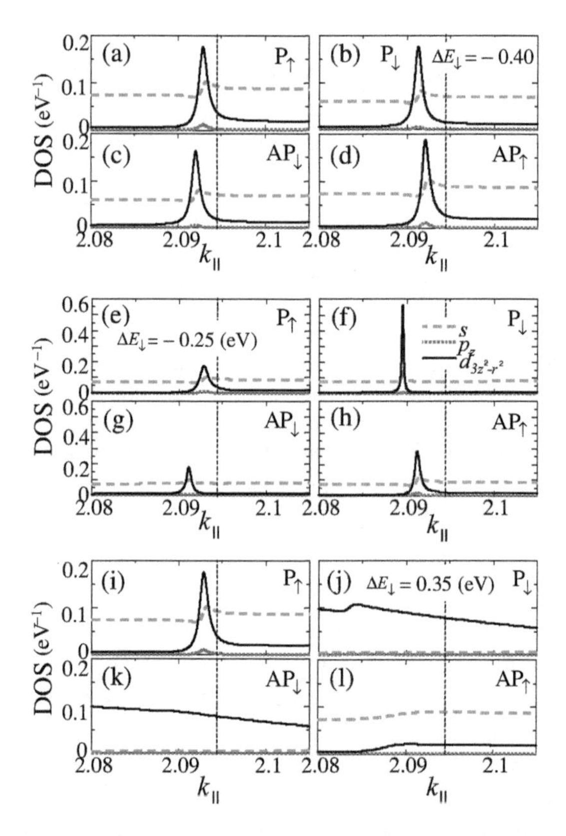


**Figure 6.36** Calculated results of momentum-resolved partial density of states at the edge site on a Ni electrode (a site denoted by a crossed circle in Fig. 6.23b). Each cluster with four panels corresponds to the momentum-resolved conductance shown in Fig. 6.35a, 6.35b, or 6.35c. Each panel includes orbital decomposed up- or down-spin density of states for parallel or antiparallel alignment. Broken lines show the position of the $\bar{K}'$ point.

The effects of band mixing at the contact may be studied by calculating the energy bands for finite-size SL/G/SL junctions with zigzag-edge contact. When little band mixing occurs at the contact, the energy bands are similar to those of zigzag-edge nanoribbons [38]; that is, the top valence band and the bottom conduction band of graphene merge at a momentum larger than $\bar{K}'$ and the edge state remains rather clear. On the other hand, when strong band mixing occurs, the edge state is strongly hybridized with the metallic bands of the electrodes to form bonding and antibonding states, and then disappears. The top valence band and bottom conduction band merge at a momentum of less than $\bar{K}'$ and form a fuzzy band as presented in Chapter 3. Comparing the calculated conductance and the energy bands, we may tentatively identify the merging point as the effective DP. Because the electronic structure of the electrodes in realistic junctions is more complicated than that of the electrodes used in SL/G/SL junctions, it is difficult to estimate the shift in the effective DP.

To reveal the underlying mechanism of the spin dependence of the $k_{\|}$ resolved conductance, we calculate the $k_{\|}$ resolved partial DOS at the site of the Ni edge, indicated by a crossed circle in Fig. 6.23b. The results given in Fig. 6.36a–d, those in Fig. 6.36e–h, and those in Fig. 6.36i–l are the $k_{\|}$ and orbital-resolved partial DOS for $\Delta E_{\downarrow} = -0.4, -0.25$, and 0.35 eV, respectively; they correspond to the results of the conductance shown in Figs. 6.35a, 6.35b, and 6.35c, respectively; Note that the partial DOSs of $P_{\uparrow}$ states in the figures are the same because only the down-spin band is shifted.

Figure 6.36a–d shows the partial DOSs of the up- and down-spin states for P and AP configurations of alloy junctions with $\Delta E_{\downarrow} = -0.4$ eV. The partial DOS of the $d_{3z^2-r^2}$ orbital shows a peak at the momentum at which the conductance also has a peak. The results shown in Fig. 6.36a–d resemble each other, and no spin dependence appears in the conductance. Unlike the results shown Fig. 6.36b, the peak of the partial DOS of the $d_{3z^2-r^2}$ orbital for $\Delta E_{\downarrow} = -0.25$ eV shown in Fig. 6.36f is quite sharp. On the other hand, the partial DOS for $\Delta E_{\downarrow} = 0.35$ eV shown in Fig. 6.36j shows no peak, although a hump structure appears for $P_{\downarrow}$ at the momentum at which the conductance $\Gamma_{P\downarrow}$ shows a peak. The $s$ component becomes small for the $P_{\downarrow}$ state. Although the partial DOSs for the $AP_{\downarrow}$ and $AP_{\uparrow}$ states with $\Delta E_{\downarrow} = 0.35$ eV are almost flat, as shown in Fig. 6.36k and 6.36l, respectively, the $d_{3z^2-r^2}$ orbital has a large component in the $AP_{\downarrow}$ state, in contrast to the large contribution of the $s$ orbital in the $AP_{\uparrow}$ state. The change in the contribution from each orbital to the partial DOS may be understood in terms of a change in the electronic state of Ni near the Fermi energy. The DOS of bulk Ni and the local DOS of the first layer of four-atomic-layer Ni are indicated in Fig. 6.37 by solid and broken curves, respectively. When $\Delta E_{\downarrow} < 0$ ($> 0$), we

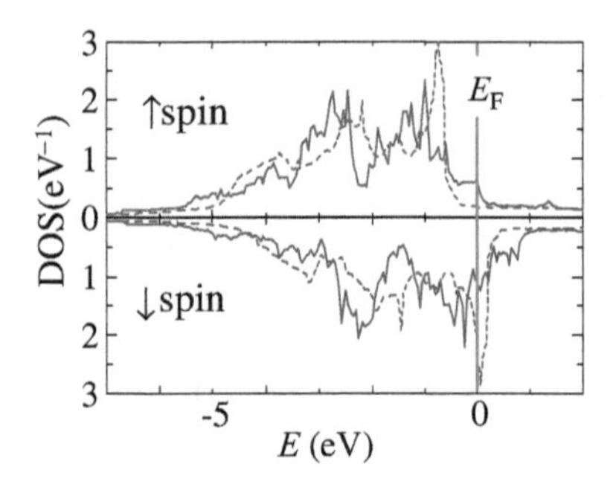


**Figure 6.37** Calculated results of density of states of bulk Ni (broken curves) and of the top atomic layer of a four-atomic-layer Ni film (solid curves).

expect the weight of the *d* component to be small (large) and that of the *s* component to be large (small).

The above results indicate that the momentum dependence of the band mixing at the contact plays an important role in the $k_{\parallel}$-resolved conductance. The sharp partial DOS of the $d_{3z^2-r^2}$ orbital in the $\mathrm{P}_{\downarrow}$ state with $\Delta E_{\downarrow} = -0.25$ eV may cause the sharpness of the $k_{\parallel}$-dependence of $\Gamma_{\mathrm{P}\downarrow}$ shown in Fig. 6.35b. Regarding $\Gamma_{\mathrm{AP}\uparrow}$ in the AP alignment, both $\mathrm{AP}_{\downarrow}$ and $\mathrm{AP}_{\uparrow}$ are relevant to the $k_{\parallel}$-dependence of $\Gamma_{\mathrm{AP}\uparrow}$, because the electronic state of $\mathrm{AP}_{\downarrow}$ on the left contact is connected to that of $\mathrm{AP}_{\uparrow}$ on the right contact. With this in mind, we find two possible sources of the broad peak of $\Gamma_{\mathrm{AP}\uparrow}$ with a maximum value much smaller than $e^2/h$ shown in Fig. 6.35c; a large difference between the shifts in the effective DPs in $\Gamma_{\mathrm{P}\uparrow}$ and $\Gamma_{\mathrm{P}\downarrow}$, and a difference in the orbital dependence of the partial DOSs for $\mathrm{AP}_{\downarrow}$ and $\mathrm{AP}_{\uparrow}$.

When there is sufficient intensity of DOS near the DP, the momentum-resolved partial DOS generally exhibits a broad peak, whereas weak or no intensity near the DP typically produces a sharp peak in the partial DOS. In the latter case, a large difference between $\Gamma_{\mathrm{P}\uparrow}$ and $\Gamma_{\mathrm{P}\downarrow}$ appears. In addition, the AP conductance depends on the shift in the effective DPs of the up- and down-spin states and on the degree of matching between the orbital components of the partial DOS at the left and right contacts. This type of mode-matching/mode-mismatching might be a generalization of the model proposed previously [4, 5, 14] and called a generalized mode-matching/mode-mismatching model.

### 6.4.6 *Discussions and Conclusions*

By adopting a realistic TB model, we studied the characteristics of the conductance and MR ratio for fcc Ni alloy/G/fcc Ni alloy junctions with zigzag-edge contact. We have clarified the necessary conditions for realizing high MR ratios in the junction: the overlapping integrals between the graphene and the electrodes should be rather weak, $1–10^{-2}$ times those in Ni metal; the width of the contact region should be larger than a few tens of zigzag chains; and Ni alloys such as Ni-Co or Ni-Cu produce high MR ratios. The mechanism has been attributed to matching/mismatching of momentum and orbital states between the electrodes and the graphene and between the left and right contacts. The mechanism might be called a generalized mode-matching/mode-mismatching mechanism. We also found that electrons or holes doped from the electrodes to the graphene may not strongly affect the conductance and MR if the potential change at the contact is less than 0.1 eV. However, the MR ratio is affected by the roughness and is suppressed by a few atomic percent of impurities.

The magnitude of the experimentally observed MR ratios is less than 10% [1–3, 39]. We speculate that the small MR ratios observed in these experiments could be attributed to junction structures that do not satisfy the conditions for high MR ratios mentioned above. Clean junctions with a suitable interlayer distance between the graphene and the electrodes should be fabricated to obtain high MR ratios in two-terminal junction structures. Such contacts may be fabricated using Pd or Pt alloys with Ni or Co. The contact resistance may be controlled by selecting an appropriate electrode material [40, 41].

We neglected the magnetism at the zigzag edges. We believe that it may not affect transport in FM/G/FM junctions with metallic contacts having a finite overlapping region because the electronic structure at the contact is strongly modified by band mixing, and because the overlapping region is wider than the region where edge magnetism is sustained. Even if edge magnetism exists at low temperature, it may disappear at room temperature [42, 43].

Here we mention briefly effects of doping and asymmetry of the junction. When electrons or holes are doped, the graphene sheet is metallic and the conductance oscillates with graphene length with a period determined by the Fermi wave vector of the doped graphene, and that the MR also oscillates with the graphene length [4]. Because the phase of the oscillation depends on spin, the MR ratio can be negative for a certain length of the graphene. Detailed calculation for fcc Ni/graphene junctions including effects of disorder, however, is left to be studied. The meaning of the asymmetry may be twofold, structural asymmetry and material asymmetry. The former has been studied for graphene junctions in a simple model by Blanter and Martin [32] but no remarkable result was presented. When the junction is asymmetrical with respect to electrode materials, the degree of the matching/mismatching at the left and right contacts may differ from each other, resulting in a complicated combination of the matching/mismatching effects on the conductance in P and AP alignments. The complication might produce negative MR for certain combinations of ferromagnetic materials for left and right electrodes.

In conclusion, we numerically simulated the conductance and MR for realistic fcc Ni alloy/G/fcc Ni alloy junctions and found that a large MR will be realized for contacts with a relatively large interlayer distance, a sufficiently wide contact region, and Ni alloy electrodes, and that a small amount of electron/hole doping at the contacts may not affect the MR. The large MR is attributed to generalized spin- and orbital-dependent mode mismatching/matching in the electronic structure near the DP. However the MR ratio decreases with increasing roughness; therefore FM/G/FM junctions must be sufficiently clean to realize high MR ratios.

## References

1. E. W. Hill, A. K. Geim, K. Novoselov, F. Schedin, and P. Blake, *IEEE Trans. Magn.*, **42**, 2694 (2006).
2. M. Nishioka and A. M. Goldman, *Appl. Phys. Lett.*, **90**, 252505 (2007).

3. M. Ohishi, M. Shiraishi, R. Nouchi, T. Nozaki, T. Shinjo, and Y. Suzuki, *Jpn. J. Appl. Phys.*, **46**, L605 (2007).
4. A. Yamamura, S. Honda, J. Inoue, and H. Itoh, *J. Magn. Soc. Jpn*, **34**, 34 (2010).
5. S. Honda, A. Yamamura, T. Hiraiwa, R. Sato, J. Inoue, and H. Itoh, *Phys. Rev. B*, **82**, 033402 (2010).
6. T. Hiraiwa, R. Sato, A. Yamamura, J. Inoue, S. Honda, and H. Itoh, *IEEE Trans. Mag.*, **47**, 2743 (2011).
7. R. Sato, T. Hiraiwa, J. Inoue, S. Honda, and H. Itoh, *Phys. Rev. B*, **80**, 004400 (2012).
8. see e.g., J. Inoue, *Nanomagnetism and Spintronics*, Elsevier (2009).
9. B. Dulbak et al., *Nat. Phys.*, **8**, 557 (2012).
10. W. Y. Kim and K. S. Kim, *Nat. Nanotech.*, **3**, 408 (2008).
11. A. Rycerz, J. Tworzydlo, and C. W. J. Beenakker, *Nat. Phys.*, **3**, 172 (2007).
12. W. Harrison, *Electronic Structure and the Properties of Solids*, W. H. Freeman (1980).
13. D. A. Papaconstantpoulos, *Handbook of the Band Structure of Elemental Solids*, Plenum, New York (1986).
14. L. Brey and H. A. Fertig, *Phys. Rev. B*, **76**, 205435 (2007).
15. D. K. Ferry, *J. Comput. Electron.*, **12**, 76 (2013).
16. Z. Wang, H. Guo, and K. H. Bevan, *J. Comput. Electron.*, **12**, 104 (2013).
17. H. Itoh and J. Inoue, *J. Magn. Soc. Jpn.*, **30**, 1 (2006).
18. H. Itoh, A. Shibata, T. Kumazaki, J. Inoue, and S. Maekawa, *J. Phys. Soc. Jpn.*, **68**, 16332 (1999).
19. H. Schomerus, *Phys. Rev. B*, **76**, 045433 (2007).
20. S. Adam, P. W. Brouwer, and S. D. Sarma, *Phys. Rev. B*, **79**, 201404(R) (2009).
21. T. C. Lin et al., *Phys. Rev. B*, **77**, 085408 (2008).
22. M. Yamamoto et al., *Phys. Rev. B*, **79**, 125421 (2009).
23. Y.-W. Son, M. L. Cohen, and S. G. Louie, *Phys. Rev. Lett.*, **97**, 216803 (2006).
24. M. Y. Han, B. Özyilmaz, Y. Zhang, and P. Kim, *Phys. Rev. Lett.*, **98**, 206805 (2007).
25. D. C. Elias, R. R. Nair, T. M. G. Mohiuddin, S. V. Morozov, P. Blake, M. P. Halsall, A. C. Ferrari, D. W. Boukhvalov, M. I. Katsnelson, A. K. Geim, and K. S. Novoselov, *Science*, **323**, 610 (2009).
26. K. Pi, K. M. KcCreary, W. Bao, Wei Han, Y. F. Chiang, Yan Li, S.-W. Tsai, C. N. Lau, and R. K. Kawakami, *Phys. Rev. B*, **80**, 075406 (2009).

27. G. Giovannetti, P. A. Khomyakov, G. Brocks, V. M. Karpan, J. van den Brink, and P. J. Kelly, *Phys. Rev. Lett.*, **101**, 026803 (2008).
28. N. Memec, D. Tomanek, and G. Cuniberti, *Phys. Rev. B*, **77**, 125420 (2008).
29. Q. Ran, M. Gao, X. Guan, Y. Wang, and Z. Yu, *Appl. Phys. Lett.*, **94**, 103511 (2009).
30. S. B-Lopez, M. Vanević, M. Kindermann, and M. Y. Chou, *Phys. Rev. Lett.*, **104**, 076807 (2010).
31. A. H. Castro Neto, F. Guinea, N. M. R. Peres, K. S. Novoselov, and A. K. Geim, *Rev. Mod. Phys.*, **81**, 109 (2009), and references therein.
32. Ya. M. Blanter and I. Martin, *Phys. Rev. B*, **76**, 155433 (2007).
33. M. Y. Han, J. C. Brant, and P. Kim, *Phys. Rev. Lett.*, **104**, 056801 (2010).
34. L. Yang, C.-H Park, Y.-W. Son, M. L. Cohen, and S G. Louie, *Phys. Rev. Lett.*, **99**, 186801 (2007).
35. E. V. Castro, K. S. Novoselov, S. V. Morozov, N. M. R. Peres, J. M. B. Lopes dos Santos, J. Nilsson, F. Guinea, A. K. Geim, and A. H. Castro Neto, *Phys. Rev. Lett.*, **99**, 216802 (2007).
36. C. E. Malec and D. Davidović, *Phys. Rev. B*, **84**, 033407 (2011).
37. H. Itoh, *J. Phys. D: Appl. Phys.*, **40**, 1228 (2007).
38. K. Nakada, M. Fujita, G. Dresselhaus, and M. S. Dresselhaus, *Phys. Rev. B*, **54**, 17954 (1996-II).
39. Z-M. Liao, H-C. Wu, J-J. Wang, G. L. W. Cross, S. Kumar, I. V. Shvets, and G. S. Duesberg, *Appl. Phys. Lett.*, **98**, 052511 (2011).
40. J. A. Robinson, M. Labella, M. Zhu, M. Hollander, R. Kasarda, Z. Hughes, K. Trumbull, R. Cavalero, and D. Snyder, *Appl. Phys. Lett.*, **98**, 053101 (2011).
41. B.-C. Huang, M. Zhang, Y. Wang, and J. Woo, *Appl. Phys. Lett.*, **99**, 032107 (2011).
42. M. Sepioni, R. R. Nair, S. Rablen, J. Narayanan, F. Tuna, R. Winpenny, A. K. Geim, and I. V. Grigorieva, *Phys. Rev. Lett.*, **105**, 207205 (2010).
43. J. Kunstmann, C. Özdoğan, A. Quandt, and H. Fehske, *Phys. Rev. B*, **83**, 045414 (2011).

# Chapter 7

# Summary

## 7.1 New Materials and New Fields of Science

It is well known that the discovery of a new material opens the way to new fields in science and in technological applications. The discovery of graphene is surely not an exception. From around 1980 many discoveries of new materials have been reported: quasi-crystals with five-rotational symmetry, high-Curie-temperature superconductivity in cuprates such as $(La\text{-}Sr)_2CoO_4$, superconductors made of Fe and As elements, magnetic multilayers that resulted in the discovery of giant magnetoresistance (GMR) and tunnel magnetoresistance (TMR), magnetic semiconductors such as (Ga-Mn)As, $C_{60}$, carbon nanotubes, and the new permanent magnet $Nd_2Fe_{14}B$ called Neomax. The establishment of fabrication techniques of Ga-N should also be highly approved.

Among these discoveries, the permanent magnet $Nd_2Fe_{14}B$ with a few dopants of Dy ions and Ga-N is now widely mass-produced to in apply various fields. Although other materials show outstanding features, some of them encounter difficulties to be overcome for technological applications. For commercialization, performance at room temperature should be guaranteed with sufficient margin. The Curie temperature of oxide superconductors could not exceed

*Graphene in Spintronics: Fundamentals and Applications*
Jun-ichiro Inoue, Ai Yamakage, and Shuta Honda

ISBN 978-981-4669-56-6 (Hardcover), 978-981-4669-57-3 (eBook)
www.panstanford.com

170 K, and the magnetic Curie temperature of (Ga-Mn)As and related magnetic semiconductors remains less than 200 K. Although the finding of suitable materials for commercial applications of these materials has not yet succeeded, the search for new materials and the study of basic physics and chemistry have expanded worldwide and have brought about rich production in the field of science.

The GMR phenomenon in magnetic multilayers has been promptly applied in the field of electronics and opened a new scientific field called *spintronics*, as mentioned in earlier chapters of this book. TMR is on the way to mass production for spintronics applications. As for carbon materials of fullerenes, carbon nanotubes, and graphene, a lot of proposals have been presented, as briefly introduced in Chapter 1.

## 7.2 GMR and TMR in Spintronics

The basic field of spintronics, started from the discovery of GMR, has expanded by taking into account various phenomena such as TMR, spin injection into semiconductors, spin transfer torque, and orbital-related phenomena, including spin–orbit interactions. In spite of successful research in chemistry and physics, the commercialization of spintronics meets a lot of difficulties to overcome the so-called "valley of death" and "river of death."

In my personal opinion, successful commercialization of the phenomena of GMR and TMR may be partly attributed to their robustness and simplicity. Robustness means that the quality of samples (magnetic multilayers) is easily controlled to obtain desirable performance, and simplicity means that the physics in the phenomena can be understood classically or semiclassically. It is also noted that high-quality magnetic trilayers and multilayers can be fabricated by the sputtering method, instead of the molecular beam epitaxy (MBE) method, which makes mass production of devices possible.

What is a more important point to be noted is that an additional technique (or physics) has been invented for commercialization. It is the effect of exchange bias for GMR. To apply GMR to magnetic sensors, the resistivity should be changed within a few Oe instead of several hundreds of Oe in the first observation of GMR. Such sensitivity could be achieved by attaching antiferromagnetically coupled layers to the GMR layers. The pinning effect of the antiferromagnetically coupled layers is decisive to the technological application of GMR. In spite of successful application of GMR using the effect of exchange bias, the physics of the latter has not been clarified yet.

Such an additional technique desirable for the application of TMR to magnetoresistive random access memory (MRAM) is the method of magnetization reversal. As shown in Chapter 5, the realistic structure of a TMR device is complicated; however, it is not a problem because the fabrication of such structures is supported by silicon technology. The first method adopted to reverse the magnetization direction of TMR junctions was to utilize the double external magnetic field, but the method was insufficient for downsizing of devices. The proposal of the method of spin transfer torque to control the magnetization direction of TMR junctions had perfect timing. Combining the method of spin transfer torque with the perpendicular magnetic anisotropy of thin magnetic films, the MRAM application of TMR has made much progress. Nevertheless, it is reported that more ingenuity would be required to suppress the current density (and therefore suppress heat production) necessary to control the magnetization direction of TMR junctions.

## 7.3 Graphene in Spintronics

The most promising applications of graphene may be high-frequency field-effect transistors (FETs) and transparent electrodes due to two-dimensionality, high mobility, electron–hole symmetry, and a single atomic layer of the graphene sheet. In spite of the outstanding performance of the physical properties of graphene, the spintronics application of graphene seems to be difficult to achieve. The largest difficulty may be the formation of an energy band gap in graphene. Realization of spin transport through graphene junctions is essential in spin FETs. Improvement of the fabrication technique of graphene sheets and the realization of suitable contact/junction structures in devices are also desirable for technological applications.

We have reported the transport properties of the structure of spin FETs made of graphene contacts with nonmagnetic and ferromagnetic metals. Most of the theoretical calculations of the spin and charge transport have been performed in the quantum regime, and favorable results have been obtained for spin FETs and magnetoresistance of ferromagnetic junctions with a graphene spacer. Because the mean-free path and spin diffusion length could be sufficiently long as compared to the length-scale graphene sheet in junctions, we expect that ballistic transport could be confirmed in graphene junctions. Thus the realization of graphene junctions in the ballistic regime would be a milestone for graphene applications in spintronics.

There could be an alternative way to achieve spintronics applications of graphene by contriving novel device structures using graphene sheets, for example, a double vertical tunnel junction of two graphene sheets reported in Chapter 5. We expect such research direction to be developed, because even in GMR and TMR applications, magnetic trilayers with nonmagnetic or insulating spacers between ferromagnetic films themselves are used combined with other materials or with other physical concepts such as exchange bias and spin transfer torque.

Furthermore, we would like to mention a possibility to find novel materials with two-dimensionality, inversion asymmetry, and possibly spin–orbit interactions. The inversion asymmetry may make magneto-electric effects possible, and therefore more interesting properties, including magnetism, could be realized. Actually it has recently been reported that transition metal dicalcogenide $WSe_2$ shows interesting properties as an FET [1].

Finally novel electronics using valley degrees of freedom and pseudo spin in graphene is also a novel direction of research [2, 3].

Summarizing, improvement of the fabrication technique of graphene sheets, novel device structures for desired properties, realization of graphene junctions that operate in the ballistic transport regime, and control of the electronic structure of graphene by adopting a bilayer contact of graphene with other materials or by fabrication of novel materials are desirable for the realization of graphene applications in the field of spintronics in the near future.

## References

1. Y. J. Zhang et al., *Science*, **344**, 725 (2014).
2. D. Gunlycke and C. T. White, *Phys. Rev. Lett.*, **106**, 136806 (2011).
3. D. Pesin and A. H. MacDonald, *Nat. Mater.*, **11**, 409 (2012).

# Appendix

## A.1 Conductance Formalism for Numerical Calculation

### A.1.1 *Kubo–Greenwood and Lee–Fisher Formalisms*

The general expression of conductivity is given by the Kubo formula [1]; however, the formula is rather complicated to be applied to numerical calculations. Instead the so-called Kubo–Greenwood formula, which is derived from the general expression of the Kubo formula by applying one-electron approximation, is used for numerical calculations. The Kubo–Greenwood formulae for the conductivity and conductance are given as

$$\sigma = \frac{\pi\hbar}{\Omega}\mathrm{Tr}\left[\mathcal{J}\delta\left(\varepsilon_{\mathrm{F}} - \mathcal{H}\right)\mathcal{J}\delta\left(\varepsilon_{\mathrm{F}} - \mathcal{H}\right)\right], \tag{A.1}$$

$$\Gamma = \frac{\pi\hbar}{L^2}\mathrm{Tr}\left[\mathcal{J}\delta\left(\varepsilon_{\mathrm{F}} - \mathcal{H}\right)\mathcal{J}\delta\left(\varepsilon_{\mathrm{F}} - \mathcal{H}\right)\right], \tag{A.2}$$

where $\mathcal{J}$ and $\mathcal{H}$ are the current operator and Hamiltonian, respectively, $\Omega$ is the system volume, and $L$ is the length of the systems. These expressions are rewritten in the following way by using Green's functions, defined as

$$\begin{aligned} G^{\pm}(E) &= [E \mp i\eta - \mathcal{H}]^{-1} \\ &= P[E - \mathcal{H}]^{-1} \pm i\pi\delta(\epsilon - \mathcal{H}). \end{aligned} \tag{A.3}$$

Here, $\eta$ is an infinitesimally small positive value and $G^+$ and $G^-$ are called retarded and advanced Green's functions, respectively. By using the relation that

$$\delta(E - \mathcal{H}) = \frac{1}{2\pi i}\left\{G^+(E) - G^-(E)\right\}, \tag{A.4}$$

conductance is expressed as

$$\Gamma = \frac{\pi\hbar}{L^2}\frac{1}{(2\pi i)^2}\mathrm{Tr}[\mathcal{J}(G^+ - G^-)\mathcal{J}(G^+ - G^-)]. \tag{A.5}$$

Here, the Fermi energy $\varepsilon_\mathrm{F}$ is omitted in the expression.

The current operator may be given as the sum of each atomic sites and expressed as

$$\mathcal{J} = \sum_l \mathcal{J}_l, \tag{A.6}$$

with

$$\begin{aligned}\mathcal{J}_l &= \frac{eta}{i\hbar}\sum_i (c^\dagger_{l+1,i}c_{l,i} - c^\dagger_{l,i}c_{l+1,i}) \\ &= \frac{ea}{i\hbar}(|l+1\rangle \boldsymbol{t}_l \langle l| - |l\rangle \boldsymbol{t}_l^\dagger \langle l+1|),\end{aligned} \tag{A.7}$$

where $l$ is a suffix for atomic layer and $i$ indicates the atomic position within the layer. Here $a$ is the lattice constant, and $\boldsymbol{t}$ indicates the hopping matrix specified by the Hamiltonian. Inserting Eq. A.6 into Eq. A.5, one obtains

$$\Gamma = \frac{\pi\hbar}{L^2}\frac{1}{(2\pi i)^2}\sum_{l,m}\mathrm{Tr}[\mathcal{J}_l(G^+ - G^-)\mathcal{J}_m(G^+ - G^-)].$$

Because of the current conservation, the summation over $l$ and $m$ gives $N^2$, where $N$ is the number of layers and $L = Na$:

$$\Gamma = \frac{\pi\hbar}{a^2}\frac{1}{(2\pi i)^2}\mathrm{Tr}[\mathcal{J}_l(G^+ - G^-)\mathcal{J}_m(G^+ - G^-)]. \tag{A.8}$$

By defining

$$\tilde{G} \equiv G^+ - G^-,$$

and inserting Eq. A.7 into Eq. A.8, we obtain

$$\begin{aligned}\Gamma &= \frac{e^2}{4\pi\hbar}\mathrm{Tr}[(|l+1\rangle \boldsymbol{t}_l\langle l|\tilde{G} - |l\rangle \boldsymbol{t}_l^\dagger\langle l+1|\tilde{G})(|m+1\rangle \boldsymbol{t}_l\langle m|\tilde{G} \\ &\quad -|m\rangle \boldsymbol{t}_l^\dagger\langle m+1|\tilde{G}) \\ &= \frac{e^2}{4\pi\hbar}\mathrm{Tr}[|l+1\rangle \boldsymbol{t}_l\langle l|\tilde{G}|m+1\rangle \boldsymbol{t}_l\langle m|\tilde{G} + |l\rangle)\boldsymbol{t}_l^\dagger\langle l+1|\tilde{G}|m\rangle \boldsymbol{t}_l^\dagger\langle m+1|\tilde{G} \\ &\quad -|l+1\rangle \boldsymbol{t}_l\langle l|\tilde{G}|m\rangle \boldsymbol{t}_l^\dagger\langle m+1|\tilde{G} - |l\rangle \boldsymbol{t}_l^\dagger\langle l+1|\tilde{G}|m+1\rangle \boldsymbol{t}_l\langle m|\tilde{G}].\end{aligned} \tag{A.9}$$

Taking the trace over atomic layers, the conductance is given as

$$\Gamma = \frac{e^2}{2h}\mathrm{Tr}\left[\boldsymbol{t}_l\tilde{G}_{l,m+1}\boldsymbol{t}_l\tilde{G}_{m,l+1} + \boldsymbol{t}_l^\dagger\tilde{G}_{l+1,m}^\dagger\boldsymbol{t}_l\tilde{G}_{m+1,l}\right.$$
$$\left. -\boldsymbol{t}_l\tilde{G}_{l,m}^\dagger\boldsymbol{t}_l\tilde{G}_{m+1,l+1} - \boldsymbol{t}_l^\dagger\tilde{G}_{l+1,m+1}\boldsymbol{t}_l\tilde{G}_{m,l}\right], \quad \text{(A.10)}$$

with

$$\tilde{G}_{i,j} = \langle i|\tilde{G}|j\rangle.$$

Because of the current conservation, the layer indices $l$ and $m$ are arbitrary, and we take $l = m$, and Eq. A.10 is expressed as

$$\Gamma = \frac{e^2}{2h}\mathrm{Tr}\left[\boldsymbol{t}_l\tilde{G}_{l,l+1}\boldsymbol{t}_l\tilde{G}_{l,l+1} + \boldsymbol{t}_l^\dagger\tilde{G}_{l+1,l}\boldsymbol{t}_l^\dagger\tilde{G}_{l+1,l}\right.$$
$$\left. -\boldsymbol{t}_l\tilde{G}_{l,l}\boldsymbol{t}_l^\dagger\tilde{G}_{l+1,l+1} - \boldsymbol{t}_l^\dagger\tilde{G}_{l+1,l+1}\boldsymbol{t}_l\tilde{G}_{l,l}\right]. \quad \text{(A.11)}$$

The last expression is called Lee–Fisher formula [2]. Here, the trace is taken over spin, orbital, and atomic sites with a layer. One may obtain conductance by calculating the local Green's functions $\tilde{G}_{l,l}$ and $\tilde{G}_{l+1,l+1}$ and nonlocal Green's functions $\tilde{G}_{l,l+1}$ and $\tilde{G}_{l+1,l}$.

## A.1.2 *Recursive Green's Function Method*

The local and nonlocal Green's functions are calculated recursively in the following way. Let us consider a half-infinite system, and let $G_{n-1,n-1}$ (which is written as $G_{n-1}$ in Fig. A.1) be a local Green's function at the surface $n-1$ of the half-infinite system. We add an isolated single layer $n$, the Green's function of which is given as $g_n$, onto the surface.

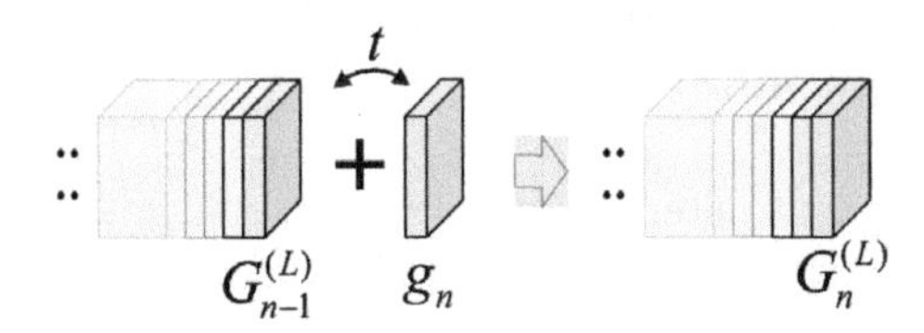


**Figure A.1** Add an isolated layer $n$ onto the surface of a half-infinite system.

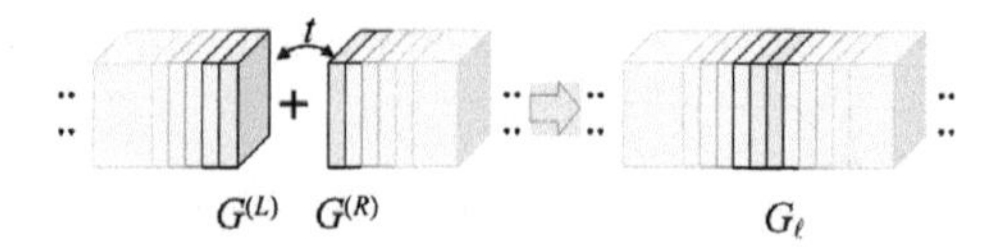


**Figure A.2** Procedure to combine left- and right-infinite systems.

The surface Green's function $G_{n,n}$ and Green's function of the isolated single layer $g_n$ are given as

$$G_{n,n} = \langle n| \frac{1}{E - \mathcal{H} + i\eta} |n\rangle, \tag{A.12}$$

$$g_n = \frac{1}{E + i\eta - \mathcal{H}_n}, \tag{A.13}$$

respectively. The new surface Green's function is denoted to be $G_{n,n}$ (Gn in the figure). By using a hopping matrix $\boldsymbol{t}$ between the layer $n$ and $n-1$, the relation between $G_{n-1,n-1}$ and $G_{n,n}$ is given recursively as

$$\begin{aligned} G_{n,n} &= g_n + g_n \mathbf{t}^\dagger G_{n-1,n-1} \mathbf{t} g_n + g_n \mathbf{t}^\dagger G_{n-1,n-1} \mathbf{t} g_n \mathbf{t}^\dagger G_{n-1,n-1} \mathbf{t} g_n + \cdots \\ &= g_n [1 - \mathbf{t}^\dagger G_{n-1,n-1} \mathbf{t} g_n]^{-1} \\ &= [E + i\eta - \mathcal{H}_n - \mathbf{t}^\dagger G_{n-1,n-1} \mathbf{t}]^{-1}. \end{aligned} \tag{A.14}$$

Here, $\mathcal{H}_n = \mathcal{H}_{n,n}$, $\vec{t} = \mathcal{H}_{n-1,n}$. Performing this procedure for left- and right-infinite systems and adding them at a certain point, we obtain local and nonlocal Green's functions $G_{n,n}$, $G_{n+1,n+1}$ and $G_{n,n+1}$, $G_{n+1,n}$. See Fig. A.2.

Green's function of the half-infinite system may be calculated by using Möbious transpormation [3] or by adopting an approximate method in which Green's functions are replaced by diagonal Green's functions.

## A.2 Alternative Method to Calculate Conductance and Its Application

Here we show an alternative method to calculate the conduction of junctions. The method is formally identical to that explained in the previous section. In the following section, we explain the method briefly and give an example of calculating the conductance and magnetoresistance (MR) of a system expressed in terms of a Rashba-type Hamiltonian. It is well known that the Rashba-type Hamiltonian caused by a strong spin–orbit interaction (SOI) gives rise to Dirac dispersion near the Γ point in momentum space in a two-dimensional electron gas (2DEG). Although many studies have been undertaken on the quantum transport for the Rashba-type Hamiltonian, the MR effect of ferromagnetic junctions involving a spacer with a Rashba-type Hamiltonian has seldom been investigated. Because the Dirac point is the only state that carries electrical transport, novel and enhanced MR properties can be expected, as discussed in Chapter 6.

### A.2.1 *Model and Formalism*

The lateral junction is two dimensional and consists of two ferromagnets separated by a nonmagnetic spacer with a Rashba-type Hamiltonian, as shown in Fig. A.3. The junctions are symmetric, and the lattice structure is assumed to be square for both electrodes and the spacer.

The Hamiltonian of the spacer is given as

$$H_S = \sum_{ij,\sigma} \varepsilon_S c^{\dagger}_{i,j,\sigma} c_{i,j,\sigma} - t \sum_{ij,\sigma\sigma'} \left[ -c^{\dagger}_{i+1,j,\sigma} c_{i,j,\sigma'} \left( i\boldsymbol{\sigma}_y \right)_{\sigma\sigma'} \right.$$

$$\left. + c^{\dagger}_{i,j+1,\sigma} c_{i,j,\sigma'} \left( i\boldsymbol{\sigma}_x \right)_{\sigma\sigma'} + \text{H.c.} \right] + \boldsymbol{\sigma} \cdot \boldsymbol{h}, \tag{A.15}$$

where the first term denotes atomic potentials, the second term is the Rashba-type hopping with a parameter $t > 0$, and the last term is a Zeeman interaction resulting from an effective magnetic field $\boldsymbol{h}$. $\boldsymbol{\sigma} = (\sigma_x, \sigma_y, \sigma_z)$ are the Pauli Matrices; $i$ and $j$ denote lattice points along the $y$ and $x$ directions, respectively; and $\sigma(= +, -)$ is the spin suffix. Because no disorder is permitted at the junction, there is translational symmetry along the interface ($x$ direction). Therefore, the electronic states are expressed in terms of the position of layer $i$ along the $y$ direction and a wave vector $k_{\parallel}$ along the $x$ direction. By using this expression, the Hamiltonians of the intra-atomic layer (actually a chain), $H^1_S$, and of the interatomic layers, $H^2_S$, of the spacer are given as

$$H^1_S = \begin{pmatrix} \varepsilon_S + h_z & 2t\sin k_{\parallel} + h_x - ih_y \\ 2t\sin k_{\parallel} + h_x + ih_y & \varepsilon_S - h_z \end{pmatrix} \tag{A.16}$$

and

$$H_S^2 = -t\begin{pmatrix} 0 & -1 \\ 1 & 0 \end{pmatrix}, \quad H_S^{2\dagger} = -t\begin{pmatrix} 0 & 1 \\ -1 & 0 \end{pmatrix}, \tag{A.17}$$

respectively. Here the atomic distance is taken as 1 and $\varepsilon_s = 0$, unless otherwise mentioned.

The Hamiltonians of the left and right electrodes, $H_{\rm L}$ and $H_{\rm R}$, respectively, are assumed to be given by a single-orbital tight-binding model with the same hopping integral $t$ as that of the spacer. By including an exchange field $\Delta_{\rm ex}$, the intralayer Hamiltonian of the electrodes can be given as

$$H_{\rm L(R)} = \begin{pmatrix} \varepsilon_0 - \frac{\Delta_{\rm ex}}{2} - 2t\cos k_\parallel & 0 \\ 0 & \varepsilon_0 + \frac{\Delta_{\rm ex}}{2} - 2t\cos k_\parallel \end{pmatrix},$$

with an on-site potential $\varepsilon_0$ of atoms in the electrode. There remains an ambiguity for the Hamiltonians at the left and right interfaces, $H_{\rm LS}$ and $H_{\rm RS}$, respectively; however, we assume that they are given by diagonal matrices as $H_{\rm LS} = -t_{\rm L}\mathbf{1}$ and $H_{\rm RS} = -t_{\rm R}\mathbf{1}$. The total Hamiltonian is given by a block tridiagonal matrix.

An expression of the conductance of the junction has been given by Caroli et al. [4] using the nonequilibrium Green's function method, in which $H_{\rm LS}$ and $H_{\rm RS}$ are treated as perturbations to infinite order. See also R. Landauer [5]. The nonequilibrium Green's function $\boldsymbol{G}$ is expressed in terms of the equilibrium Green's function $\boldsymbol{g}$ in which $H_{\rm LS} = H_{\rm RS} = 0$. In the zero-bias limit, the conductance is given as

$$\Gamma = \frac{1}{N_{k_\parallel}} \sum_{k_\parallel} \Gamma(k_\parallel), \tag{A.18}$$

with

$$\Gamma(k_\parallel) = \frac{|e|^2}{h}(2\pi t_{\rm L} t_{\rm R})^2\, {\rm Tr}\, \boldsymbol{D}_{Lk_\parallel} \boldsymbol{G}_{1n}^R(k_\parallel) \boldsymbol{D}_{Rk_\parallel} \boldsymbol{G}_{n1}^A(k_\parallel). \tag{A.19}$$

Here, $\Gamma(k_\parallel)$ is the momentum-resolved conductance, $\boldsymbol{D}_{{\rm L(R)}k_\parallel}$ a local density of states (DOS) of the semi-infinite left (right) electrode, and $\boldsymbol{G}_{1n}^{\rm R(A)}(k_\parallel)$ is a nonlocal retarded (advanced) Green's function between the left and right edges of the spacer (layer 1 and layer $n$). Note that $\boldsymbol{G}_{1n}^{\rm R(A)}(k_\parallel)$ fully includes the effects of the electrodes. The trace is taken over spin.

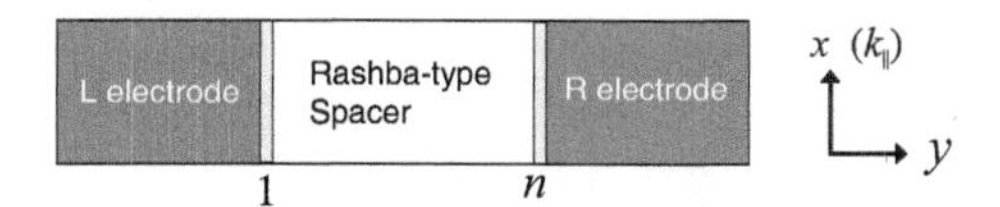


**Figure A.3** Schematic figure of a lateral junction made of ferromagnetic electrodes and a spacer with a Rashba-type hopping term. The junction is periodic along the $x$ direction, and 1 and $n$ are layer indices of the left and right edges of the spacer.

The nonlocal Green's function $\boldsymbol{G}_{1n}$ and local Green's function $\boldsymbol{G}_{11}$ are given, respectively, as

$$\boldsymbol{G}_{1n} = \left[(\mathbf{1} - \boldsymbol{V}_n \boldsymbol{g}_{nn}) \boldsymbol{g}_{1n}^{-1} (1 - \boldsymbol{g}_{11} \boldsymbol{V}_1) - \boldsymbol{V}_n \boldsymbol{g}_{n1} \boldsymbol{V}_1\right]^{-1} \quad \text{(A.20)}$$

and

$$\boldsymbol{G}_{11} = [\boldsymbol{g}_{11} + \boldsymbol{G}_{1n} \boldsymbol{V}_n \boldsymbol{g}_{n1}] (\mathbf{1} - \boldsymbol{V}_1 \boldsymbol{g}_{11})^{-1}. \quad \text{(A.21)}$$

Here $\boldsymbol{V}_1 = t_{\mathrm{L}}^2 \boldsymbol{g}_{\mathrm{L}}$ and $\boldsymbol{V}_n = t_{\mathrm{R}}^2 \boldsymbol{g}_{\mathrm{R}}$, in which $\boldsymbol{g}_{\mathrm{L(R)}}$ is the local Green's function at the edge of the left (right) electrode, and $\boldsymbol{g}_{1n}$ is the nonlocal Green's function of the isolated spacer. The latter is easily calculated recursively, as shown in the appendix. Here the superscripts $R$ and $A$ have been omitted.

The nonlocal Green's function may be calculated recursively. Hamiltonians of the two-dimensional model in a mixed presentation may be given in terms of a block tridiagonal matrix with a size of $n \times n$. Green's function is expressed as

$$\boldsymbol{G} = [(\omega \pm i0)\,\mathbf{1} - \boldsymbol{H}]^{-1}. \quad \text{(A.22)}$$

The matrix of $(\omega \pm i0)\,\mathbf{1} - \boldsymbol{H}$ has the form

$$\boldsymbol{F} = \begin{bmatrix} \boldsymbol{f}_1 & -\boldsymbol{t} & & & & \\ -\boldsymbol{t}^\dagger & \boldsymbol{f}_2 & -\boldsymbol{t} & & & \\ & -\boldsymbol{t}^\dagger & \boldsymbol{f}_3 & \cdot & & \\ & & \cdot & \cdot & \cdot & \\ & & & \cdot & \cdot & -\boldsymbol{t} \\ & & & & -\boldsymbol{t}^\dagger & \boldsymbol{f}_n \end{bmatrix}, \quad \text{(A.23)}$$

and the matrix element of Green's function is given as

$$\boldsymbol{G}_{1n} = \left(\boldsymbol{F}^{-1}\right)_{1n}, \quad \text{(A.24)}$$

which is obtained recursively as

$$\begin{aligned} \left(\boldsymbol{F}^{-1}\right)_{1n} &= \boldsymbol{g}_1 \boldsymbol{t} \bar{\boldsymbol{g}}_2 \boldsymbol{t} \bar{\boldsymbol{g}}_3 \boldsymbol{t} \cdots \bar{\boldsymbol{g}}_{n-1} \boldsymbol{t} \bar{\boldsymbol{g}}_n \\ &= \boldsymbol{g}_1 \Pi_{i=2}^{n} (\boldsymbol{t} \bar{\boldsymbol{g}}_i), \end{aligned} \quad \text{(A.25)}$$

with

$$\bar{g}_i = \left[f_i - t^\dagger \bar{g}_{i-1} t\right]^{-1}, \tag{A.26}$$

$$\bar{g}_2 = \left[f_2 - t^\dagger g_1 t\right]^{-1}, \tag{A.27}$$

$$g_1 = f_1^{-1}. \tag{A.28}$$

Here,

$$f_i = \begin{bmatrix} z - \varepsilon_0 & -2t \sin k_\parallel \\ -2t \sin k_\parallel & z - \varepsilon_0 \end{bmatrix}, \tag{A.29}$$

$$t = -t \begin{pmatrix} 0 & -1 \\ 1 & 0 \end{pmatrix}, \tag{A.30}$$

$$t^\dagger = -t \begin{pmatrix} 0 & 1 \\ -1 & 0 \end{pmatrix}. \tag{A.31}$$

The Zeeman term has been neglected. The expression for $G_{n1}$ is given by $\left(F^{-1}\right)_{1n} = g_1 t^\dagger \bar{g}_2 t^\dagger \bar{g}_3 t^\dagger \cdots \bar{g}_{n-1} t^\dagger \bar{g}_n$. Because $t^\dagger = -t$, $\left(F^{-1}\right)_{1n} = \left(F^{-1}\right)_{1n}$, that is, $g_{1n} = g_{n1}$ for odd values of $n$, and $g_{1n} = -g_{n1}$ for even values of $n$.

## A.2.2 *Calculated Results*

To treat the noncollinear alignment of the magnetizations of the left and right electrodes, the spin axis of the left electrode is canted by an angle $(\theta, \varphi)$. We first give briefly results of the conductance for junctions with nonmagnetic electrodes, and then we present the main MR results for junctions with ferromagnetic electrodes. We also give results of conductance for junctions with a magnetic field $h = (h_x, h_y, h_z)$ and for junctions with doped spacers. The number of atomic layers, $n$, of the spacer is $10^3$–$10^5$ (typically about $10^3$).

We make a distinction between results with even and odd numbers of atomic layers, $n$, because a difference in the sign of the left-going and right-going hopping of electrons in the Rashba Hamiltonian gives $g_{1n} = +(-)g_{n1}$ for odd (even) $n$, as shown in the appendix. In the following we take $t_L = t_R = t \equiv 1$ and introduce an infinitesimally small imaginary energy $\eta = 10^{-6}$ for convenience of the numerical calculations.

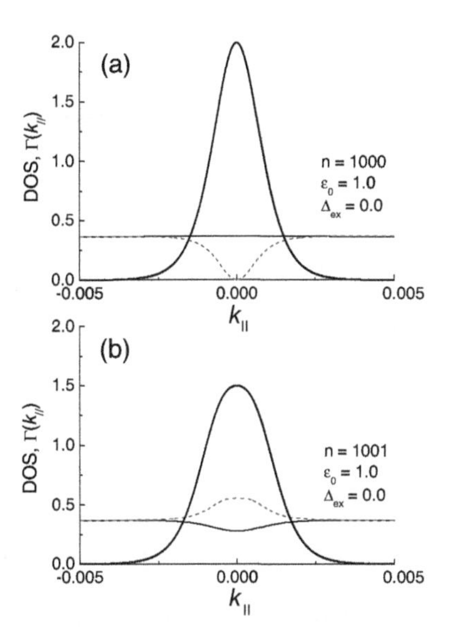


**Figure A.4** Calculated results of momentum-resolved conductance $\Gamma(k_{\|})$ (thick curves) for (a) $n = 1000$ and (b) $n = 1001$. Thin curves show the momentum-resolved DOS at the edge of the spacer, and broken curves show the contribution to the local DOS from the electrode.

### A.2.2.1 Nonmagnetic Electrodes

Figures A.4a and A.4b show the calculated results of the momentum-resolved local DOS at the spacer interface and momentum-resolved conductance $\Gamma(k_{\|})$ for $n = 1000$ and 1001. The units of $\Gamma(k_{\|})$ are $e^2/h$. The calculated values of $\Gamma(k_{\|} = 0)$ coincide with those obtained semi-analytically.

We find that the conductance $\Gamma(k_{\|} = 0)$ for $n = 1000$ is quantized, whereas that for $n = 1001$ is smaller than the quantized value. As shown in Fig. A.4a the contribution to the local DOS from the electrodes (DOS0) is zero for even $n$. In other words, contributions from the $V_1$ and $V_n$ terms in Eq. A.21 vanish at $k_{\|} = 0$, and the local DOS is determined only by the Dirac point of an isolated space. In contrast, reflection and transmission of the electron wave function occur at the interface for odd $n$, and the conductance becomes smaller than the quantized value. The difference between the conductances for even and odd $n$ values is caused by a characteristic

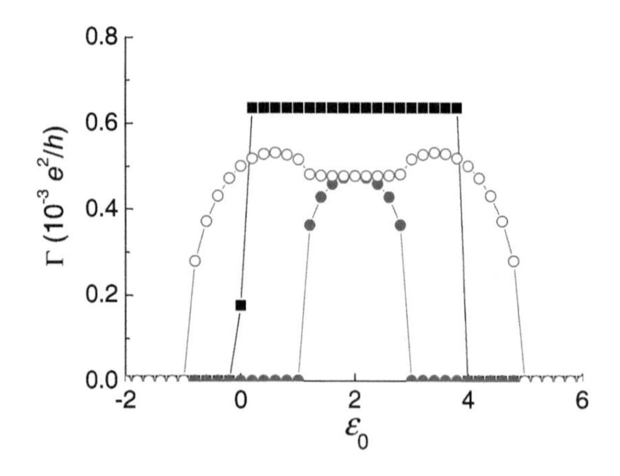


**Figure A.5** Calculated results of conductance of ferromagnetic junctions as a function of the atomic potential $\varepsilon_0$ of the electrodes. Black squares are the results for $\Delta_{ex} = 0$ with $n = 1000$, and closed and open circles are those for $\Delta_{ex} = 2$ with $n = 1000$ and 1001, respectively.

feature of the Rashba-type hopping term. The difference between the integrated values of $\Gamma(k_\parallel)$ is, however, sufficiently small, as expected.

In addition to the contribution of the Dirac point at $k_\parallel = 0$ to $\Gamma(k_\parallel)$, $\Gamma(k_\parallel)$ has nonzero values for $k_\parallel \neq 0$. This is caused by tunneling transport through the energy gap near the Dirac point of the spacer [6, 7]. Because the energy gap is infinitesimally small near the Dirac point, electron tunneling survives even for large $n$, but the half-width of $\Gamma(k_\parallel)$ decreases with increasing $n$. As a result, the integrated conductance $\Gamma$ decreases weakly as $\Gamma \propto 1/n$. Note that the numerical error becomes large when $n \times \eta \gtrsim 1$.

### A.2.2.2 Ferromagnetic electrodes

For junctions with ferromagnetic electrodes, the magnetization alignment of the left and right electrodes may be controlled by a weak external magnetic field. When $\theta = 0$ $(\pi/2)$ and $\varphi = 0$, the magnetization is parallel (P). Without an external magnetic field, the magnetization alignment might be antiparallel (AP), and it changes to P with an external magnetic field. When the resistivity is increased (decreased), the MR is called positive (negative) MR.

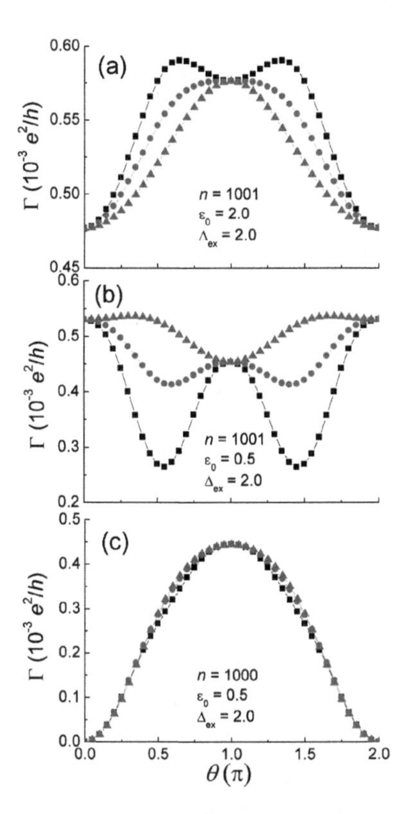


**Figure A.6** Calculated results of Γ of ferromagnetic junctions as a function of the relative angle $\theta$ between two magnetizations of electrodes with $\varphi = 0$ (squares), $\pi/4$ (circles), and $\pi/2$ (triangles) for (a) $n = 1001$ and $\varepsilon_0 = 2.0$, (b) $n = 1001$ and $\varepsilon_0 = 0.5$, and (c) $n = 1000$ and $\varepsilon_0 = 0.5$.

We first note that the Dirac dispersion of the spacer may give rise to a half-metallic-like state at the interface. Because of the square lattice, the energy dispersion of the intralayer of the electrode is given as $\varepsilon_{k_\parallel \sigma} = \varepsilon_0 - \sigma\,\Delta_{\text{ex}}/2 - 2t\cos k_\parallel$. The band is further broadened by $\pm 2t$ by the interlayer hopping term. Because the Dirac point appears at $\varepsilon = 0$ with $k_\parallel = 0$ when $\boldsymbol{h} = 0$, matching of the conduction state of the electrodes is subject to the relation $0 < \varepsilon_0 - \sigma\,\Delta_{\text{ex}}/2 < 4t$. When the relation does not hold, no ballistic conduction occurs even though the DOS is not zero at $\varepsilon = 0$. This is a half-metallic-like state for ballistic conduction via the Dirac point.

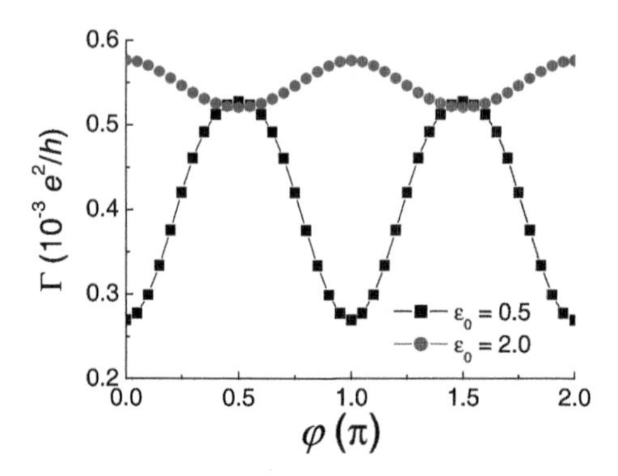


**Figure A.7** Calculated results of $\Gamma$ for ferromagnetic junctions as a function of the angle $\varphi$ between two magnetizations with $\theta = \pi/2$. The number of atomic layers of the spacer is $n = 1001$ and $\Delta_{ex} = 2.0$.

Figure A.5 shows the $\varepsilon_0$ dependence of the conductance in symmetric junctions with $\varepsilon_s = 0$. Results for $\Delta_{ex} = 0$ (paramagnetic electrodes) are nonzero only when $0 < \varepsilon_0 < 4$ ($t = 1$), as explained above. There is almost no difference between results for $n = 1000$ and those for 1001. For ferromagnetic electrodes with $\Delta_{ex} \neq 0$ a further complication appears in the dependence of $\Gamma$ on $\varepsilon_0$. The calculated $\Gamma$ results of a junction with $\Delta_{ex} = 2.0$ are shown by red circles in Fig. A.5. Filled and open circles indicate results for $n = 1000$ and 1001, respectively. The relation shown in the previous paragraph should give nonzero conductance of up- and down-spin electrons for $-1 < \varepsilon_0 < 3$ and for $1 < \varepsilon_0 < 5$, respectively, and the total conductance might be finite for $-1 < \varepsilon_0 < 5$. This explanation holds only when $n$ is odd. When $n$ is even, $\Gamma = 0$ in regions $-1.0 < \varepsilon_0 < 1.0$ and $3.0 < \varepsilon_0 < 5.0$. The results are attributed to Rashba-type hopping, which is nonzero only between opposite spin states. When $n$ is even, electrons that enter into the down-spin state should exit the spacer via the up-spin state. Because a half-metallic-like state appears in the regions $-1 < \varepsilon_0 < 3$ and $1 < \varepsilon_0 < 5$ for down- and up-spin states, respectively, $\Gamma$ is finite only in the region $-1 < \varepsilon_0 < 1$.

Thus the conductance of ferromagnetic junctions is strongly affected by the matching of the electronic states of the electrodes with the Dirac point and by the matching of the spin state at the left interface with that of the right interface. When a noncollinear magnetization alignment of the electrodes is introduced, this matching or mismatching may produce interesting MR effects as described below.

Figure A.6 shows the dependence of $\Gamma$ on the relative angle between two magnetizations in ferromagnetic junctions, that is, the MR effect. Values of $\Gamma$ are plotted as a function of $\theta$ for fixed angles $\varphi = 0, \pi/4$, and $\pi/2$. Values of the parameters are taken as (a) $n = 1001$, $\varepsilon_0 = 2.0$, and $\Delta_{ex} = 2.0$, (b) $n = 1001$, $\varepsilon_0 = 0.5$, and $\Delta_{ex} = 2.0$, and (c) $n = 1000$, $\varepsilon_0 = 0.5$, and $\Delta_{ex} = 2.0$ in Figs. 6a, 6b, and 6c, respectively. As can be seen in Fig. A.5, a half-metallic-like state occurs for $\varepsilon_0 = 0.5$ but not for $\varepsilon_0 = 2.0$.

The $\theta$ dependence of $\Gamma$ shown in Fig. A.6a indicates that the conductance tends to increase in AP alignment of electrode magnetizations and that $\Gamma$ is weakly dependent on $\varphi$. The larger conductance in AP alignment than in P alignment indicates a positive MR, which is much different from the usual tunnel magnetoresistance (TMR) and giant magnetoresistance (GMR) effects in which a negative MR usually occurs [8]. This feature is attributed to the Rashba-type hopping, which occurs between opposite spin states, and to the matching or mismatching at the interface between the two spin states. Because no half-metallic-like state exists at the interface for $\varepsilon_0 = 2.0$, the results for $n = 1001$ are almost the same as those for $n = 1000$ (not shown).

The $\theta$ dependence of $\Gamma$ shown in Fig. A.6b indicates a strong effect of the angle $\varphi$. Combined with the matching or mismatching between spin states, the half-metallic-like state realized at the interface for $\varepsilon_0 = 0.5$ may give rise to a complicated dependence on angle. The $\varphi$ dependence of $\Gamma$ is also related to anisotropic-like MR, as will be mentioned later.

Figure A.6c shows the $\theta$ dependence of $\Gamma$ for $\varepsilon_0 = 0.5$ and $n = 1000$. When $\theta = 0$, a half-metallic-like state appears at both interfaces, as shown in Fig. A.5, and the even number of atomic layers in the spacer disconnects the conduction path between the left and right electrodes. As a result, the conductance vanishes for $\theta = 0$. With increasing relative angle between the two magnetizations, the conduction path becomes connective, and the conductance increases, resulting in an infinite positive MR. This is completely

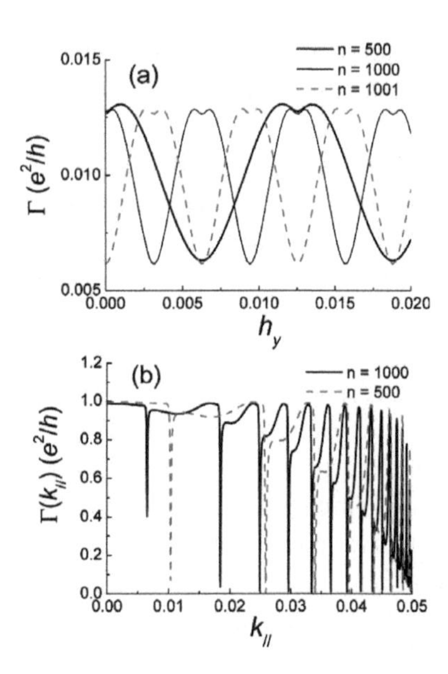


**Figure A.8** (a) Calculated results of conductance as a function of an external magnetic field $h_y$ for a hole-doped spacer with $n = 500$, 1000, and 1001. Parameter values are $\varepsilon_0 = 0.5$, $\Delta_{\text{ex}} = 2.0$, $\varepsilon_s = -0.1$, and $\theta = \pi$ (antiparallel alignment). (b) Momentum-resolved conductance for a hole-doped spacer with $n = 500$ (broken curve) and 1000 (solid curve) with the same parameter values as those in (a) but $h_y = 0$.

opposite to the infinite negative MR of tunnel junctions that occurs with ideal half-metallic ferromagnets.

Figure A.7 shows the dependence of $\Gamma$ on the angle $\varphi$ with a fixed angle $\theta = \pi/2$ (in-plane magnetization). The value of $\Delta_{\text{ex}} = 2.0$, and $\varepsilon_0 = 0.5$ or 2.0. The number of atomic layers in the spacer is $n = 1001$. The value of $\Gamma$ oscillates with a period $\pi$. Because the magnetization of the left electrode is rotated in the plane, the conductance oscillation may be anisotropic magnetoresistance (AMR) [9]. Because the present MR effect appears in junctions made of nonmagnetic spacers, the effect could be more similar to tunneling anisotropic magnetoresistance (TAMR) [10] rather than to the usual AMR in ferromagnetic metals. When $\varepsilon_0 = 2.0$, the interface is not half-metallic-like and the oscillation amplitude of $\Gamma$ is weak, whereas for junctions with $\varepsilon_0 = 0.5$ the interface is half-metallic-like and the oscillation amplitude becomes large.

### A.2.2.3 Effects of Magnetic Field

Here the electrodes are assumed to be nonmagnetic. The effect of the magnetic field $\boldsymbol{h} = (h_x, h_y, h_z)$ on $\Gamma$ is found to be rather weak. For $\boldsymbol{h} = (h_x, 0, 0)$, the momentum-resolved conductance $\Gamma(k_\parallel)$ shifts by $h_x$. Because $\Gamma(k_\parallel)$ has finite values near $k_\parallel = 0$, the integrated conductance $\Gamma$ is unaffected by $h_x$. The effect of $\boldsymbol{h} = (0, h_y, 0)$ is not obvious. Although $\Gamma(k_\parallel)$ varies around $k_\parallel = 0$ and depends nonmonotonically on $h_y$, the integrated value $\Gamma$ is independent of $h_y$.

For $\boldsymbol{h} = (0, 0, h_z)$, the Dirac point at $k_\parallel = 0$ splits off, resulting in an insulating state. Therefore, the conductance decreases exponentially with increasing $h_z$. The effect of $h_z$, however, is weaker for odd values of $n$ than for even values. The reason for this may be explained as follows. $\Gamma(k_\parallel)$ for even $n$ is larger than for odd $n$ in regions with larger values of $k_\parallel$. However, the effect of $h_z$ is strong only near $k_\parallel = 0$. Therefore the contribution from the larger $k_\parallel$ region to $\Gamma$ is stronger for even $n$ than for odd $n$.

### A.2.2.4 MR for Doped Junctions

The conductance of an electron-doped spacer may be calculated by shifting the atomic potential by $\varepsilon_s < 0$ and that for a hole-doped spacer by shifting the atomic potential by $\varepsilon_s > 0$. Numerical calculations show that the conductance increases linearly with increasing doping rate, as expected from the linear dispersion near the Dirac point. When the interface is not half-metallic-like, there is almost no difference between the conductance for $n = 1000$ and $n = 1001$. However, the difference becomes large for half-metallic-like interfaces. The MR effect is also enhanced.

Figure A.8a shows the conductance of hole-doped junctions ($\varepsilon_s = -0.1$) calculated as a function of external magnetic field $\boldsymbol{h} = (0, h_y, 0)$ for several values of $n$. The magnetization alignment of the electrodes is AP, that is, $\theta = \pi$, and the interfaces are half-metallic-like ($\Delta_{\text{ex}} = 2.0$ and $\varepsilon_0 = 0.5$). We see that the conductance oscillates with $h_y$ and the oscillation period becomes large with deceasing $n$. Note that the extrema of $\Gamma$ are almost independent of $n$ because the spacer is now metallic. The oscillatory feature results from the quantum well states of the hole-doped spacer. Figure A.8b shows the momentum-resolved conductance for $n = 500$ and 1000 without a magnetic field. The vanishingly small conductance for certain values of $k_{\parallel}$ suggests the existence of quantum well states, which are probably caused by the half-metallic-like interfaces of the junction.

### A.2.3 *Discussion and Summary*

We have studied MR effects of ferromagnetic junctions involving a nonmagnetic spacer with a Rashba-type hopping Hamiltonian under the assumption of ballistic transport. Large positive MR effects in these junctions are attributed to two characteristic features of the spacer with Rashba-type hopping. One is the appearance of a half-metallic-like state at the interface, owing to the strong momentum filtering effect by the Dirac point. The other is the strong matching or mismatching of the spin states at the left and right interfaces caused by electron hopping between opposite spin states in the Rashba-type Hamiltonian. The matching or mismatching condition between spin states depends on the magnetization alignment, and therefore large MR effects appear. Furthermore, the MR effects depend on the number of atomic layers of the spacer because of the difference in sign between the left- and right-going hopping integrals of the Rashba-type Hamiltonian. In-plane rotation of magnetization in the left or right electrodes produces another type of MR that is anisotropic. This type of MR is attributed to the combined effect of the strong momentum filtering by the Dirac point and Rashba-type hopping. We should, however, note that these characteristic features of the MR are predicted for the ballistic transport regime in clean samples. Disorder in electrodes, fluctuations of spacer length, and other factors may weaken these characteristic features. It is also noted that Rashba-type hopping in graphene occurs between two sublattices, and therefore such MR would not occur at graphene junctions. However, half-metallic-like states at the interfaces do produce large MR, as demonstrated for realistic graphene junctions [6, 7, 11, 12]. The large MR effects studied in this work could be observed in junctions involving surfaces of topological insulators in which Rashba-type hopping occurs between opposite spin states.

Experimental studies of MR for junctions made of surfaces of topological insulators would be quite interesting.

## A.3 Spin–Orbit Interaction

The Hamiltonian of the SOI is in general given by

$$H_{\mathrm{SO}} = \frac{e\hbar}{4m^2c^2}\left(\boldsymbol{\sigma}\cdot[\mathrm{E}\times\mathrm{p}]\right) \tag{A.32}$$

$$\simeq -\frac{e}{2m^2c^2}\left(\frac{dV}{rdr}\right)(\boldsymbol{s}\cdot\boldsymbol{l}), \tag{A.33}$$

where $c$ is the light velocity, $E$ is an electric field induced by a potential gradient $dV/dr$, and $\boldsymbol{l}$ is the orbital angular momentum. There are several origins of the internal electric field:

- Inversion asymmetry of the lattice such as the zinc-blende structure. The SOI is called Dresselhaus-type SOI [13].
- Inversion asymmetry of the structure such as the 2DEG. The SOI is called Rashba-type SOI [14, 15]. The Rashba Hamiltonian for a 2DEG is given as

$$H_0 = \begin{pmatrix} \frac{\hbar^2}{2m}k^2 & i\lambda\hbar k_- \\ -i\lambda\hbar k_+ & \frac{\hbar^2}{2m}k^2 \end{pmatrix}, \tag{A.34}$$

  where $k_{\pm} = k_x \pm ik_y$ with $\boldsymbol{k} = (k_x, k_y)$ the electron momentum in the 2DEG plane, $\lambda$ is the spin–orbit coupling. The eigenvalues are given as

$$E_{k\pm} = \frac{\hbar^2k^2}{2m} \pm \lambda\hbar k. \tag{A.35}$$

- $\ell - s$ coupling on atoms.
- Potential gradient caused by imputiry potential and $\ell - s$ coupling of atoms.

The first three types of SOIs are called intrinsic SOIs since they reside uniformly in the system, while the last one is called an extrinsic SOI as it is caused by impurity potentials. The SOI in semiconductors is usually caused by the Dressehaus- and/or the Rashba-type SOI, while that in metals and alloys may be related to the $\ell - s$ coupling.

The SOI shown above may be recasted in different ways to clarify the role in transport properties. Since Eq. A.32 may be written as $H_{SO} \propto \boldsymbol{\sigma} \cdot \boldsymbol{h}_{\text{eff}}$, one may consider that the SOI subsists a momentum-dependent effective magnetic field. It is also rewritten as $H_{SO} \propto [\sigma \times \boldsymbol{E}] \cdot \boldsymbol{p}$, from which one can derive a velocity $\upsilon = dH_{SO}/d\boldsymbol{p} \propto [\sigma \times \boldsymbol{E}]$ for conduction electrons. The velocity is called anomalous velocity. It is also noted that the SOI results in complex wave functions of electrons, from which a phase called Berry phase appears. As shown below, Berry phase may be interpreted to be an effective magnetic field in the momentum space.

## References


1. R. Kubo, *J. Phys. Soc. Jpn.*, **17**, 975 (1957).
2. P. A. Lee, and D. S. Fisher, *Phys. Rev. Lett.*, **47**, 882 (1981).
3. A. Umerski, *Phys. Rev. B*, **55**, 5266 (1997).
4. C. Caroli, R. Comberscot, P. Nozieres, and D. Saint-James, *J. Phys. C*, **4**, 916 (1971); **5**, 21 (1971).
5. R. Landauer, *IBM J. Res. Dev.*, **1**, 223 (1957).
6. A. Yamamura, S. Honda, J. Inoue, and H. Itoh, *J. Magn. Soc. Jpn.*, **34**, 34 (2010).
7. S. Honda, A. Yamamura, T. Hiraiwa, R. Sato, J. Inoue, and H. Itoh, *Phys. Rev. B*, **82**, 033402 (2010).
8. See, for example, the article by J. Inoue in *Nanomagnetism and Spintronics*, edited by T. Shinjo (Elsevier, Amsterdam, 2009).
9. I. A. Cambell and A. Fert, in *Ferromagnetic Materials*, Vol. 3, edited by E. P. Wohlfarth (North-Holland, Amsterdam, 1982), p. 747.
10. C. Gould, C. Ruster, T. Jungwirth, E. Girgis, G. M. Schott, R. Giraud, K. Brunner, G. Schmidt, and L. W. Molenkamp, *Phys. Rev. Lett.*, **93**, 117203 (2004).
11. T. Hiraiwa, R. Sato, A. Yamamura, J. Inoue, S. Honda, and H. Itoh, *IEEE Trans. Magn.*, **47**, 2743 (2011).
12. R. Sato, T. Hiraiwa, J. Inoue, S. Honda, and H. Itoh, *Phys. Rev. B*, **85**, 094420 (2012).
13. G. Dresselhaus, *Phys. Rev.*, **100**, 580 (1955).
14. E. I. Rashba, *Sov. Phys. Solid State*, **2**, 1109 (1960).
15. Y. A. Bychkov and E. I. Rashba, *J. Phys. C*, **17**, 6039 (1984).

# Index

An environmentally friendly book printed and bound in England by www.printondemand-worldwide.com

PEFC Certified

This product is from sustainably managed forests and controlled sources

www.pefc.org

PEFC

PEFC/16-33-415

This book is made of chain-of-custody materials; FSC materials for the cover and PEFC materials for the text pages.